ARBURG

Practical Guide to Injection Moulding

Edited by

V. Goodship

Rapra Technology Limited
Shawbury, Shrewsbury, Shropshire, SY4 4NR, UK
Tel: +44 (0)1939 250383 Fax: +44 (0)1939 251118 http://www.rapra.net

ISBN: 1-85957-444-0

Typeset, printed and bound by Rapra Technology Limited.

Contents

ARBURG GmbH & Co.
Postfach 11 09
72286 Lossburg
Germany

Tel.: +49 (0) 74 46 33-00
Fax: +49 (0) 74 46 33 3365
Website: http://www.arburg.com
E-mail: contact@arburg.com

ARBURG Ltd.
Tachbrook Park Drive
Warwick
CV34 6RH
United Kingdom

Tel.: +44 (0) 1926 457 000
Fax: +44 (0) 1926 457 020
Website: http://www.arburg.co.uk
E-mail: uk@arburg.com

Dear Reader,

ARBURG have been manufacturing injection moulding machines since 1954. The ARBURG ALLROUNDER range of injection moulding machines that are specified and supplied today can be traced back to an original model designed and built by one of the two sons of our founder Arthur Hehl.

Karl Hehl introduced a machine for small scale production of camera flash bulb leads encapsulated in polymer. At that time they were typically produced using metal shrouds that could short out in damp and humid conditions. His innovative idea was to manufacture them with plastic, encapsulating the plug and thus avoiding the problem occurring. Karl, together with his brother Eugen are still actively involved in the company to this day.

Since that time ARBURG have become one of the major global players in the injection moulding field. Our state of the art manufacturing facility is based in the town of Lossburg in the Black Forest, Southern Germany. We pride ourselves on being an innovative and forward thinking company that set very high standards in the quality of our machine build, integrating optimum specifications for machines and robots to meet the increasingly demanding requirements our customers seek. We also actively look to develop long-term partnerships in offering our customers technical assistance to ensure that they remain competitive in their chosen marketplace. This may be in multi-component moulding, liquid silicone rubber, ceramic, thermoset or elastomer processing.

ARBURG has over forty years of experience in the field of multi-component injection moulding. We are renowned as being one of the leading companies in this field and customers worldwide look to us to provide a wide range of services to assist in their projects. With this breadth and depth of knowledge we are well equipped to support this 'ARBURG practical guide to injection moulding'.

We trust that you find this guide both informative and enjoyable.

Yours sincerely

Colin Tirel
Sales Director
ARBURG Ltd

Preface

This book is designed as a guide to the process of injection moulding. Whilst it is imagined to be primarily of interest to engineers and engineering students, many of the chapters should be accessible to non-specialist readers. It concentrates mainly on the moulding of plastics but also provides a basic guide to the moulding of ceramics and metal powders.

It can be basically split into three sections. Chapters 1-4 form the underlying basics of the rest of the book. Chapters 5-8 are the heart of the book, the processing guides. Chapter 9 is a troubleshooting guide. Finally Chapter 10 is an introduction to some of the more specialist injection moulding processes such as multi-material and gas assisted moulding.

This book should equip the reader with the knowledge to understand the relationship of material, machine and mould tool. The materials covered include common thermoplastic and thermoset materials, as well as liquid silicone rubber (LSR). It also provides information on the machinery required for successfully moulding these materials.

Acknowledgements

Thanks to Colin Tirel and Marcus Vogt at ARBURG for their assistance and help in gathering the material for this guide and the many people who contributed to the original source materials. This book was based on the ARBURG Plastics Technology Course KT1 and their associated technical literature.

Thanks also to Sally for letting me loose on another project.

Special mentions to Sammi for his thoughts and input, and Deb for her help with the drawings.

Dedication

This book is dedicated to my family and friends, who make it all worthwhile, and especially to Big D., Cider Deb, Jake (Rastus-face), Evie and Gandalf.

'The future belongs to those who believe in the beauty of their dreams'
Eleanor Roosevelt

Dr Vannessa Goodship
WMG
University of Warwick

1 Introduction

1.1 The Big Picture

When the Hyatt brothers, John and Isaiah, built and patented the first injection moulding machine in 1872, it is doubtful if they could possibly have imagined the impact this invention would have on the world. It spawned a worldwide industry employing approximately 0.5 million workers in the US alone.

'Miscellaneous plastics products' which also includes other plastic manufacturing industries such as extrusion, is the 4th largest manufacturing industry in the US. In 2001, it was estimated that there were 21,000 operating plastics industry establishments in the US, generating approximately $321 billion in shipments. If upstream, supplying industries are also included, this total annual shipment is nearly $409 billion. The smaller UK plastics sector by comparison is worth £18 billion annually.

In Western Europe the consumption of thermoplastic materials alone in 2002 was around 33 million tonnes. Of this, engineering resins reached a market size of around 2.5 million tonnes. The country to country usage of materials varies widely, for example Italy is the largest consumer of unsaturated polyester resins in Europe consuming 23% of the European market demand in 2002. 27% of all plastic consumed is used by injection moulders. The biggest injection moulding industry of all the European countries is found in Germany which not surprisingly therefore uses the most raw material.

The packaging sector accounts for about 24% of this market, with the automotive industry 18% and the electrical equipment industry 18%. Other important markets for plastics materials are the construction industry and consumer products.

A major factor for the entire injection moulding industry has been the globalisation of the market. The manufacturing base for many plastic injection mouldings has moved to low cost plants such as those found in China, India and Eastern Europe. To adapt to the increased competition, Western manufacturers have moved into more technically difficult products and mouldings, as well as developing the markets into more advanced processes such as those that will be discussed in Chapter 10.

From these few facts and figures it can be seen that injection moulding is an extremely large and important manufacturing industry, but what is injection moulding?

1.2 Introduction to Injection Moulding

Injection moulding is one of the most common processes used to produce plastic parts. It is a cyclic process of rapid mould filling followed by cooling and ejection. A variety of materials both plastic and non-plastic can be used as feedstock. However, the machine must be configured for the type of material used.

The material, which is generally available as grains or powder, is plasticised in an injection unit and injected into a clamped mould under high pressure (500-1500 bar). The main advantage of injection moulding is that it is a very economical method of mass production. Ready parts with tight tolerances can be produced in one step, often completely automatically. In general after-processing is not necessary. It is also possible to integrate different functions into one part to avoid the formation of different components that would be more expensive, e.g., the base of a typewriter with integrated guidance and fixing elements, the springy components of a printer element, a lens with integrated prisma to stop down a beam of light.

To guarantee a high quality in the injection moulded parts the following points have to be considered:

- The material has to be plasticised and injected carefully to avoid negative effects on the material properties.
- The process settings (such as pressures and temperatures) concerning the machine and mould have to remain constant with regard to time and space.
- Basic parts of an injection moulding machine.

An example of a commercially available injection moulding machine is shown in Figure 1.1. The basic parts that make up a machine are shown in Figure 1.2.

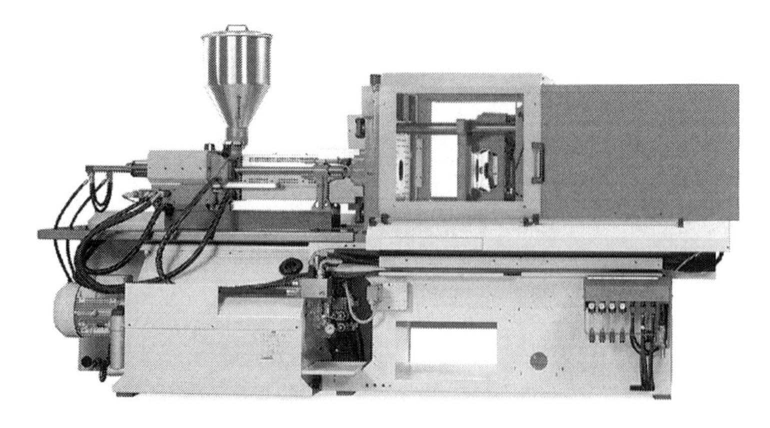

Figure 1.1 Injection moulding machine

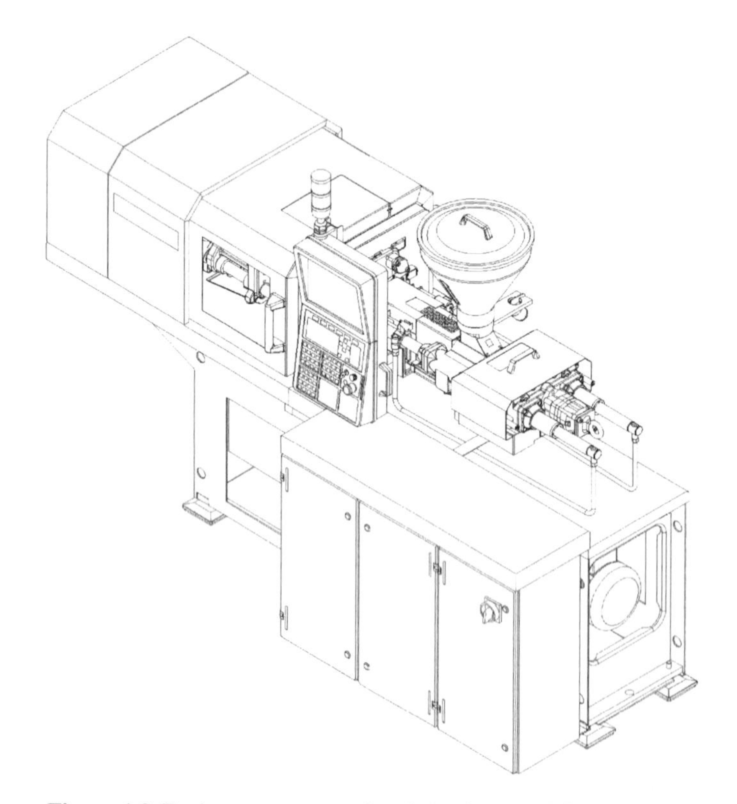

Figure 1.2 Basic components of an injection moulding machine

The control unit and control cabinet allow the machine operator to control and monitor the process. Control systems will be discussed in Chapter 5.

The injection unit is there to melt and meter the material into the tool. Machinery will be discussed in Chapter 3.

The clamp unit holds the injection moulding tool and gives the required clamp force to hold the two sides of the mould tool together. Tooling will be discussed in Chapter 4.

The machine base with hydraulics houses the systems that drive the movement of the machine. Each machine has a certain 'footprint'. This is the floor space it requires for operation.

1.3 The Injection Moulding Process

In injection moulding the mould and the plasticising area are separated from each other. The plasticising area, i.e., the plasticising cylinder temperature, is kept at the level of the processing temperature. The mould on the other hand, is kept cold enough for demoulding of the injection moulded part (thermoplastics) or warm enough for crosslinking (thermosets). The plasticised material is injected into the clamped mould. In an injection moulding machine, the clamping unit which contains the mould and the injection unit are integrated. Completely automated production is possible if the mould is installed with a vertical parting line. This enables the parts to fall down and out of the mould after demoulding. Injection moulding machines are typically used for the processing of thermoplastics. There are two types of injection unit available: a piston injection unit and a screw piston injection unit (reciprocating). The reciprocating screw method is the most common. For the processing of thermosets only screw piston machines can be used. This is because without the screw, the dwell time would be too long and the risk of early crosslinking would be too high. The injection sequence for both types of machine now follows.

1.3.1 Piston Injection Unit

Injection sequence:

1. Injection starts (clamped mould, start of piston movement) (Figure 1.3)
2. Injection and dosage
3. Holding pressure (to balance the solidification shrinkage) (Figure 1.4)
4. Ejection (Figure 1.5)

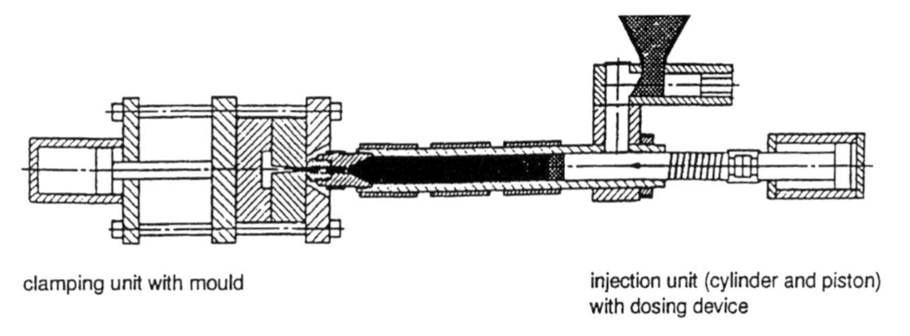

Figure 1.3 Piston injection unit (1)

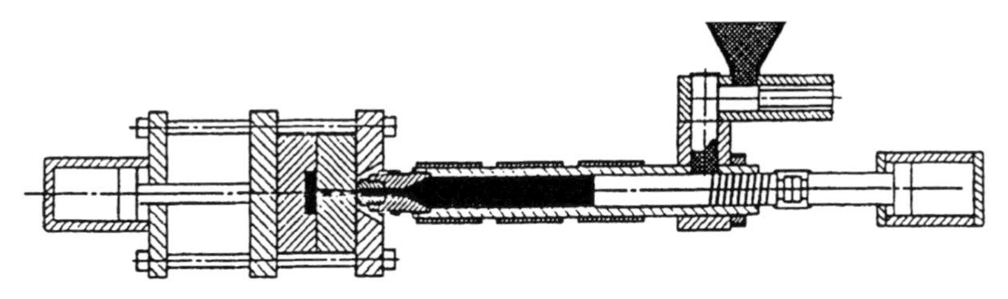

Figure 1.4 Piston injection unit (2)

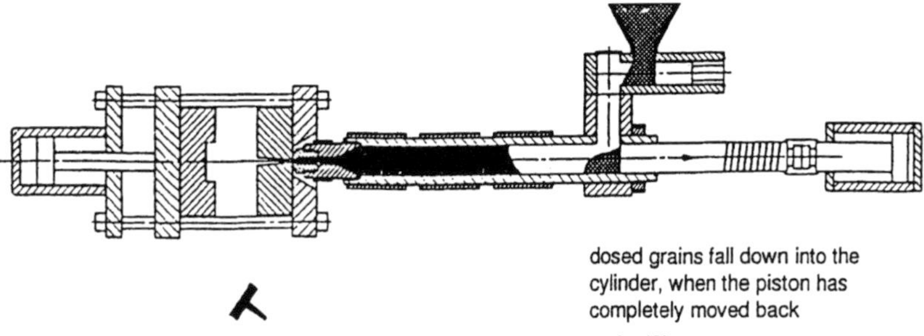

Figure 1.5 Piston injection unit (3)

In the piston injection unit, the material is transported step by step through the heated plasticating cylinder until the required temperature is reached. A dosage device replaces the consumed plastic by new material. As the dwell time in the cylinder is long, this method is not suitable for processing heat sensitive materials such as rigid polyvinyl chloride (PVC) and thermosets. Figures 1.3-1.5 illustrate the machine cycle.

1.3.2 Reciprocating Screw Machine

For a reciprocating screw machine the process cycle can be split into five stages:

1. In stage one, as shown in Figure 1.6, material is injected into the tool.

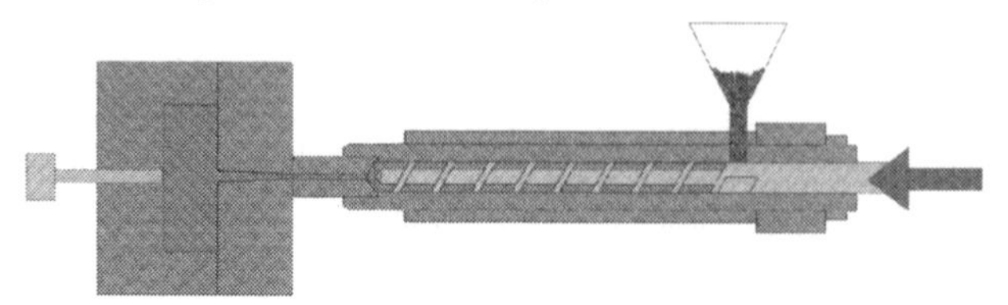

Figure 1.6 Injection moulding: injection

2. In stage 2 (Figure 1.7), the screw begins to turn and retract, metering a specified weight of molten material for the next shot. The previous shot is now cooling in the closed tool.

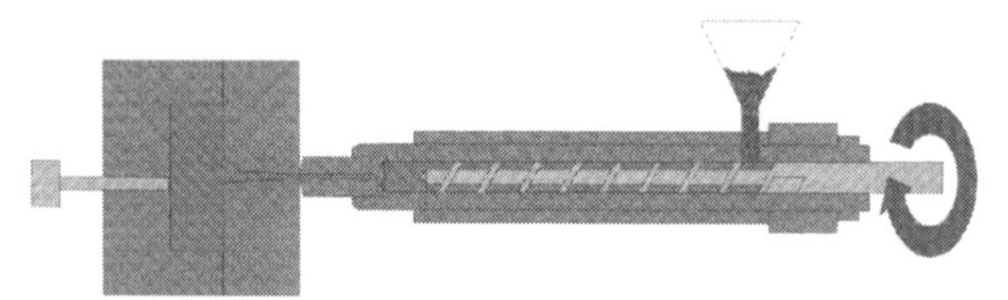

Figure 1.7 Injection moulding: metering

3. In stage 3, the injection unit moves back from the clamping unit as shown in Figure 1.8.

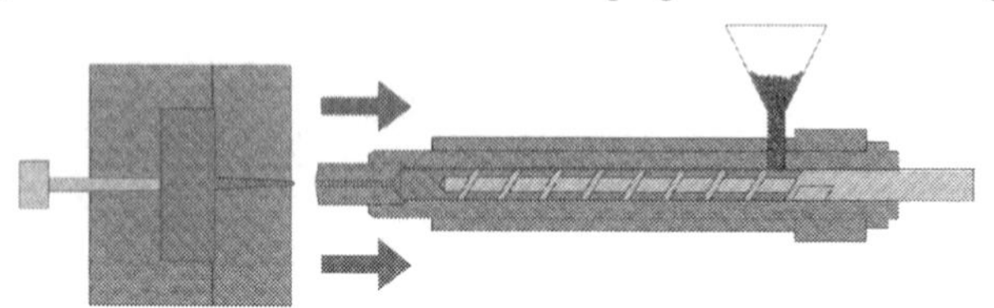

Figure 1.8 Injection moulding: injection unit retracts

4. Stage 4 is shown in Figure 1.9. In this stage the tool opens to reveal a cooled injection moulded component.

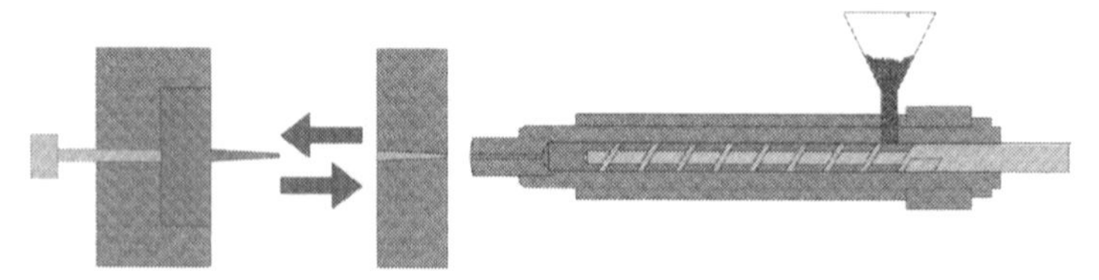

Figure 1.9 Injection moulding: mould open

5. Stage 5 is ejection of the part as shown in Figure 1.10. The injection unit will then move forward to the clamp unit to start a fresh cycle as shown in stage 1.

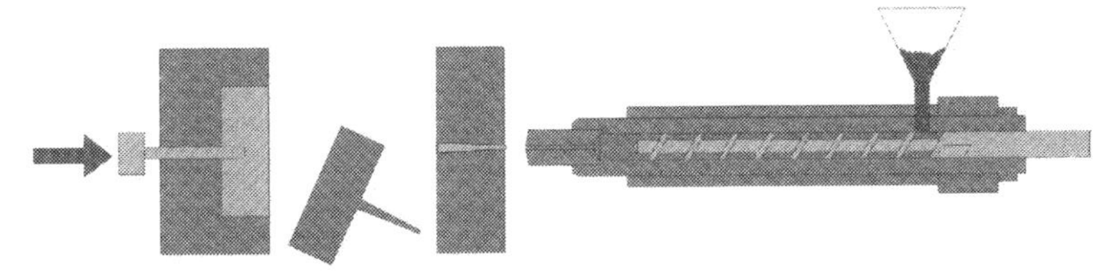

Figure 1.10 Injection moulding: ejection

Thermoplastics as well as thermosets and classic elastomers can be processed with screw injection units. These material types will be discussed in more depth in Chapters 6 and 7.

Thermoplastics are injected into a cold mould. The temperature of the mould must be sufficiently below the melting temperature of the material for it to solidify. This is because solidification is a physical process.

Thermosets and classical elastomers are injected into a hot mould to make the crosslinking of the material possible. Crosslinking is a chemical process.

In the screw piston injection unit, the material is dosed and plasticised simultaneously. The material is kneaded thoroughly by a rotating, axially movable screw. It is heated up to the processing temperature by the heat transfer of the hot cylinder wall and by friction. The material is transported by the screw to the screw tip. As the nozzle opening (cylinder opening) is still closed, the screw moves backwards. As soon as enough material is in the area in front of the screw tip, the screw is stopped. This is controlled by a limit switch or by a stroke measure device. This is the end of the plasticising and dosage stage.

In screw piston injection machines, the material is plasticised more homogeneously and has to stand less thermal stress than in piston injection units, as the plasticising itself happens just shortly after the injection.

1.3.3 Breaking Down the Injection Moulding Cycle

A single injection moulding cycle can be broken down into three distinct stages: plastication, mould filling and cooling with solidification.

1.3.3.1 Plastication

This stage is carried out in the injection unit and is similar to the process of extrusion.

The polymer flow rate is governed by the material processing conditions of the plastication stage: a combination of material rheology, barrel temperature and shear, back pressure and screw speed. The basic aim is to produce a homogeneous melt for the next stage where the material enters the mould. Moulding parameters which control the plastication stage are cylinder temperature, screw back temperature and back pressure.

1.3.3.2 Filling

Here the injection unit delivers a pre-set amount of molten polymer to the mould tool.

The parameters of mould filling are of great importance to the end result especially when considering factors such as warpage (orientation effects) and surface finish (skin formation). Filling dynamics are also thought to be the major factor in affecting the levels of residual stress. It is important that injection speeds are reproducible as slight changes can cause variations in the end product. Injection speeds that are too high can cause jetting and degradation and thus affect mechanical properties. A low speed may cause an increase in pressure requirements due to a thicker frozen layer and short shots (incomplete filling of the mould).

Thinner sections will generally need faster injection speeds than thick walled parts, mainly because of the decrease in the importance of the relationship between mould filling time and cooling time with a thicker section. The important thing is that the speeds are reproducible from one shot to the next. Important moulding parameters for filling are the injection speed and injection pressure.

1.3.3.3 Packing and Solidification

Once the material is in the tool, filling must be completed (tool packing), the part cooled and finally ejected. The purpose of the packing stage is to add extra material to compensate for the shrinkage caused by the decreasing density of the solidifying polymer. If the additional polymer were not injected the component would shrink and warp due to nonuniform cooling.

Ideally the packing and cooling stages should be such that the final dimensions are maintained as close as possible to design tolerances. Variables during this stage are packing pressure, packing time and the mould temperature. Bad mould design can lead to inconsistent cooling along the dimensions of the mould surface which can cause increased residual stresses. Once the material has cooled sufficiently, the component can be injected and the injection cycle continues. The cycle does not occur sequentially, while one part is cooling, plastication of the next cycle has already begun.

A breakdown of the cycle and the relative time for each stage is shown in Figure 1.11.

The injection moulding process and the parameter effects will be further discussed in Chapter 8.

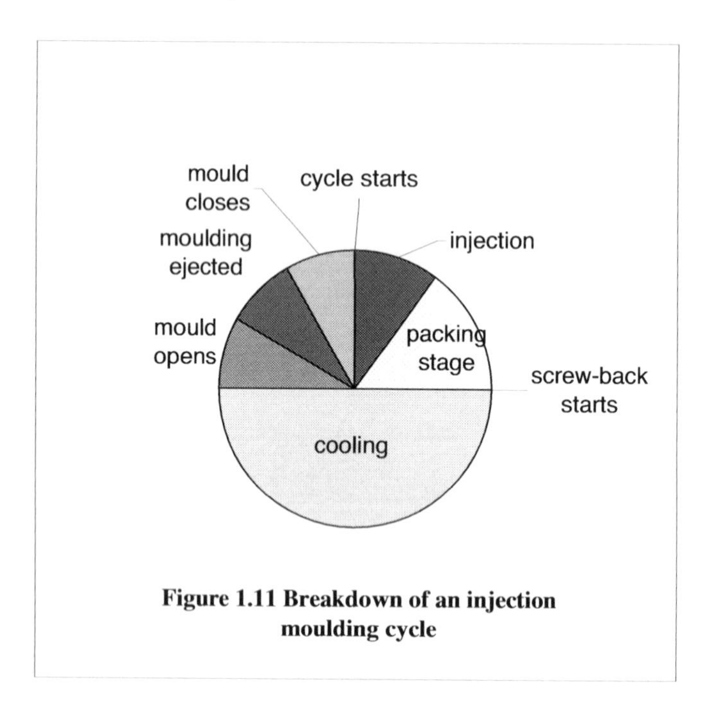

Figure 1.11 Breakdown of an injection moulding cycle

A brief introduction to injection moulding has now been given. Before moving on to consider it in greater depth, a quick summary of other commercial methods of polymer processing will now be considered for comparison.

1.4 Comparison with Other Moulding Methods

1.4.1 Extrusion

Extrusion is used for the production of intermediate products such as profiles, boards and sheets. Generally screw extruders are used. For the processing of thermosets, piston extruders are also suitable. The material is kneaded thoroughly and plasticised by friction and additional cylinder heating. It is fed through an opened mould (die) with a diameter corresponding to the desired profile (Figure 1.12).

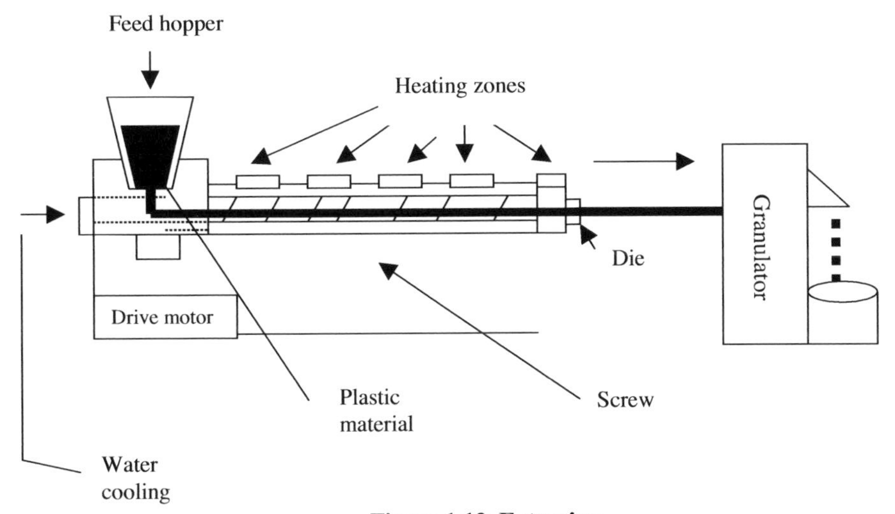

Figure 1.12 Extrusion

The extruder is very similar in design and purpose to an injection unit on an injection moulding machine. However, there are certain differences in both design and purpose. Whilst the material is simply fed through an extruder, an injection unit must also have the ability to rapidly inject the molten polymer into the mould tool cavity.

1.4.2 Compression Moulding

With compression moulding a weighed amount of material is placed in an open mould tool and then compressed (Figure 1.13). During the clamping movement, the material is heated up to its processing temperature by heat transfer from the hot mould allowing it to form to the shape of the tool. After sufficient pressure and temperature has been applied the tool is opened. Thermosets are demoulded from the hot mould when crosslinking is complete. Thermoplastics have to be cooled until the part is rigid enough for demoulding. Therefore the cycle times for thermoplastics are very long (12-20 minutes). In consequence, this method is only used for thick wall parts or for boards.

The compression moulding process exerts a much lower level of shear on the material than injection moulding, which is a high shear method.

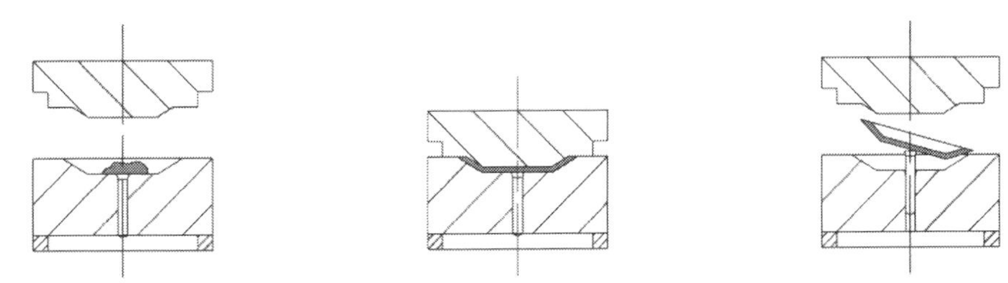

Figure 1.13 Compression moulding

1.4.3 Transfer Moulding

In a transfer mould, there is a special plasticising area (cylinder) as well as cavities.

With an open mould a weighed amount of material is placed into the plasticising area (Figure 1.14a). After mould clamping (Figure 1.14b), the material is pressed through the sprue channels into the cavities by the piston (Figure 1.14c). The material is warmed by heat transfer from a hot mould. Frictional heat also warms the material, which occurs in the tool but especially when it passes though the sprue channels.

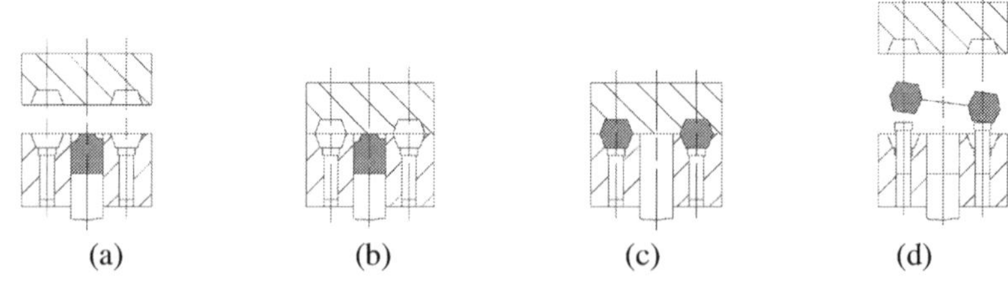

Figure 1.14 Transfer moulding

Practically, transfer moulding is used for the processing of thermosets only, as there are no advantages with the processing of thermoplastics compared to the injection moulding method.

1.4.4 Blow Moulding

Blow moulding is the third most commercially important process for plastics production after extrusion and injection moulding. It is used to produce a range of hollow articles for example bottles, fuel tanks and other large containers. There are two main variations, injection blow moulding and extrusion blow moulding. The process sequence for both is the same.

- The material is either extruded or injected to produce a tube like preform.
- The preform is blown out to the shape of the mould and then cooled.

Injection blow moulding is most commonly employed for the production of transparent soft drinks containers. However, extrusion blow moulding is the one most commonly employed for mouldings such as shampoos and detergent containers, plastic drums and milk bottles and is described below.

The material is fed through a transfer screw (which is very similar to an extruder), into a die head where the material is melted and passes out through a die as a tube like extrudate termed a parison as shown in Figure 1.15. This process can be either continuous or with larger articles intermittent. The parison extrudes down vertically and relies on the hot strength of the plastic to hold the parison weight in shape. For these reason blow moulding uses far more viscous materials than would normally be employed for the injection moulding process. A low viscosity material would simply pour out of the die onto the floor or split off before the parison had formed.

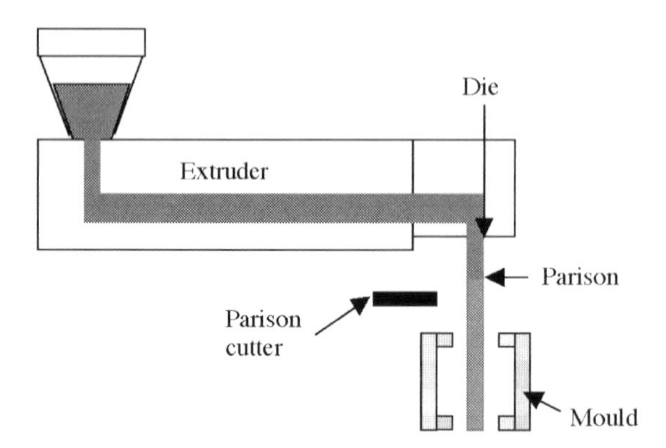

Figure 1.15 Blow moulding stage 1

The parison will continue to extrude until it reaches the base of the mould. At this point the mould will close and the parison will be cut above the mould with a hot knife. The mould is now moved away from the parison, taking the cut off slice of parison closed in the mould with it. The next parison continues to extrude. The inflation of the parison can now commence. A blow pin comes down into the top of the mould and blows air in, to inflate the hot parison against the sides of the mould as shown in Figure 1.16. The mould is cooled with water and this aids heat transfer to help solidify the newly formed article. The blow pin is removed and the tool opens to eject the part. The mould can then return to collect the next parison and start a new moulding cycle.

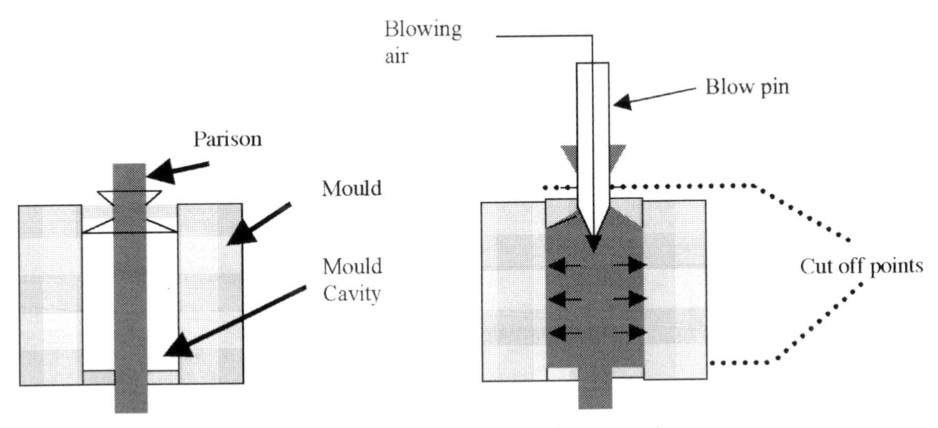

Figure 1.16 Blow moulding stage 2

The criteria for a blow moulding material are quite specific:

- Suitable viscosity
- High melt strength
- Be extendable when inflating (this is often given as an inflation ratio for the material)
- Be able to seal at the base of the moulding.

During inflation the material expands at a constant rate to give a constant wall thickness across the body of the moulding.

1.4.5 Film Blowing

In film blowing the plastic material is fed through an extruder to an annular die opening. The cylindrical molten tube is inflated from the inside by blowing air, creating a bubble of material that can be fed and collected onto rollers. Cooling is achieved by blowing air through a cooling ring above the die. The process is shown in Figure 1.17.

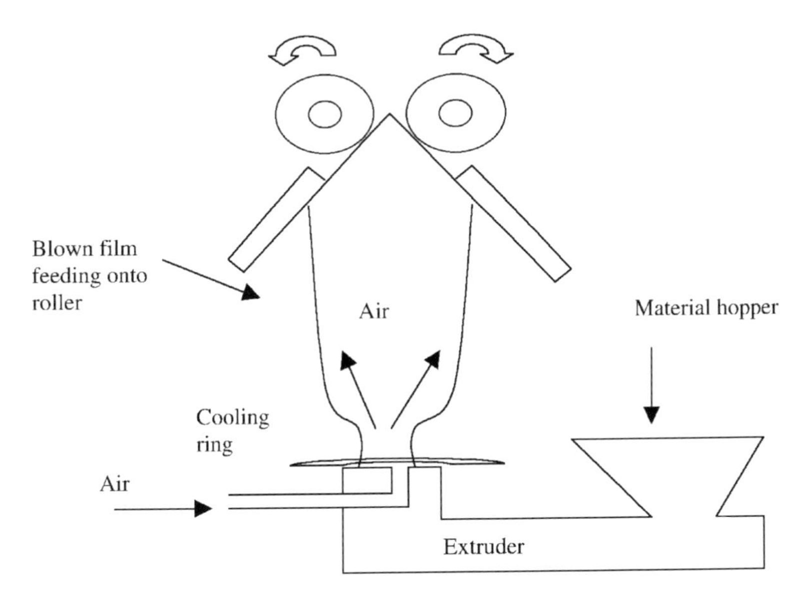

Figure 1.17 Film blowing

Large amounts of film scrap are available due to the short lifetime associated with packaging and industrial film materials such as carrier bags, dustbin liners and plastic sacks. A typical lifetime of products of this type is only two years. 100% recyclate material may be used in low grade applications such as bin liners, other products such as carrier bags may incorporate scrap with the virgin material.

Again, like blow moulding, film blowing has specific material requirements in terms of melt strength, viscosity and inflation characteristics. Generally film blowing is limited to polyolefin materials, the majority of usage being of low density polyethylene (LDPE), linear low density polyethylene (LLDPE) and high density polyethylene (HDPE).

1.4.6 Intrusion Moulding

The intrusion process is suited to mixed plastics. Generally they are finely ground before processing to aid dispersion. The process has elements of extrusion in that a plasticating unit is used to homogenise the mixture, which is then fed into an open mould. Once the plastic has cooled it can be ejected or removed depending on the type of mould that is used. The process is tolerant to contaminants such as mixed plastics, sand, glass, wood and paper and providing a minimum polyolefin fraction of around 40% is present. Other contaminants become embedded in this low melting fraction. This process is generally used to produce large, geometrically simple shapes such as profiles and panels for wood replacement applications such as fencing, posts and scaffolding.

1.4.7 Injection-Compression Moulding

The injection-compression moulding process combines elements of both the injection moulding process and the compression moulding process.

Compared with injection moulding, there is a big reduction in the filling pressures required. There is also less orientation of the material, which can improve properties and reduce anisotropy. This can be of special importance in moulding thin-walled components where dimensional stability is important, and with transparent mouldings where good optical properties are required.

Figure 1.18 shows the formation of an injection-compression moulded product. A pre-set amount of material is injected into a partly closed mould (a). The mould then closes to the size of the desired final component, squeezing the material by compression to complete cavity filling (Figure 1.18b).

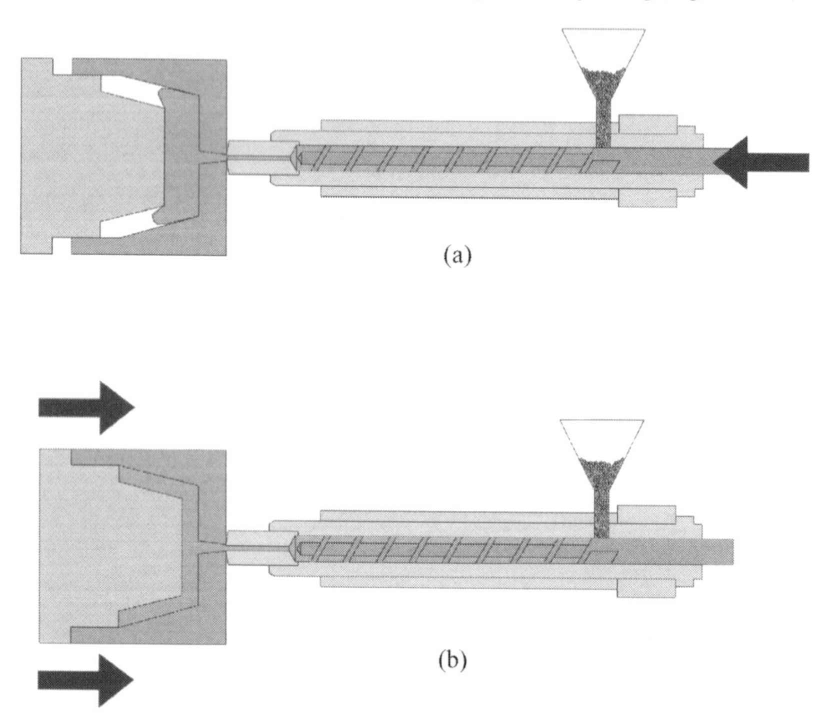

Figure 1.18 Injection-compression moulding

It can be seen that injection-compression is a variation of the injection moulding process. Other variations on the injection moulding process such as this will be discussed in Chapter 10.

Other processes for thermoset and thermoplastics also exist which are not discussed here. The suggested reading at the end of this chapter provides sources of further information.

1.5 Conclusion

This chapter has briefly introduced the fascinating subject of injection moulding. A comparison with the various processing methods also discussed in this chapter is given in Table 1.1. For some components, there may be more than one suitable process route. The final choice of process will depend on both economics and the component design. For the rest of this book however, only the injection moulding process will be discussed.

Table 1.1 Comparison of processes

Process	Complexity of parts	Example of parts	Forming action	Mould
Extrusion	Fairly simple profiles	Window profiles, tubes, granules	None	None
Injection moulding	Complex (solid)	TV housing, car handles	Injection	Closed
Blow moulding	Complex (hollow)	Bottles	Inflation	Closed
Film blowing	Very simple films	Carrier bags	Inflation	None
Compression moulding	Simple	GMT components, structural parts	Compression	Closed
Transfer moulding	Simple	Thermoset mouldings	Compression	Closed
Intrusion moulding	Simple	Low strength profiles and panels	Compression	Open
Injection-compression moulding	Complex	Thin walled mouldings	Compression	Closed

Further Reading

V. Goodship, *Introduction to Plastics Recycling*, Rapra Technology Limited, UK, 2001

W. Michaeli, *Plastics Processing: An Introduction*, Carl Hanser Verlag, Germany, 1995.

C. Rauwendaal, *Polymer Extrusion*, 3rd Edition, Carl Hanser Verlag, Germany, 1994

D.V. Rosato and D.V. Rosato (Eds.), *Blow Molding Handbook*, Carl Hanser Verlag, Germany, 1989.

J.F. Stevenson (Ed.), *Innovations in Polymer Processing: Molding*, Carl Hanser Verlag, Germany, 1996.

2 Introduction to Plastics

2.1 Introduction

Plastics were originally seen as substitute products for traditional materials such as metal and wood. However, now they have become as irreplaceable as the classic materials themselves. Plastics have managed this achievement because of their unique versatility and the ability to tailor their properties, which other materials cannot match. Our modern everyday life would be inconceivable without plastics. The use of plastics enables us to solve problems that are insoluble with the classic materials, whether it be – to name only a few examples – in electronics, light engineering, medical technology, space technology or machine and vehicle manufacture.

Plastics are made up of polymers and other materials that are added to them to give the desired characteristics. Natural polymeric materials such as rubber, shellac and gutta percha have a long history as raw materials for man. The first thermoplastic, celluloid, was also manufactured from a natural product, from cellulose. Even today, there are still some cellulose based plastics, i.e., the cellulose acetates (CA). Cellulose is already composed of the large molecules that are characteristic of plastics (macromolecules). However, to manufacture CA plastics, they still have to be 'prepared' with acetic acid. The first injection moulding machine was built and patented in 1872 in order to mould cellulose materials.

Today the vast majority of plastics are manufactured artificially, i.e., the macromolecules are built up from smaller molecules (predominantly from carbon and hydrogen). Basically, plastics can also be manufactured from their basic constituents, carbon and hydrogen (coal and water). For economic reasons, however, similar to petrol manufacture, plastics are nowadays almost exclusively manufactured from products generated by the fractionated distillation of crude oil.

We can, therefore, divide plastics into:

1. Plastics made from natural substances, e.g., Celluloid, cellulose acetate, vulcanised fibre, casein plastics (galalith)
2. Artificial plastics, e.g., polyethylene, polystyrene, polyamide

However, the origin of plastics, whether obtained from naturally occurring large molecules or synthetically prepared from smaller molecules, makes no difference to the subsequent processing.

The first synthetically developed polymer was called Bakelite after its inventor, Leo Baekeland in 1907. The material, phenol-formaldehyde, is a thermoset phenolic resin. However, up until 1924, and the work of Herman Staudinger, there was no real understanding of the chemical structure of polymers. Staudinger proposed the concept of linear molecular chains and macromolecules, which once accepted by the scientific community (this actually took several more years), allowed the doors to open on the synthesis and development of new polymeric materials. This new understanding of the structure of polymers allowed the development of plastics such as polyvinyl chloride and cellulose acetate in the 1920s. The 1930s saw the introduction of polyamides, polystyrene and acrylics as well as the introduction of single and twin screw extruders for polymer processing. New polymeric materials continued to be introduced, their development fuelled by the Second World War. The 1940s saw the introduction of epoxies, polyethylene and acrylonitrile-butadiene-styrene (ABS) to name but three. The 1950s saw the birth of the polypropylene industry as well as polyethylene terephthalate (PET) and polycarbonate (PC). American companies developed a number of engineering materials in the 1970s including polyphenylene sulphide and a number of fluoropolymers such as DuPont's (US) ethylene-tetrafluoroethylene copolymer Tefzel and perfluoralkoxy plastics under the trade name Teflon PFA. In 1973 Dynamit Nobel (Germany) introduced polyvinylidene fluoride (Dyflon) into the market and the 1980s saw the development of liquid crystal polymers (LCP).

The development of new polymers has now slowed due to the expense and difficulty of synthesising new materials, however new plastics are still being developed by mixing existing materials together. These materials are called polymer alloys and blends. An alloy has a single glass transition temperature (this will be explained later), and generally has better properties than the individual components. A blend has more than one glass transition temperature and has properties between those of the original materials. An example of a commercially successful blend is ABS.

2.2 Structure and Typical Properties of Polymers

The word polymer derives from the Greek word *poli*, which means many and the word *meros*, which means parts. This is because polymers are made up of a number of smaller repeated units called monomers. The simplest and most commonly used monomer is ethylene. Chemically it consists of two carbon atoms (C) and four hydrogen atoms (H). It can be represented in the two ways shown in Figure 2.1. The lines in this diagram represent bonds that exist between the atoms to form a molecule.

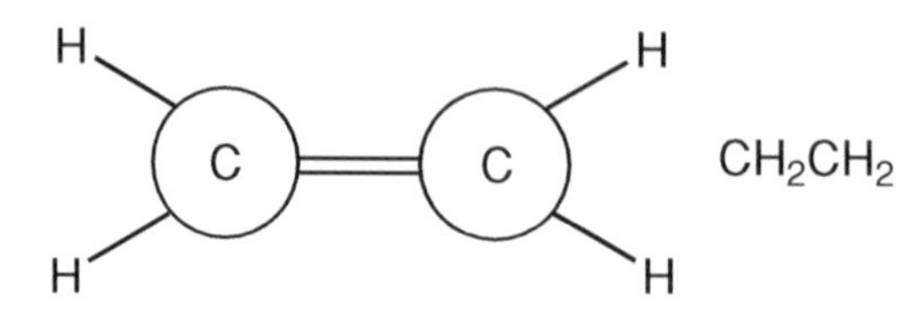

Figure 2.1 An ethylene molecule

It is the existence of the double bond between the carbon atoms in ethylene, which allows the creation of polyethylene. This happens when the monomers are combined by a process called polymerisation to form a chain such as the one shown in Figure 2.2. A chain of useful polymer may consist of 200-2000 monomers joined together. This particular type of polymerisation is called addition polymerisation.

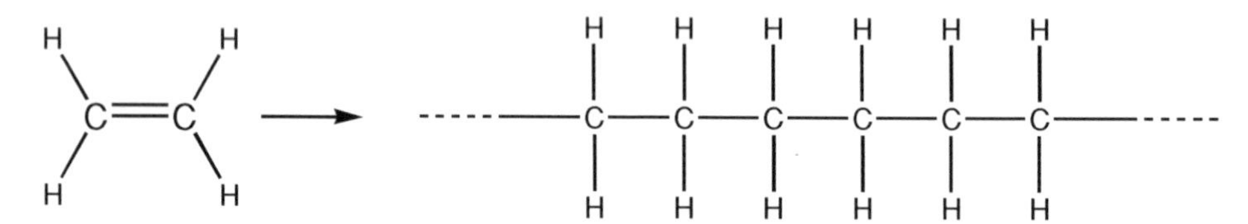

Figure 2.2 Polyethylene molecule

It can be seen from Figure 2.2, that carbon atoms form the backbone of the polymer. Many polymeric systems are made of long chains of carbon atoms such as this. But polymers are not confined to carbon forms and injection moulding materials such as liquid silicone rubbers (LSR) have different chemical structures as the backbone of the polymer chain. This will be illustrated later in Chapter 7. The repeat unit structures of a number of common polymers are shown in Table 2.1.

2.3 Formation of Macromolecules

There are three separate types of formation reactions for macromolecules. In **polymerisation**, the polymers merge in lines and form macromolecules without any kind of by-product being produced. This has already been illustrated in Figure 2.2. The number of monomers contained in a macromolecule is also referred to as the degree of polymerisation. If the macromolecules are made up of only one type of monomer, they are called homopolymers, one example being standard polystyrene. If different monomers form part of the structure of a macromolecule, then they are called copolymers (mixed polymers). Thus, for example, SB consists of styrene and butadiene monomers. ABS comprises three different types of monomers: acrylonitrile, butadiene and styrene.

In **polycondensation** macromolecules are formed from monomers under the splitting action of another substance, usually water. If polymerisation takes place at a temperature of over 100 °C, steam is formed. The macromolecules generated by polycondensation can be made up of one type of monomer, (e.g., PA 6), or 2 different monomers, (e.g., PA 66), depending on the shape of the monomers. Further examples of polycondensates are polycarbonate (PC) and linear polyesters such as polyethylene terephthalate (PET).

In **polyaddition** macromolecules are formed from monomers, without the generation of any cleavage products. Two different monomers are always required for polyaddition. The monomers undergo slight changes during this process. A few atoms change places between the different monomers. Examples are polyurethane (PU) and epoxy resins (EP).

The type of formation reaction plays no part in the subsequent injection moulding of thermoplastics. The molecules are already complete before injection moulding begins. They are merely melted, and then solidify in the mould to form the component.

In the processing of thermosets however, the type of formation reaction can have effects, if **crosslinkage** is brought about by polycondensation. Indeed, the locking pressure of the mould must then be great enough to prevent the mould halves being forced apart by steam production, thus making the component porous or blistered. Particular attention should be paid to this during the compression moulding or transfer moulding processing of thermosets. In injection moulding, the mould clamp force selected for the mould must be high enough to avoid any danger of its being forced open by steam, because of the high injection pressures required.

Table 2.1 Characteristic structures of common polymers

Polymer	Repeat unit(s)
Polyethylene (PE)	$-\!\!\left(CH_2-CH_2\right)_n\!\!-$
Polypropylene (PP)	$-\!\!\left(CH_2-CH(CH_3)\right)_n\!\!-$
Polystyrene (PS)	$-\!\!\left(CH_2-CH(C_6H_5)\right)_n\!\!-$
Polyamide 6 (PA 6)	$-\!\!\left(C(=O)-(CH_2)_5-NH\right)_n\!\!-$
Polyamide 66 (PA 66)	$-\!\!\left(C(=O)-(CH_2)_4-C(=O)-NH-(CH_2)_6-NH\right)_n\!\!-$
Polyethylene terephthalate (PET)	$-\!\!\left(CH_2-CH_2-O-C(=O)-C_6H_4-C(=O)-O\right)_n\!\!-$
Polyvinyl chloride (PVC)	$-\!\!\left(CH_2-CH(Cl)\right)_n\!\!-$
Epoxy resin (EP)	$-\!\!\left(O-C_6H_4-C(CH_3)_2-C_6H_4-O-CH_2-CH(OH)-CH_2\right)_n\!\!-$
Melamine-formaldehyde resin	$H_2C=O$; melamine (2,4,6-triamino-1,3,5-triazine, ring of C and N with three NH_2 groups)

2.4 Molecular Weight

A polymeric material may have many macromolecular chains, all of various lengths or repeat units. The molecular weight distribution is used to describe this variation and the average size of these chains determines the polymer molecular weight. As well as the backbone, polymers may also have side chains of varying lengths. This branching from the main chain also affects the properties of the polymer. Branching restricts the ability of the polymer chains to pack together. Therefore branching affects the density of a polymer. For example with high density polyethylene there is very little branching hence giving the term high density, low density polyethylene in contrast has many branches. An illustration of this effect on both density and melting point can be seen in Table 2.2.

Table 2.2 Influence of branching on properties of polyethylenes

Polyethylene	Melting point (°C)	Density (g/cm^3)	Tensile strength (MPa)	Number/type of branches
LDPE	110-120	0.91-0.93	17-26	Long branches
LLDPE	122-124	0.92	13-27	10-35 short branches (per 1000 carbon atoms)
HDPE	130-135	0.94-0.97	21-38	4-10 short branches (per 1000 carbon atoms)

It can be seen that the properties of the polymer are strongly linked to molecular weight and molecular weight distribution, and it is necessary to match the molecular weight, material properties and flow characteristics in order to shape the material during processing and give the desired final product.

2.5 Plastics

Plastics are made up of polymers and other materials that are added to increase the functionality. The actual polymer content within a plastic can vary widely from less than 20% to nearly 100%. Those plastics consisting of virtually 100% polymer are termed 'prime grades'. The level and type of the other additives used depends on the application for which the plastic is intended. There is a vast range of materials available in the market, in the USA alone over 18,000 different grades of plastic material are available.

Plastics can be subdivided into three main categories, thermoplastics, thermosets and elastomers. This distinction is based on both the molecular structure and the processing routes that can be applied. These three classes of materials will now be introduced.

2.5.1 Thermoplastic

These materials melt and flow when heated and solidify as they cool. On subsequent re-heating they regain the ability to flow. This means they can be reprocessed and hence recycled by re-melting them. Thermoplastics are used to make consumer items such as drinks containers, carrier bags and buckets. The most common thermoplastic materials and their applications are shown in Table 2.3.

When thermoplastics solidify they can take one of two molecular structures: an amorphous structure or a semi-crystalline structure which are both illustrated in Figure 2.3.

When semi-crystalline materials are cooled the molecular structure tends to become highly ordered and crystals are formed. The size of these crystalline regions varies according to both the structures of the chains themselves and the cooling rate. These materials displays sharp melting points unlike amorphous materials that soften. Semi-crystalline materials also tend to shrink more due to this molecular rearrangement, some materials may shrink by as much as 20%. This shrinkage will be more in the direction of flow due to the molecular realignment caused by the process of injection moulding.

Table 2.3 Common thermoplastics and their applications

Thermoplastic polymer	Applications
High density polyethylene (HDPE)	Packaging, pipes, tanks, bottles, crates
Low density polyethylene (LDPE)	Packaging, grocery bags, toys, lids
Polypropylene (PP)	Caps, yoghurt pots, suitcases, tubes, buckets, rugs, battery casings
Polystyrene (PS)	Mass produced transparent articles, yoghurt pots, fast food foamed packaging, cassettes
Polyamide (PA)	Bearings, gears, bolts, skate wheels, fishing lines
Polyethylene terephthalate (PET)	Transparent carbonated drink bottles
Polyvinyl chloride (PVC)	Food packaging, shoes, flooring

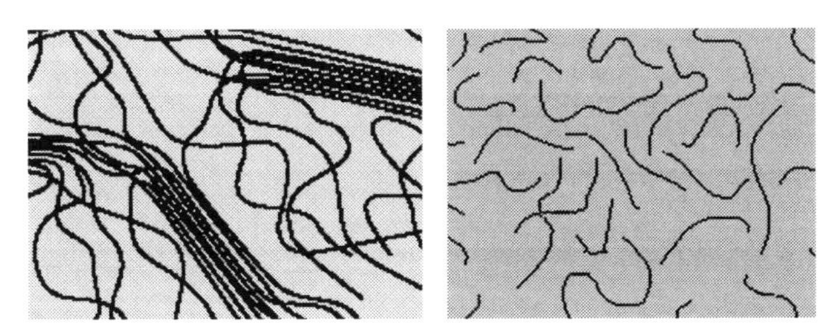

Figure 2.3 Semi-crystalline (left) and amorphous (right) materials

2.5.2 Thermosets

Thermoset injection moulding compounds change their structure when injected. Before injection moulding, they still consist of thread-shaped molecules similar to thermoplastics. However, during a process termed 'curing' the molecules crosslink forming a highly dense network of bonds. This makes the material stiff and brittle and the thermoset moulded parts can then no longer be melted. Thermoset materials decompose before they can melt, therefore, they cannot be reprocessed in the same way as thermoplastics. The differences in the arrangement of molecules between thermoplastics and thermosets can be seen in Figure 2.4. Thermosets are often used where their strength and durability can be utilised, some common thermosets are shown in Table 2.4.

Table 2.4 Common thermoset materials

Thermoset polymer	Application
Epoxy	Adhesives, electrical insulation
Melamine	Heat resistant laminate surfaces, i.e., kitchen worktops
Phenolics	Heat resistant handles for pans, irons, toasters
Polyurethane (PU)	Rigid or flexible foams for upholstery and insulation
Unsaturated polyesters	Partitions, toaster sides, satellite dishes

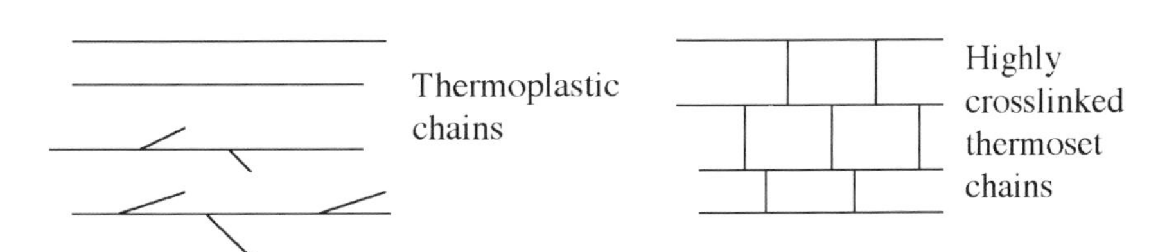

Figure 2.4 Arrangements of thermoplastic and thermoset molecular chains

Different properties arise from the divergent forms of the molecules. When thermoplastic compounds are injection moulded, the structure of their molecules is not changed. Melting only increases the mobility of the molecules. In contrast once thermoset plastics have crosslinked they can no longer be melted. Thermoset materials can thus be deformed only elastically and, in contrast, cannot be deformed plastically. The rigidity of the thermoset materials depends on how narrow or wide the spaces between the network of crosslinking of the molecules are. A material with a wide gap can be deformed elastically to a large degree. Such thermosets are also known as elastomers.

2.5.3 Elastomers

According to the American Society for Testing and Materials (ASTM), a polymeric elastomer is defined as 'a material that, at room temperature, can be stretched repeatedly to at least twice its original length, and upon immediate release of the stretch, will return with force to its approximate original length'. Which means, to put it in layman's terms, it's rubbery!

2.5.3.1 Thermoplastic Elastomers (TPEs)

For a long time elastomer materials were limited to thermoset type materials with permanent crosslinks, such as polyisoprene (natural rubber). However one major drawback with these materials was that they could not be recycled, which led to the development of thermoplastic elastomers (TPEs). One common example is styrene-butadiene-styrene (SBS). Since styrene and butadiene do not mix, the styrene breaks up when the SBS is heated allowing the material to be re-melted. TPEs offer considerable advantages over thermoset type elastomers:

- Lower part costs due to a lower material density
- Faster cycle times
- Recyclable scrap and parts
- Can be coloured.

The market for TPEs has also benefited from advanced processing techniques such as over-moulding (see Chapter 10). The ability to use TPEs either alone or in combination with other materials has enabled them to be used in applications such as

- Gaskets, seals and stoppers
- Shoe soles/heels, wrist straps
- Handles, grips, knobs.

The properties of thermoplastic elastomers that have made them so commercially successful are their lower modulus and flexibility. The ability to recover from stress and return to their original shape make them suitable for applications such as sealing rings etc. Of course thermosetting rubbers materials have long been available and it is these materials that TPES are replacing.

Whilst newer material types have emerged in recent years, primary TPE types can be categorized into two generic classes, block copolymers (styrenics, copolyesters, polyurethanes and polyamides) or thermoplastic/elastomer blends and alloys (thermoplastic polyolefins and thermoplastic vulcanisates). These TPE types are known as two-phase systems as essentially, a hard thermoplastic phase is coupled mechanically or chemically with a soft elastomer phase. The result is a TPE that has the combined properties of the two phases.

Thermoplastic elastomers usage has increased significantly in recent years and is expected to continue to rise. Worldwide, consumption was estimated at 1,400,000 metric tonnes/year in 2000. Some examples of common thermoplastic elastomers are shown in Table 2.5. As well as adhesion considerations, the suitability of a TPE will also depend on properties such as its hardness and compression ratio. Hardness can be defined as the resistance of the material to indentation and is usually measured on a durometer using a Shore Hardness scale. TPEs tend to be rated on a Shore A scale, the softest materials ranging from around 3 Shore A and the hardest to 95 Shore A. Out of interest, thermoplastic materials are measured on a different, Shore D scale. This gives an indicator of the differences in properties. TPE material grades are available with a range of hardness levels, however it must be considered that hardness is also a function of the thickness of the material and the substrate beneath. In multi-shot applications, the thinner the layer of elastomer on the substrate the harder it will feel due to the effect of the harder substrate beneath it.

Table 2.5 Examples of TPEs and substrates

Type	Elastomer description	Subgroup
TPE-O	Polyolefin blends	PP/EPDM PP/EPDM crosslinked
TPE-V	Polyolefin alloys	Various
TPE-S	Styrene	SEBS SBS SEBS/PPE
TPE-A	Polyamide	PA 12 based PA 6 based
TPE-E	Polyester	Polyesterester Polyetherester
TPE-U	Polyurethane	Polyester urethane Polyether ester urethane Polyether urethane
SEBS = styrene-ethylene/butylene-styrene terblock copolymer		

The compression set (CS) is often specified for sealing type applications and is a measure of the deformation after compression for a specified deformation, time and temperature. The usual test method is ASTM D395 or ISO 815. Therefore a compression set of 100% would represent a material that did not recover and a compression set of 0%, a material that behaved completely elastically and a value of 30% means the material regained 70% of its original thickness.

$$CS(\%) = [(h_i - h_f)/(h_i - h_c)]*100$$

where h_i=initial height, h_f=final height, h_c=height during compression.

The usual method calls for a 25% deformation, the larger the deformation, the lower the compression set will be.

2.5.4 The Formulation of Plastics

It was stated earlier in the chapter that plastics are mixtures of polymers and other materials. There are many additives commercially available that can be mixed with polymers. For example, glass or carbon fibre reinforcement gives increased strength. Flame retardants can be added for flame resistance. They can be coloured with pigments for aesthetic or technical purposes, or they can be made more heat and light resistant by the use of stabiliser additives. Those are just a few examples of the numerous possibilities. In addition, the levels of each of these additives can also be varied. It is easy to see how so many grades of plastic have come to exist. It is also hopefully now apparent how they have been able to compete so successfully with other materials as diverse as glass, metal and wood. A list of some of the more common plastic additives is given in Table 2.6.

Plastics, which incorporate reinforcing materials such as glass fibre or clay, are called polymer composites. Composites can be made from both thermoplastic and thermoset polymers, with materials such as glass and carbon fibres, which increase the tensile stiffness and strength of the resultant materials. Polymer composites are widely used for a variety of applications where their high strength to weight ratio can be utilised. The automotive industry especially has taken advantage of these properties to make weight savings over metal components to improve fuel efficiency.

Table 2.6 Common additives for plastics	
Additive	**Purpose**
Reinforcement, e.g., glass fibre	Increased strength and stiffness
Extenders, e.g., calcium carbonate	Cost reduction: much cheaper than polymer
Conductive fillers, e.g., aluminium powder	Improved thermal and electrical conductivity
Flame retardant	Increased fire resistance
Light stabilisers	Increased resistance to degradation from daylight
Heat stabilisers	Increased resistance to degradation from elevated temperature exposure
Pigments	Give colour, improved aesthetic properties
Plasticisers	Improved flow properties and increased flexibility
Coupling agents	Improved interface bonding between the polymer and a reinforcing agent
Foaming agents	Lightness and stiffness
Mould release agents	Processing aids
Antistatic additives	Prevent dust build up on consumer items

2.5.5 The Binding Structure of Plastics

What is the relevance of the binding structure of polymers?

The characteristics of most interest to the injection moulder are those which are needed in order to melt the injection moulding compound and inject it into the mould. It is important to know the temperature at which the individual plastic becomes molten, since different plastics react differently to heating depending on the holding power of the chains of molecules. We also need to know the decomposition temperature, where the molecule chains break apart into smaller chains or into the initial materials. This process is described as gasification, and indicates severe damage to the material.

Generally, the forces which keep the molecule in the material compound are less powerful than the chemical bonds present between atoms. Therefore materials with chemical bonds such as crosslinked thermosets are much harder to break down than materials such as thermoplastics. Polymers with a polar component, (e.g., CN, NH, CO), generate greater intermolecular forces than those without. Polar groups, however, cause higher dielectric losses, because of their dipole character.

2.6 The Effects of Processing on Thermoplastics

In order to understand what happens to polymers during injection moulding, the effects of processing on thermoplastic materials will now be considered. Processing in its simplest form is the act of melting, forming and solidifying the melt. During this process three classes of properties need to be considered.

- Deformation processes, which enable the product to be formed
- Heat and heat transfer, which enable the polymer to melt, flow and then solidify
- Changes which occur to the structure and properties as a result of processing.

In order to study deformation processes it is necessary to consider rheology.

2.6.1 Rheology

Rheology is the study of deformation and flow. Consider the following flows:

1. Squeezing toothpaste through a tube
2. Tipping honey from a spoon
3. Pouring water into a glass.

The toothpaste, honey and water all have different viscosities. The toothpaste is the most viscous, the water the least. Viscosity represents the resistance to flow. In viscous flow, a material continues to deform as long as a stress is applied. To put it more simply, in the case of the toothpaste tube, we can squeeze out the toothpaste for as long as we squeeze the tube.

Consider the toothpaste again, when it is squeezed and then released it does not return to the original shape. A rubber ball however, does return to its original shape when released and likewise polymer melts will try to return to their original shape when stress is removed. So polymers combine both viscous and elastic properties and demonstrate what is called a viscoelastic response to stress.

The interaction between viscosity and elasticity frequently determines the success of any processing operation. Processing conditions must take into account not only how the polymers flow in their molten state but also how they change as the temperature goes up and down and the polymers melt and solidify. During injection moulding the switch from mould packing to holding pressure causes the material viscosity to drastically change. It has been vividly described 'as changing from flowing like honey to flowing like modelling clay'. This explains why consistently switching at the same point can make the difference between producing consistent parts as opposed to producing rejects.

Polymer melts have viscosities in the range 2-300 Pa.S (for comparison the viscosity of water is 10^{-1} Pa.S). Two very common materials in commercial use are LDPE and PA.

As LDPE is heated, it turns from a solid, to a viscous gummy liquid and then to a mobile fluid as temperature is increased. PA on the other hand turns quite suddenly from a solid to a low viscosity (watery) fluid. It is important to understand that polymers may react differently to the heat and stress applied to them, in order to find the most suitable processing conditions for moulding.

2.6.2 Heat and Heat Transfer

A thermoplastic cannot flow when it is in its solid state. To enable it to flow it needs to be heated to either

- Above its melting point (T_m), if it is a crystalline material

or

- To its glass transition temperature (T_g), if it is an amorphous thermoplastic.

Some melting points of common polymers are shown in Table 2.7. The melt temperature is also sometimes referred to as a flow temperature or a crystalline melting point for amorphous and crystalline polymers respectively. Amorphous materials do not have a clearly defined melting point like crystalline materials, but tend to soften and flow instead. At the melt temperature, which may be a fairly sharp point or a range, the polymer behaves as a viscous liquid and flows as the molecules are free to slide over one another. Highly crosslinked polymers do not have a melting temperature. The chemical bonds between the chain form a rigid structure.

Table 2.7 Melting points of common polymers

Polymer	**Tm (°C)**
Polyethylene (PE)	135
Polypropylene (PP)	170
Polyethylene terephthalate (PET)	245
Polyamide 6 (PA6)	233

From these examples, a wide range of melting points can be clearly seen. Once above the melting point the viscosity will decrease as the temperature increases. The rate of this change varies, according to the particular type of material. Eventually the polymers will reach a point where they become thermally unstable and start to degrade. Generally, this shows itself as a discolouration of the melt turning yellow or brown. On excessive heating, polymers burn or decompose.

2.6.3 Physical and Chemical Change

As well as the effects of heat on the polymer, the effects of shear also need to be considered. Shear is a type of force that involves exerting stress across the surface of the polymer, while in effect, the base of the polymer is untouched. This action is shown in Figure 2.5. The action of shear, like temperature, causes a decrease in viscosity. This is due to the shearing action causing mechanical damage and breakage to the polymer chains. At very high levels of shear this leads to degradation.

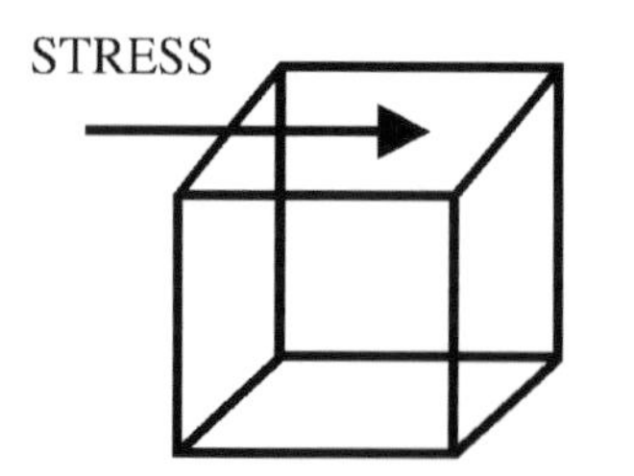

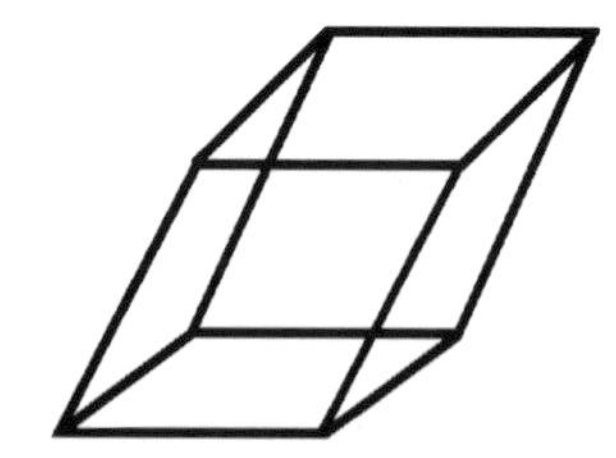

Figure 2.5 Shearing flow

With this in mind, good thermal stability is a requirement for most polymer processing operations, as the combined action of heat and shear can produce degradation. The effect of this is chain breakage. The length of the chain is related to the molecular weight, the molecular weight to the properties. Therefore any change in chain length will affect:

- Molecular weight and viscosity
- Tensile and impact properties.

Combined heat and shear can also result in:

- Change in colour (yellow or brown discolouration)
- Reduced fibre length in glass filled materials.

Consideration must be given to the residence time of the material in the processing machine as a long residence time may cause thermal deterioration especially in heat sensitive materials such as PVC.

2.6.4 Fountain Flow

Before injection the material in the injection unit is plasticated. Once the material is molten and homogeneous it is ready for injection. In order for injection to occur the pressure in the nozzle must be high enough to overcome the flow resistance of the melt. The flow is then pressure driven as the material moves from a region of high pressure to a region of low pressure. The larger the pressure and pressure gradient the faster the material will flow. In order to mould with an increased flow length, an increased entrance pressure is required. The flow length of a material under any particular processing conditions and wall thickness will depend on the thermal properties and shear properties of the material.

In the process of injection moulding, polymers come under both shear stress and to a lesser extent elongational stresses. Elongational flow occurs when the melt encounters abrupt dimensional changes such as in the gate region however it is shear flow that dominates the mould filling process. This is because injection moulding tools fill by a 'fountain flow' effect. This is shown in Figure 2.6. The first material to enter the mould is under very little pressure as it flows into the channel, but as it proceeds towards the extremities of the mould, it is progressively cooled. The material that follows is hotter and begins to pressurise that already there. As the mould has filled, a skin of solid plastic has formed on the wall. This causes a reduction in the effective gap, through which later molten material passes and so increases the shear rate. The extra shear stress causes orientation of the molecular structure and the melt is generally too viscous to allow for relaxation of all the stress. As a result it is often found that shrinkage in the direction of flow during filling is larger than that in the circumferential direction, this is due to orientation effects. Material will be orientated in the direction of flow and will shrink more in the direction of flow than perpendicular to it. It is not always possible to prevent this from occurring but it can be alleviated by differential cooling of the mould tool.

The parameters of mould filling, packing and cooling have been shown to be more important to final part quality than the plastication stage. All of these processes will be described in greater detail in Chapter 8.

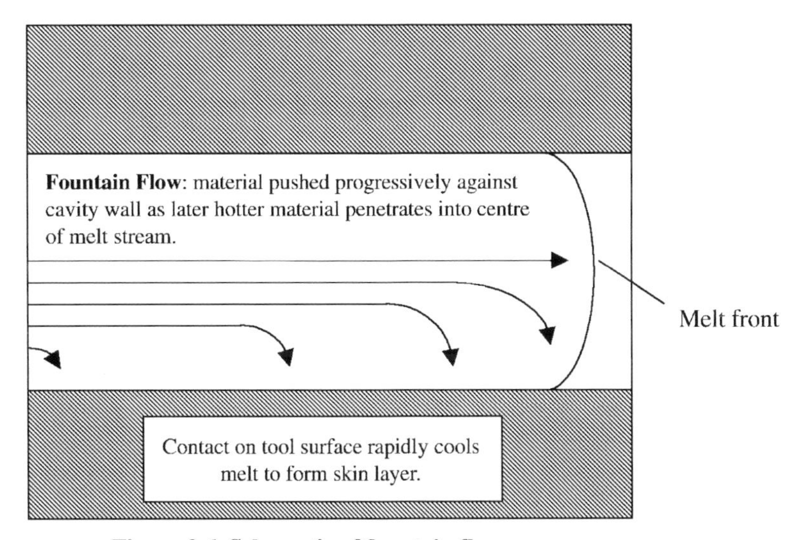

Figure 2.6 Schematic of fountain flow

2.7 Conclusion

This chapter has presented a general introduction to polymers and plastics and some of their important properties that relate to injection moulding. There are more detailed chapters on thermoplastics (Chapter 6) and thermosets (Chapter 7) later in this book.

3 Injection Moulding Machinery

In Chapter 1, the basic parts of an injection moulding machine were identified as:

- Injection unit
- Machine base with hydraulics
- Control unit and control cabinet
- Clamping unit with mould

This chapter begins with a breakdown of the purpose and the components of an injection unit. A glossary of common injection moulding machinery terms can be seen below.

Clamping unit	The part of the machine in which the mould is mounted. It provides both force and movement
Daylight	The maximum distance between the stationary and moving platens
Ejector	Ejects the moulded parts from the tool. The mechanism is activated through the clamping unit. The ejection force may be hydraulic, pneumatic or mechanical.
Fixed platen (stationary platen)	The inner face of the clamping unit that does not move during mould closing. It contains mould mounting holes.
(Full) hydraulic clamp	A clamping unit triggered by a hydraulic cylinder which is directly connected to the moving platen. The mould is opened and closed by application of direct hydraulic fluid pressure. This also provides the clamping force required to keep the mould closed during injection
Injection cylinder	Part of the injection unit which includes the screw, nozzle, hopper, heaters.
Injection screw	Transports, mixes, plasticises and injects the material from the hopper to the machine nozzle.
Injection unit	The part of the machine which feeds, melts and injects the material into the mould.
Mould (tool)	Contains the cavity to inject into. Consists of two halves, a stationary and moving half which attach onto the platens. Halves are linked via the tie bars.
Moving platen	The inner face of the clamping unit that moves during mould closing. The moving half of the mould is bolted onto this platen. This platen also contains the ejection mechanism and mould mounting holes.
Plunger unit	Injects and plasticises material by heating it in a chamber between the mould and a plunger. The heating is done by conduction. The plunger forces material into the chamber which in turn forces the melt already there into the mould.
Tie bars (tie rods, tie beams)	Bars which link and align the stationary and moving platens together.
Toggle clamp	A clamping mechanism with a toggle directly connected to the moving platen.
Reciprocating screw	A screw which both plasticises, and injects material

3.1 Injection Units

The first aim of the plastication stage is to produce a homogeneous melt for the next stage where the material enters the mould. A second important function of the injection unit is the actual injection into the mould. Here, it is important that injection speeds are reproducible as slight changes can cause variations in the end product. An injection unit is shown in Figure 3.1.

Figure 3.1 An injection unit

There are two different injection units available.

3.2 Piston (Plunger) Injection Unit

The design of this unit was based on a method used to mould rubber. Material is metered by a dosage device and transported through the heated plasticising cylinder until it is in front of the plunger at the correct temperature. The material residence time in the cylinder is very long, making this method unsuitable for heat sensitive material such as rigid PVC and thermosets. This type of machine is not widely used as it was replaced by the reciprocating screw piston injection unit.

3.3 Reciprocating Screw Piston Injection Unit

This is the most common type of unit and will be the basis for further discussion in this section. Thermoplastics as well as thermosets and classical elastomers can be processed with screw piston injection units with the process cycle as described in Chapter 1.

In the screw piston injection unit, the material is plasticised and dosed simultaneously as previously described in Section 1.4.2. The design of a plasticising screw has several advantages over a piston type mainly in the ability to produce a homogeneous melt as a result of mixing. The flow of the material is also improved as shear from the screw lowers the viscosity of the material. The long residence times present in the piston type machines are eliminated allowing heat sensitive materials such as PVC to be processed. The screw is also easier to purge and less prone to degradation or material hang-ups. Important parameters for these screws are:

- The diameter of the screw and the ratio of the diameter to the length (L/D ratio see Section 3.5)
- Shot capacity
- Plasticising rate
- The amount of material that can be melted at any one time (plasticising capacity).

3.3.1 Shot Capacity

The shot capacity is the full amount as a weight or volume of material injected during moulding from the screw. This is usually given as a shot capacity for polystyrene, and will vary with material. The shot size is the amount of material required to fully fill a moulding tool.

3.3.2 Plasticising Capacity

This is the maximum rate at which the injection unit can deliver polymer melt. In extrusion this is a continuous process. However, it should be remembered that injection is an intermittent process, therefore the plasticising rate will be lower. To calculate the melting rate consideration should be given to the overall cycle time.

The effectiveness of plastication depends on the shot size, cylinder capacity, screw design, screw speed and heater band power. It will also vary from material to material.

3.3.3 The Feeding Hopper

Material is placed in the hopper prior to plastication. It must be designed to avoid material bridging in the throat and to let gravity feed the material. Material hold up spots must be avoided. Additives, especially when they are different weights to the polymer, may tend to accumulate and be fed inconsistently. This can lead to variations in melt quality. The hopper may contain magnets to collect metal contamination, which must be prevented from entering the feed system. It may also contain grids to prevent large particulates from entering and blocking the feeding system, especially important if using recyclate materials. Keeping the feed system cool is also important, if material begins to melt in the throat of the feeding system it may stick to the sides of the throat and in extreme cases block the machine completely.

3.3.4 The Injection Cylinder

Once the material has passed through the hopper, it enters the injection barrel. The barrel will consist of a number of separately controlled heating zones as can be seen in Figure 3.2. The heat is generated from conduction of heat from the cylinder and also the heat generated by the shearing action of the screw on the material feedstock. Polymers are not particularly good conductors of heat; therefore the polymer thickness in any section of the screw tends to be kept low. The amount of shear is material dependent, mainly viscosity related and controlled by the machine screw back and back pressure.

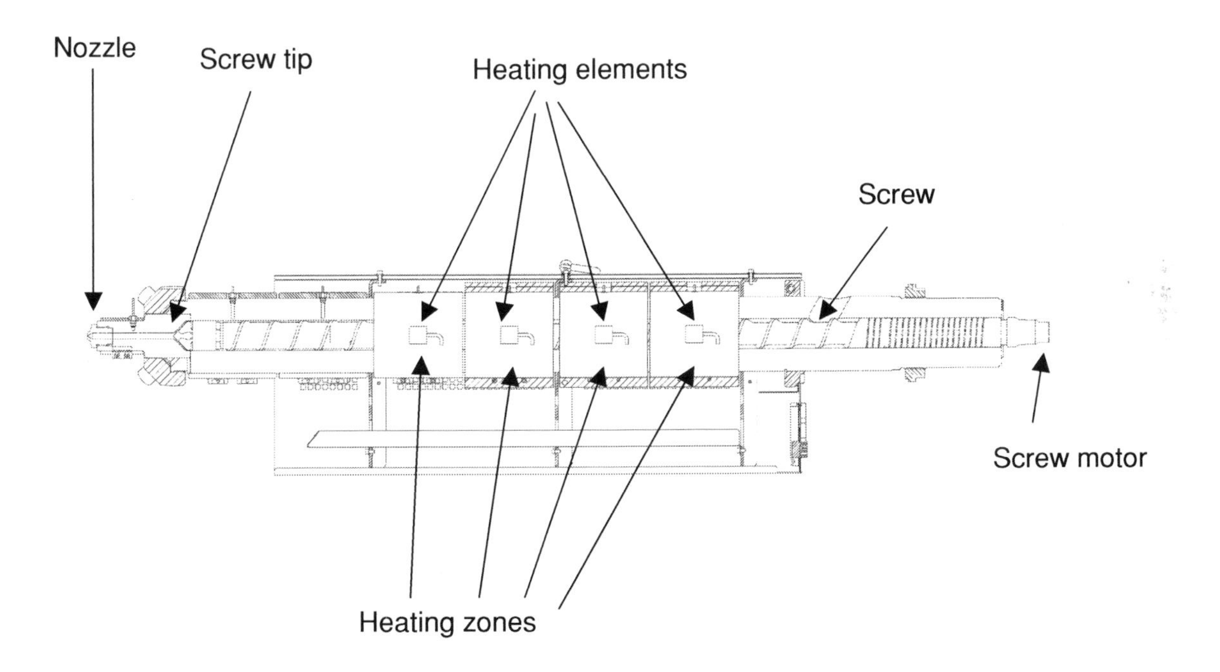

Figure 3.2 Schematic of an injection unit

3.4 Selection of the Injection Unit

To select the correct configuration of injection unit for a particular material or material range, consideration must be given to the following factors:

- The selection of the correct nozzle type
  - flat?
  - radius?
  - shut-off nozzle?
- Screw and cylinder outfitting must be adapted to suit the raw material being processed.
  - The geometry of the screw must be correct (Figure 3.3 shows some of the variety available)
  - The screw should be of a suitable corrosion resistance, (e.g., nitrided, Arbid (boronising heat treated), bimetallic)
- The dosage volume should be around 20-80% of the total shot capacity. It should be ensured that the residence time is not excessive for sensitive materials.
- Dosage capacity and melt capacity must be sufficient for processing requirements
- Whether a mixing cylinder (screw) is required for use with colour additives
- Whether a hydraulic accumulator is necessary for moulding with long lines of flow.

Figure 3.3 Various injection screws

Some typical designs for injection cylinders are shown in Figure 3.4.

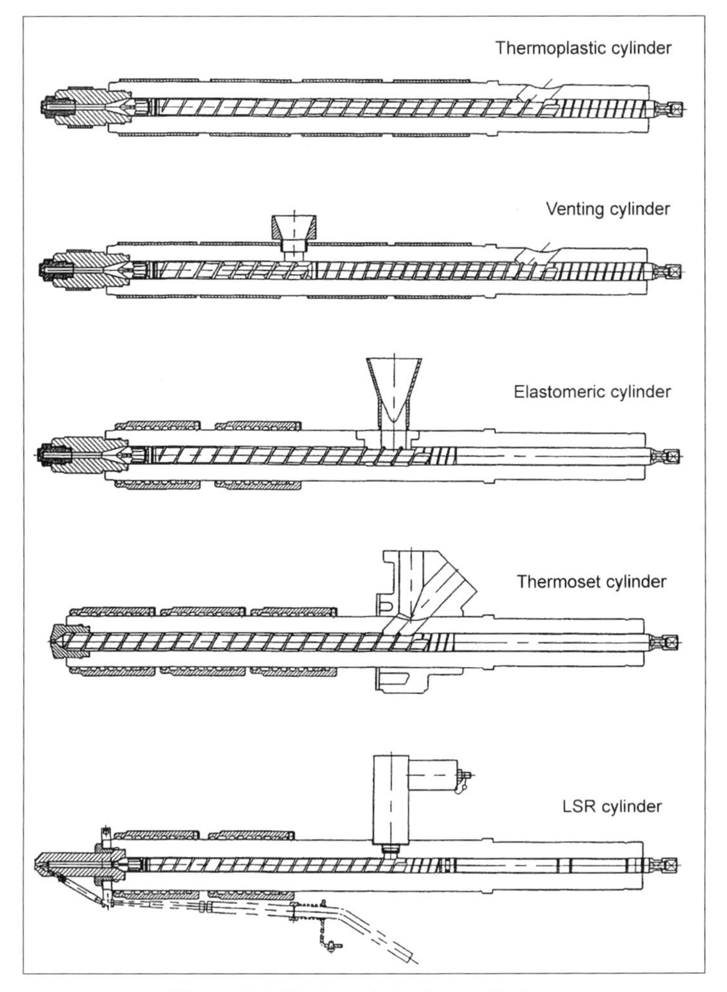

Figure 3.4 Various injection cylinders

The design of the screw is very important for plastication. Some typical screw configurations will be shown in the next section.

3.5 L/D Ratio

Perhaps the most important specification for the screw is the L/D ratio. This is the ratio of the length of the screw (the flighted length) to its diameter. For thermoplastics this ratio will be a minimum of about 20:1. With thermosets, elastomers and LSR the L/D is approximately 14:1. For extended plasticising screws the L/D may be 24:1. This is usually used for thermoplastics with colour additives, especially with PP and PE. This enables better mixing of the colourant. On fast cycle machines with increased capacity, a higher L/D ratio may also be beneficial. This ratio has also been found to improve melt performance with compact disc manufacturing machines. More information on mixing can be found in the next section.

From Figure 3.5 it can be seen that the design of the screw along its length is not constant but varies. Generally screws are designed with three distinct regions: (1) a feed section, (2) a melting transition region (2-compression region) and (3) a metering section. The size of these regions will vary dependent upon the characteristics of the material it was designed for. A passive screw has low shear and compression and acts only to melt and transfer the melt from one end to the other. A standard screw has regions of shear mixing and compression.

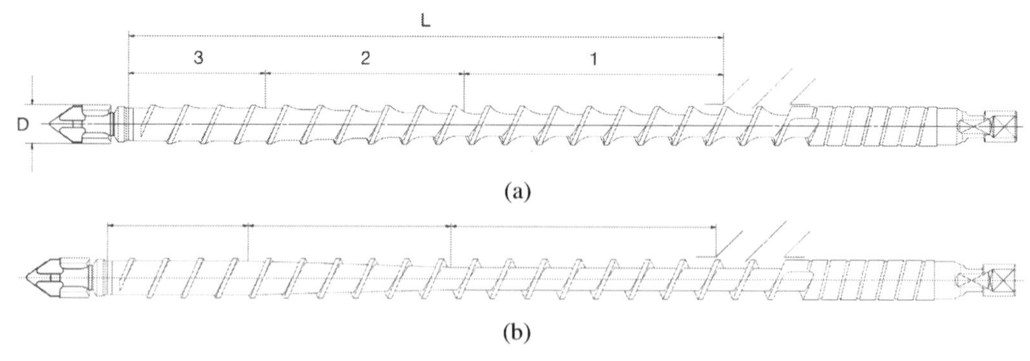

Figure 3.5 Sections of an injection screw, (a) passive screw, (b) standard screw

The compression ratio can be defined as the ratio of the flight depth in the feed section to that in the metering section. As an example, a screw for a polyamide material may have a compression ratio of 3:1 on a 20L/D screw with a 30 mm diameter.

A standard compression ratio is roughly 2:1 for thermoplastics, for sensitive thermoplastics such as PVC or with metal/ceramic powder, this may drop to 1.6:1. Compressionless screws with a compression ratio of 1 are used for processing materials such as elastomer and LSR. The design of screws for various materials is illustrated in Figure 3.6.

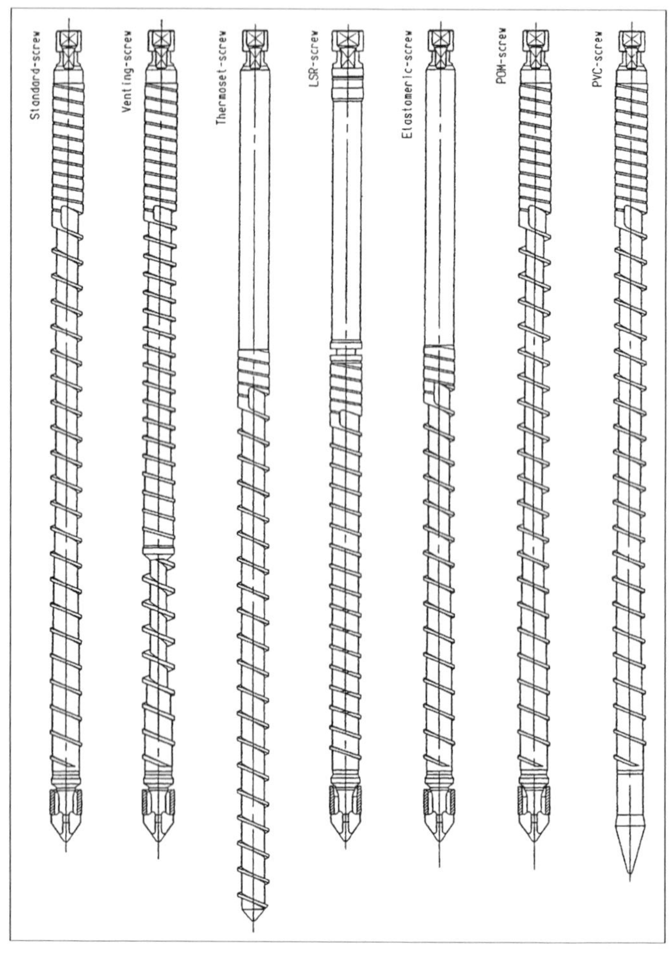

Figure 3.6 Injection moulding screws
(Note: POM is polyoxymethylene)

3.5.1 Mixing Screws for Additives and Fast Cycling Machines

In cases where an increased mixing ability in the screw is required, there are two potential solutions: to add mixing elements or to use an extended screw design.

3.5.1.1 Mixing Elements

Where higher levels of mixing are desired, mixing elements help create a more homogeneous melt as far as the distribution of temperature and filler is concerned.

The quality of the melt depends above all on two criteria:

- The retention time of the melt in the mixing element
- The number of melt partitions.

The loss of pressure in mixing elements should be kept as small as possible, as every loss of pressure leads to a reduction in plasticising quality. A mixing element must therefore be designed as follows:

- The free volume of the raw material in the mixing element must be as large as possible
- The melt deflection must be compulsory
- The flow cross-section must be as wide as possible.

One solution is to use Rhombus mixers. They are similar to slot disk mixing elements that are used in extrusion techniques. However, the Rhombus elements also have a certain conveying effect. An important advantage of these mixing elements is the compulsory deflection of the melt. On its way to the screw tip the melt flows through several cut-offs in the slot disks. Mixing elements with one or more channels and flights with cut-offs do not have compulsory deflection, as a certain percentage of the melt flows along the screw channels without being impeded; so there is only a limited mixing effect.

3.5.1.2 Extended Plasticising Cylinder

An alternative is to use an extended plasticising cylinder. This is suitable in the following circumstances:

1. If a high melting capacity is required (raw material with a high specific thermal capacity, i.e., PE, PP, PA)
   - short cycle times
   - medium cycle times and high product weight (dosage volume over 50%)
   - moulding compounds that tend to shearing (→ increase in the heating energy/overall heat energy ratio)
2. If masterbatch is used for colouring purposes and the homogeneity of the colours is of great importance.
3. If the temperature level during the moulding process must be reduced or peak temperatures must be eliminated.
   - reduction of cycle time (shorter cooling time)
   - avoiding decomposition of heat sensitive raw material

The combination of cylinder and screw depends on the specific case it is needed for, for example:

- Case 1. Extended screw with standard geometry (the separate zones of the screw are extended according to a constant factor)
- Case 2. Standard screw + Rhombus mixing element + metering zone extension

The cylinder is heated with an additional heater band or extended heater bands with increased heating power. This is shown in Figure 3.7.

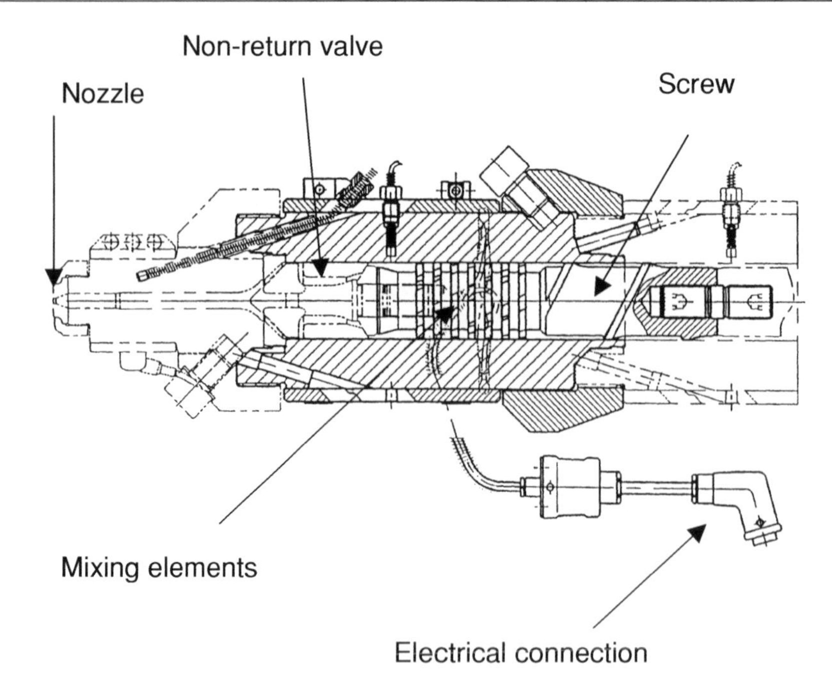

Figure 3.7 Cylinder extension module with extended plastication, additional heating circuit and mixing port

3.5.2 Non-Return Valve

Many materials require the use of a valve with a check ring to be fitted to the end of the screw to prevent backflow. They also help to ensure that a constant cavity pressure is maintained. The most important design consideration is that they should avoid flow restrictions or hold up of the melt flow. Non-return valves are more prone to wear than other components, so it must be ensured that suitably toughened materials are used in manufacture. Various designs of non-return valves are shown in Figure 3.8.

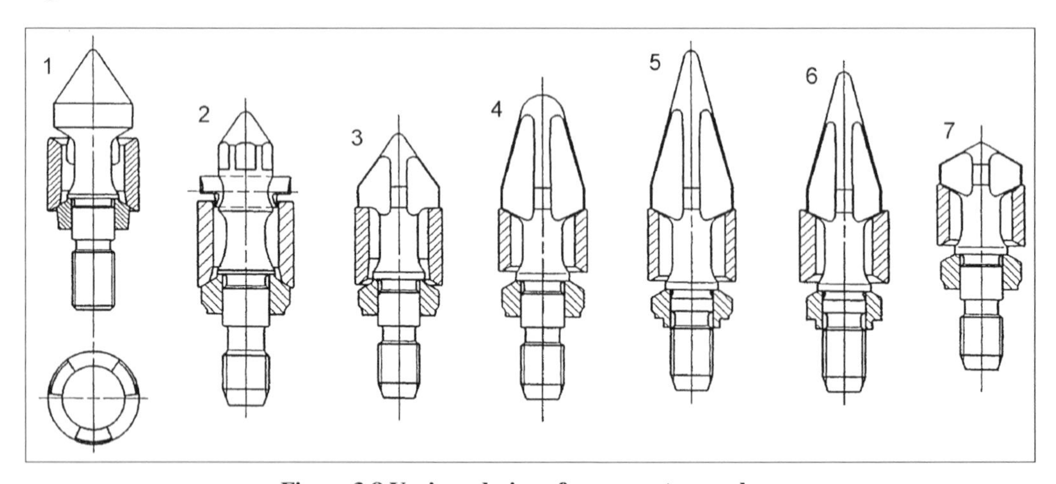

Figure 3.8 Various designs for non-return valves

1. For polyester paste
2. Low abrasion type with M-C-V series machines
3. Standard type on M-C-V series machines
4. Standard type on ARBURG ALLROUNDER
5. Old type only on hydronical unit from 35 mm upwards
6. Old type with smaller flow cross-section
7. Spare part tips for combination of new screw with old nozzle

3.6 General Information on Wear and Tear

There are several causes of wear and tear which include:

- Incorrect adjustment of process parameters, e.g., back pressure too high, dosage speed too high, no dosage delay, incorrect adjustment and setting of temperatures for plasticating cylinder and feed yoke.
- Wear and tear generated by raw materials, e.g., mechanical wear caused by fibre glass, glass spheres, stone powder, metallic powder, ceramic powder
- Chemical corrosion, e.g., with additives, flame resistant materials, materials containing fluorides

There are several ways to determine any mechanical wear and/or chemical corrosion. Mechanical wear can be seen by grooves and surface abrasion in one direction. Chemical corrosion can leave large and small holes in different areas and directions as well as surface deposits. To determine the wear of the screw and barrel consideration must be given to the original heat treatment method used. With nitride and Arbid methods the surface thickness can be measured. With bimetal outfitting the surface can be examined. Generally if the heat treated surface has worn down this signifies to the manufacturer that the units have worn out.

Measurements of the thickness can be carried out according to the barrel and screw dimensions:
- by measurement with a micrometer or callipers, etc., e.g., manufacturer's dimensions
- barrel 30 mm (30.00 mm - 30.05 mm)
- screw 30 mm (29.85 mm - 30.15 mm)
- bushing 30 mm (29.97 mm ± 0.01 mm)

- Desired gap between barrel and bushing
  - Standard 0.04 mm
  - LSR 0.02 mm

There are ways to minimise wear and tear on the injection units by use of proper process parameter adjustment, the correct selection of barrel and screw for the job and suitable heat treatment outfitting (see hot combinations and applications)

3.7 Unit Hardening Treatment

Examples of the resistance of various plasticising screw and cylinder outfitting options are shown in Table 3.1. Table 3.2 lists the classifications of the various treatments and Table 3.3 gives recommendations for the type of outfitting for some common materials. From these three tables it should be relatively straightforward to ensure that the correct equipment is being used and wear and tear can be minimised.

Table 3.1 Plasticising screw and cylinder outfitting

Metal treatment type	Plasticising		Resistant to abrasion	Resistant to corrosion	Thickness (mm)	Hardness	Temperature range up to
	Cylinder	Screw					
Nitride	+	+			0.30	800 V	250 °C
Arbid	+	+	*	*	0.12	1800 V	450 °C
BMA	+		**	*	2.5-3.0	63 HRC	450 °C
BMK	+		*	**	2.5-3.0	56-58 HRC	450 °C
VSX	+		*	*	3.00	54 HRC	450 °C
PH		+	**	*	2.00	57-58 HRC	450 °C
PK		+	*	**	2.00	54-56 HRC	450 °C

Note:
* resistant
** very resistant
generally all nozzles are in nitride execution with different steel types
non-return valves in nitride execution on screws with Arbid
Nitride = low abrasion
Arbid = abrasion proof
BMA = ultra low abrasion

Table 3.2 Abrasion classifications	
Abrasion classification	**Cylinder/screw**
Low abrasion	Nitride/Nitride
Abrasion proof	Arbid/Arbid BMA/Arbid[1] VSX/Arbid[2]
Ultra low abrasion	BMA/PH
Corrosion proof	BMK/PK
[1] only V-series from 370 V up, 470/520 °C and aggregate 50 [2] only 270 V	

Table 3.3 Recommendations for use	
Abrasion classification	**Material types**
Low abrasion	Thermoplastics without corrosive or abrasive components (LDPE, HDPE, PP, PS, CA)
Abrasion proof	Thermoplastics with corrosive and/or abrasive components (ABS, PVC, POM, PBT, PET, PA); low to middle content of filling material (e.g., glass, French chalk, TiO_2)
Ultra low abrasion	Thermoplastics with strongly abrasive and corrosive components, high temperature material (PC, PBT, PET, PA, POM, PE with high share of glass, PES, PSU, PEI, PEEK, PPS, PPA)
Corrosion proof	Thermoplastics with highly corrosive components (CPVC, FEP, ETFE, PFA, E/CTFE, PVDF), also flame protected types

Notes:

CA	=	cellulose acetate	PEI	=	polyetherimide
CPVC	=	chlorinated PVC	PES	=	polyether sulfone
ETFE	=	ethylene-tetrafluoroethylene	PFA	=	perfluoro (alkoxyalkane) copolymer
FEP	=	tetrafluoroethylene hexafluoropropylene	PPA	=	polyphthalamide
			PPS	=	polyphenylene sulfide
PBT	=	polybutylene terephthalate	PSU	=	polysulfone
PEEK	=	polyetherether ketone	PVDF	=	polyvinylidene fluoride

3.8 The Nozzle

The nozzle provides the connection between the injection cylinder and the mould tool. Its job is to convey the material with minimal pressure or heat change. There are two common types of nozzle.

- Open nozzle
- Nozzle shut-off valve.

Various designs of nozzle are shown in Figures 3.9-3.11. The nozzle itself may not necessarily be made of just one piece. A tip that is screwed into the nozzle body can be replaced or repaired. This may need to be an abrasion and corrosion resistant tool steel tip. For optimum flow conditions, there must be no material hang-ups. Therefore the flow must be streamlined. The land length is generally kept to a minimum dictated by the strength requirements. For high pressure applications an increased flange diameter may be required.

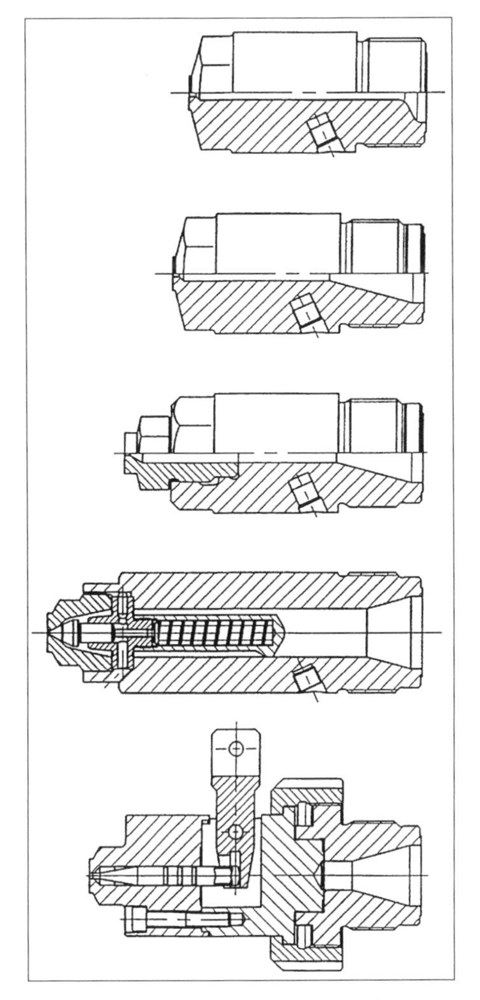

Figure 3.9 Various nozzle designs

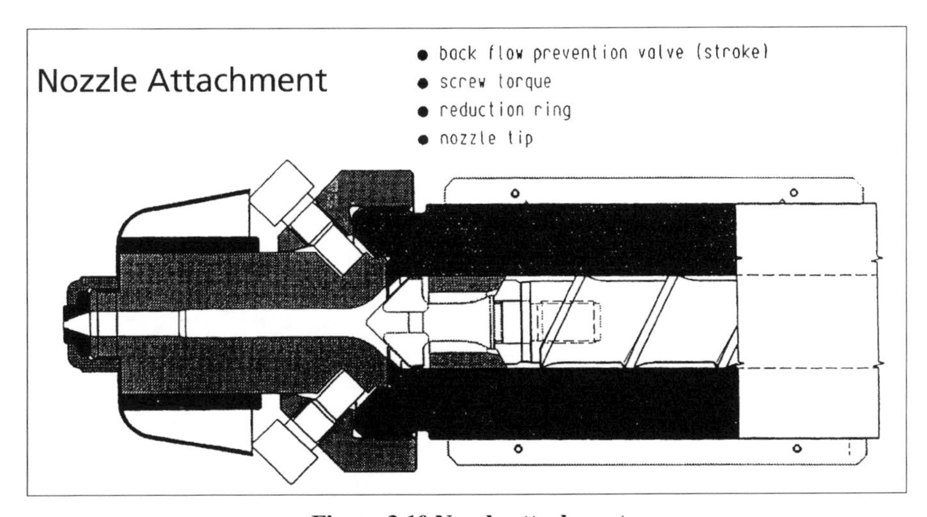

Figure 3.10 Nozzle attachment

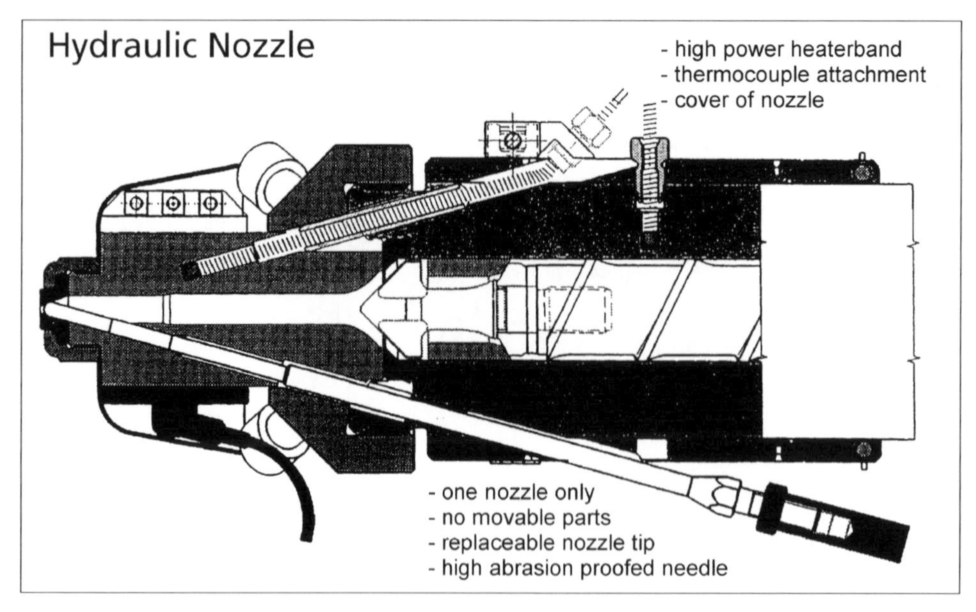

Figure 3.11 Hydraulic nozzle replaceable

It is essential that the temperature of the nozzle be controlled. The location of the heating and control is equally important else material degradation or premature material freezing (cold slugs) may occur. A thermocouple can be used close to the gate and heater. Thermocouples may extend into the melt rather than measuring the temperature of the nozzle, some measurement locations are shown in Figure 3.12.

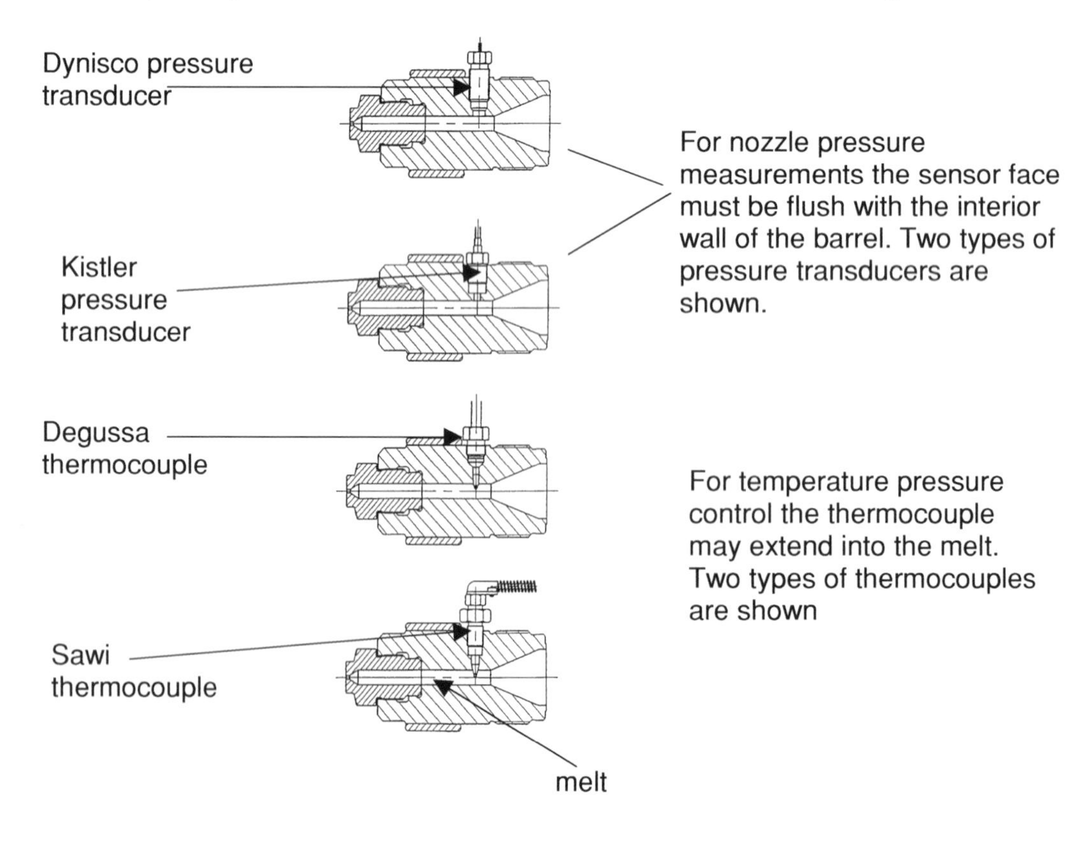

Figure 3.12 Measurement locations in nozzle body

3.9 Clamping Units

The clamping units of injection machines are described and rated separately to the injection unit. The clamping units are required to enable mounting and holding of the two mould halves. They must also provide sufficient clamping force during injection and cooling to enable effective moulding. The mould halves must also open and close accurately and smoothly to enable part injection and begin the next process cycle. Injection machines can be run by hydraulics, a hydraulic and toggle combination or by electrical power. The clamping units on injection moulding machines use hydraulic force.

Figure 3.13 shows a clamping unit. The stationary platen is attached to the machine with four tie rods connecting it to the movable platen. Figure 3.14 shows a direct hydraulic clamping system The clamp ram moves the moving platen until it reaches the stationary platen and the pressure begins to build up. The ejectors are fitted onto the moving platen and can be activated once the tool is opened and the moving platen retracted.

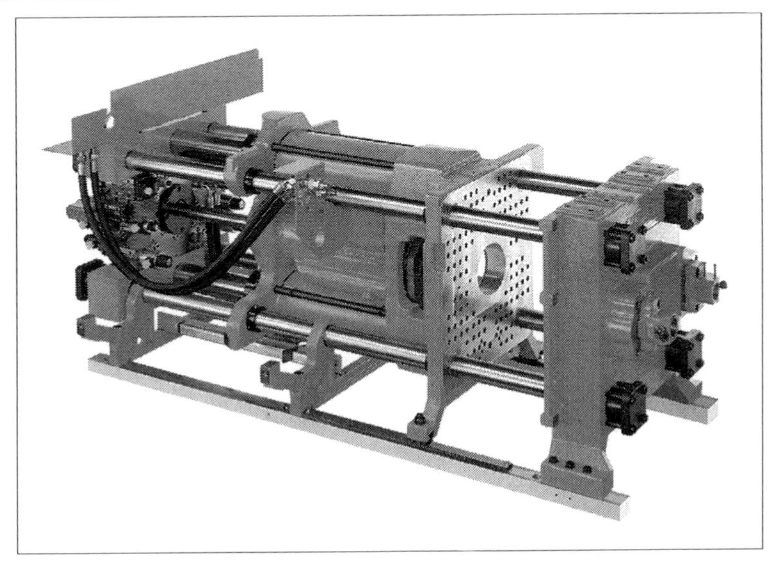

Figure 3.13 A clamping unit

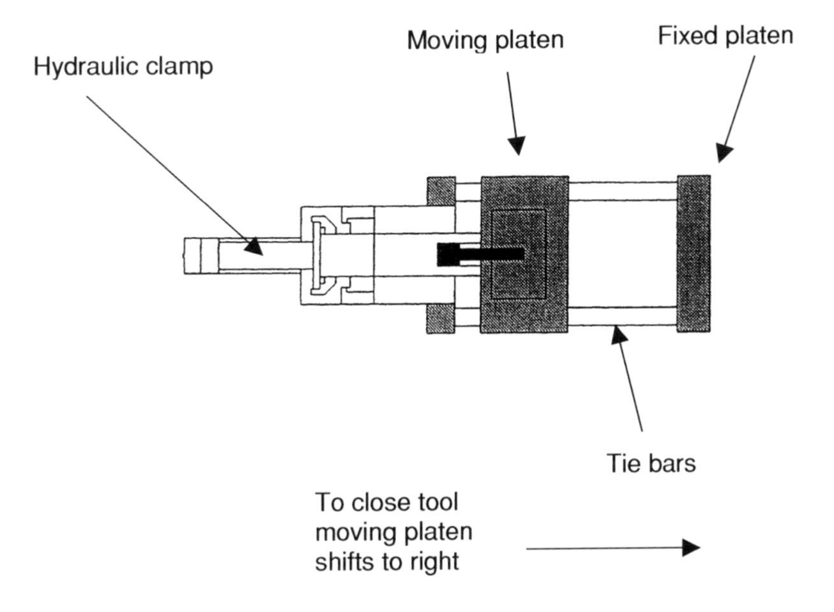

Figure 3.14 Direct hydraulic clamping unit

A toggle type clamping unit is shown in Figure 3.15. This design enables the force to be amplified. It is basically two metal bars attached by a pivot. One end is attached to the stationary plate, the other to the movable platen. When open it forms a distinctive 'V' configuration and when closed the bars form a straight line. The advantage in this design is that a much smaller force from the hydraulic cylinder is required, the size of this advantage varies but can be as high as 50:1. A further advantage of the toggle machines is that once extended the toggles remain there until retracted making them self-locking. The hydraulic system on the other hand requires the application of constant line pressure. The disadvantages are that it is more difficult to control the speed and force of a toggle mechanism. It must also be adjusted for different depths of mould tool to ensure that the toggle is fully extended.

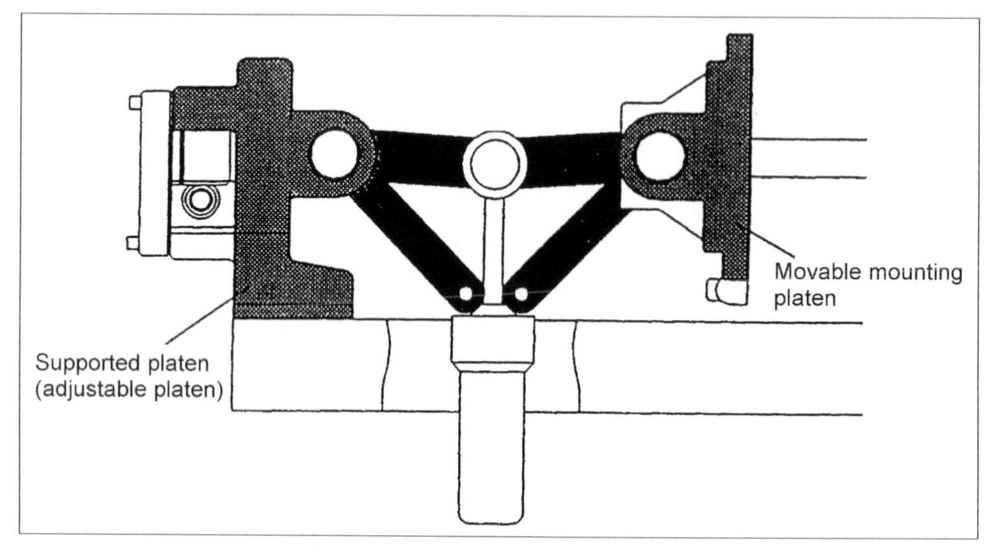

Figure 3.15 Toggle type clamping unit

3.9.1 Differential Piston System

For the opening and closing movements of the mould tool a minimal volume of oil is required. The oil volume required from the pump results from the differential surface and stroke of the piston (approx. 7% of the clamping cylinder volume). The rest of the oil flows through the borings in the main piston as a result of the piston stroke. The pressure cycle is shown in Figures 3.16-3.18.

When opening with increased opening force (high pressure opening) the control piston closes the main piston. The main piston and opening piston now open the mould with 50 bar pressure.

When the injection unit is in the vertical position and 'braking' is selected, the borings in the main piston are closed shortly before the end of the opening motion, This ensures an exact positioning of the movable platens in their lower-most end position. Sinkage of the movable platern on an idle machine is also avoided.

3.9.2 Mould Weights

Each machine will have a maximum permitted mould weight for the movable mould halves. These values should not be exceeded for any reason as production problems and premature wear would be the result. Examples of some commercially available machines from the manufacturer ARBURG are shown in Table 3.4, the type of supports are shown in Figure 3.19.

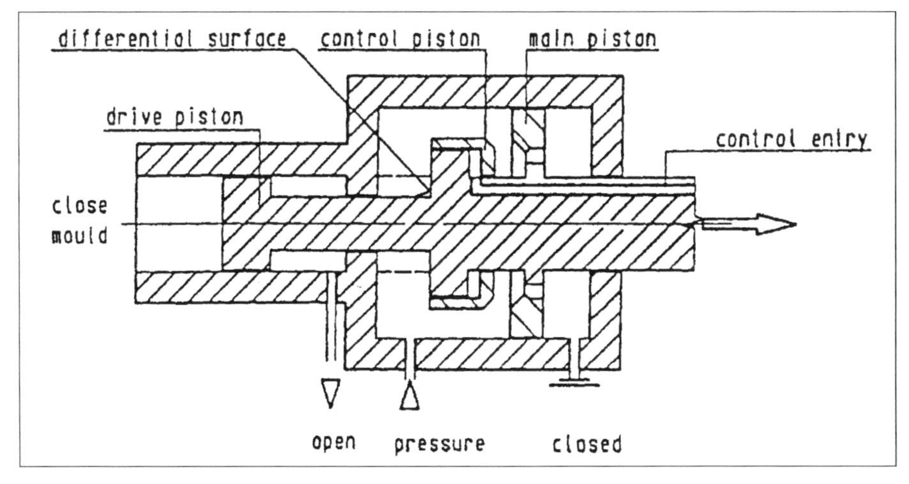

Figure 3.16 Mould closing

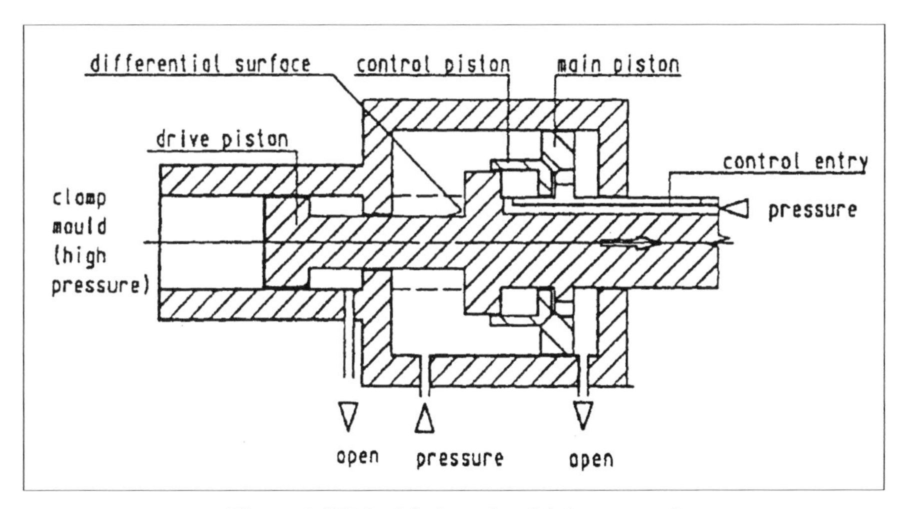

Figure 3.17 Mould clamping (high pressure)

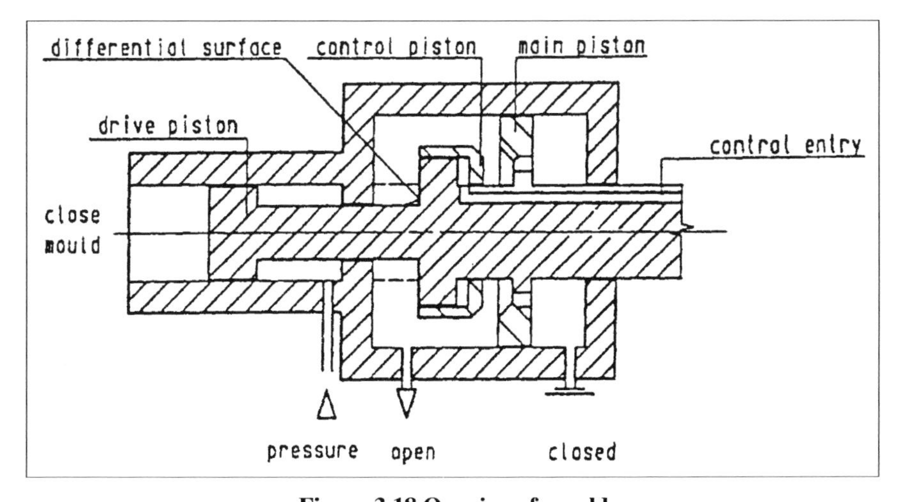

Figure 3.18 Opening of mould

Table 3.4 Maximum weights of movable mould half (kg)								
	Without support		**Vertical support**		**Vertical and horizontal support**		**Support of the tie bars**	
Machine type Allrounder®	S	75	-		-			
221 K/M	S	160	-		-			
305 K	S	50	-		-			
170 CMD	S	100	-		-			
220 H/M	-		-		-			
220 S	S	160	-		-			
270 H/M	-		-		S	400		
270 C/V	-		-		-		S	400
270 S	S	250	-		-			
320 H/M	-		-		S	400		
320 C/V	S	250	-		O	600	S	400
370 M	-		-		S	600		
370 C/V	S	250	-		O	600	-	
420 M	-		-		S	600		
420 C/V	-		S	1000	-			
470/520 M	-		-		S	1000		
470/520 C/V								
S = standard outfitting, O = option, - = not available								

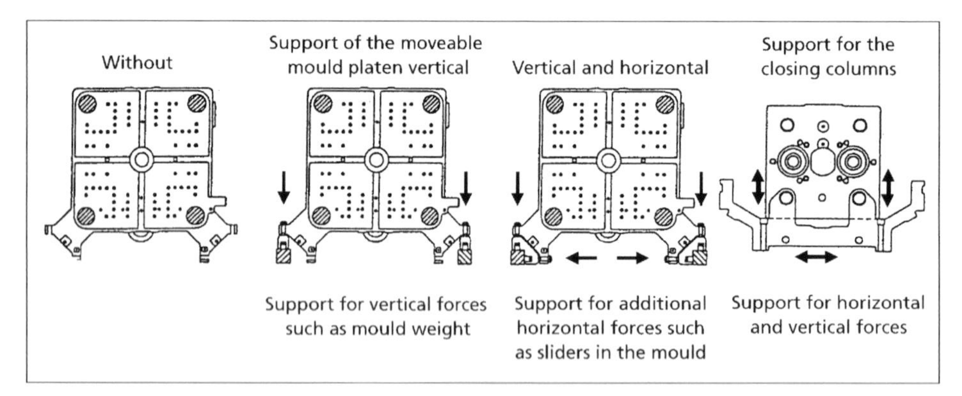

Figure 3.19 Support of the movable platen

3.10 Selection of the Clamping Unit

To select a clamping unit consideration must be given to the following factors:

- Injection mould size
  - dimensions
  - centering
  - ejector coupling
  - mould weight

- Projected surface area (amorphous and semi-crystalline materials require different clamp forces)
  - table
  - amorphous cm^2 x 4 = kN
  - semi-crystalline cm^2 x 6-7 = kN

- Select higher clamping force with split fallower moulds
  - with thin walled parts and thermoset materials

The next section will consider issues relating to the mould clamping force requirements.

3.11 Mould Clamping Force

3.11.1 Mould Clamping Force Level

The mould clamping force must be set high enough to prevent flash. This is caused by the swelling of the mould under the compound force during initial injection resulting in the compound coming out of the mould cavity. The mould clamping force required depends on the size of the moulded component surface projected onto the parting plane, and on the internal mould pressure. When the machine is started up with a new mould, the internal mould pressure required is not yet known. Naturally, from experience, it may be possible to extrapolate a figure, which depends on the type of plastic and the component format. However, practice has continually shown that it is initially quite sufficient to use a guide value of 2.5-5 kN/cm^2 to calculate a projected moulded component surface. For settings during the start-up period, the calculations are performed using the upper limiting value, even with easy-flowing compounds, such as PS or PE, for which 2.5-3.5 kN/cm^2 is adequate throughout.

Example

Size of component moulded surface projected onto the parting plane = 52 cm. Guide value for mould clamping force = 52 x 5 = 260 kN. In the course of the test injections, the mould clamping force can be reduced until just above the point at which leaks begin. The mould clamping force should be as low as possible, i.e., only as high as necessary to save energy on the toggle clamping units in order to keep wear as low as possible. Another reason to keep the mould clamping force as low as possible is the air extraction from the mould cavity required during initial injection.

3.11.2 Mould Clamping Force and Mould Rigidity

It is found in most cases, that flash (see Chapter 9 for a description) is not caused by inadequate mould clamping force or locking pressure, but by an insufficient rigidity of the mould in the areas concerned.

The mould plates carrying the moulding nest can be thoroughly bent in an area of the central bores and, above all, in the area of the ejector system, under the effect of the internal mould pressure. Thus, the mould plates must have sufficiently high flexural strength, if the formation of scratches is to be avoided. This danger is especially great for the mould plate on the ejector side. So, if possible, additional support columns, with an oversize of 0.03-0.05 mm, by comparison with the external supports, should be incorporated.

If the moulding nests are directly built into the mould plate, there is a further measure available to reduce the tendency to leak and/or scratch formation. Only one sealing edge is left standing round the moulding cavities, and the remaining surface of the parting plane is free ground to approximately 0.05 mm. This means that the sealing pressure increases, for the same mould clamping pressure, as a result of a reduction in the sealing faces. The free grinding also favours air extraction from the mould cavity during filling. Rigid moulds thus require mould clamping forces which are not so high; that means the machine can be improved and energy saved in this way – under certain circumstances, it can even mean using a smaller machine.

3.11.3 Setting Mould Closing and Clamping Force

The mould closing force is the force required to physically shut the tool. The clamping force is the force required to hold the tool shut when the material is injected and is much higher than the closing force.

3.11.3.1 Mould Closing Force

Before setting mould closing force the mould protection functions need to be set to prevent any damage from the moulds hitting one another. A low value is generally sufficient, i.e., 5.0 kN.

If the value is not sufficient to close the mould the force should be increased little by little, however, first check if there is anything hindering smooth closing. The desired closing force can then be set, e.g., 20 kN.

3.11.3.2 Clamping Force

The size of the clamping force depends on the moulded part's projected area in the parting line. The projected area is measured in cm^2 or square inches (1 sq in = 6.451 cm^2). Table 3.5 relates the specific clamping force needed per cm^2 to various plastic materials.

Table 3.5 Clamping force requirements for cm^2

Material	PS, SB, SAN, ABS, CA, CAB, PVC soft	PMMA, PPO mod., PC, PSU/PES, PVC hard
	PE soft (LDPE), PE rigid (HDPE)	PP, PA, POM, PET, PBT, PPS
Specific clamping force	2.5 to 5.0 kN/cm^2 (16.1 to 32.3 kN/sq in)	5.0 to 7.0 kN/cm^2 (32.3 to 45.2 kN/sq in)
Notes: SAN = styrene-acrylonitrile copolymer, PPO = polyphenylene oxide, CAB = cellulose acetate butyrate		

To calculate, multiply:

projected area x specific clamping force = clamping force

Example:

Material: SAN
Specific clamping force: 2.5 to 5 kN/cm^2, average value 3.5 kN/cm^2
Moulded part's projected area: 90 cm^2
Calculation of clamping force: 3.5 x 90 = 315 (kN)

3.11.3.3 Second Clamping Force

At the end of the holding pressure phase the clamping force can be reduced. Generally this will be half of the value calculated previously.

The second clamping force should last to the end of the holding pressure phase.

3.12 Data for Mould Closing Force

Table 3.6 gives recommended closing force values (clamping force).

The clamping force (closing force) to be set can be calculated from the specific closing force and the projected injected part surface in the parting line.

Example

Specific closing force = 3.5 kN/cm^2 = average value for SAN from the table above.

A projected injection part surface of
20 cm^2 results in a closing force of = 3.5 x 20 cm^2 = 70 kN
90 cm^2 results in a closing force of = 3.5 x 90 cm^2 = 315 kN.

The closing forces (clamping forces) in the fill phase are higher with thin walled parts.

$$\text{Closing force (clamping force)} = z \times A_{proj} \times pw$$

where z = number of cavities, A_{proj} = projected parting line part surface and pw = mould cavity pressure

According to Figure 3.20, pw is dependent on the flow path/wall thickness relationship (L/s) and the wall thickness (s).

Table 3.6 Recommended closing force values (clamping force)		
Injection material	**Recommended values for specific closing force** (kN/cm^2)	**Practical values for the mould cavity pressure, based on the recommended closing force values** (bar)
PS	1.5-3.5	150-350
SB	2.0-4.0	200-400
SAN	2.5-4.5	250-450
ABS	3.0-5.5	300-550
PVC rigid	2.5-5.0	250-500
PVC soft	1.5-3.0	150-300
CA	2.5-4.5	250-450
CP	2.0-3.5	200-350
PMMA	3.5-5.5	350-550
PPE mod. (PPO mod.)	3.5-6.0	350-600
PC	3.5-6.5	350-650
PSU/PES	4.0-6.0	400-600
PEI	3.5-6.5	350-650
PE soft	2.0-6.0	200-600
PP	3.0-6.5	300-650
PA 4.6	4.5-7.5	450-750
PA6	3.5-5.5	350-550
PA6.6	4.5 –7.5	450-750
PA6.10	3.0-5.0	300-500
PA11, PA12	3.5-5.5	350-550
PA amorph	3.5-4.5	350-450
POM	5.5-10.5	550-1050
PET	4.5-7.5	450-750
PBT	4.0-7.0	400-700
PPS	3.5-6.5	350-650
FEP	3.0-6.0	300-500
PAA	3.0-7.0	300-700
LCP	3.0-8.0	300-800
Thermosets/Elastomers		
Classic elastomers	2.0-6.0	200-600
PE-U	2.0-4.5	200-450
LSR	0.8-2.5	80-250
Notes: PMMA = polymethyl methacrylate, PPE = polyphenylene ether, PAA = polyacrylic acid, CP = cellulose propionate, PE-U = thermoplastic elastomer (urethane grades)		

Example 1 Coffee Cup

Flowpath (L) = 80 mm
Wall thickness (s) = 0.5 mm
L/s = 80/0.5 = 160
Cavity count (z) = 1
A_{proj} = 51 cm^2
pw (according to Figure 3.20) = 650 bar ≈ 6500 N/cm^2
Closing (clamping) force = 1 x 51 x 6500 = 331500 = 332 kN

Example 2 *Marmalade Container*

Flow path (L)	=	85 mm
Wall thickness (s)	=	0.5 mm
L/s	=	85/0.5 = 170
Cavity count (z)	=	1
A_{proj}	=	71 cm^2
pw (according to Figure 3.20)	=	750 bar ≈ 7500 N/cm^2
Closing (clamping) force	=	1 x 71 x 7500 = 532500 N = 533 kN

Example 3 *Flower Pot*

Flow path (L)	=	120 mm
Wall thickness (s)	=	0.55 mm
L/s	=	120/0.55 = 218
Cavity count (z)	=	1
A_{proj}	=	113 cm^2
pw (according to Figure 3.20)	=	800 bar = 8000 N/cm^2
Closing (clamping) force	=	1 x 113 x 8000 = 904000 N = 904 kN

It should be noted in connection with Figure 3.20 that higher cavity pressures occur with high viscosity materials.

IMPORTANT: The relationship between flow path and wall thickness should not exceed 250/1 with high viscosity materials; multiple injections are required if this is the case.

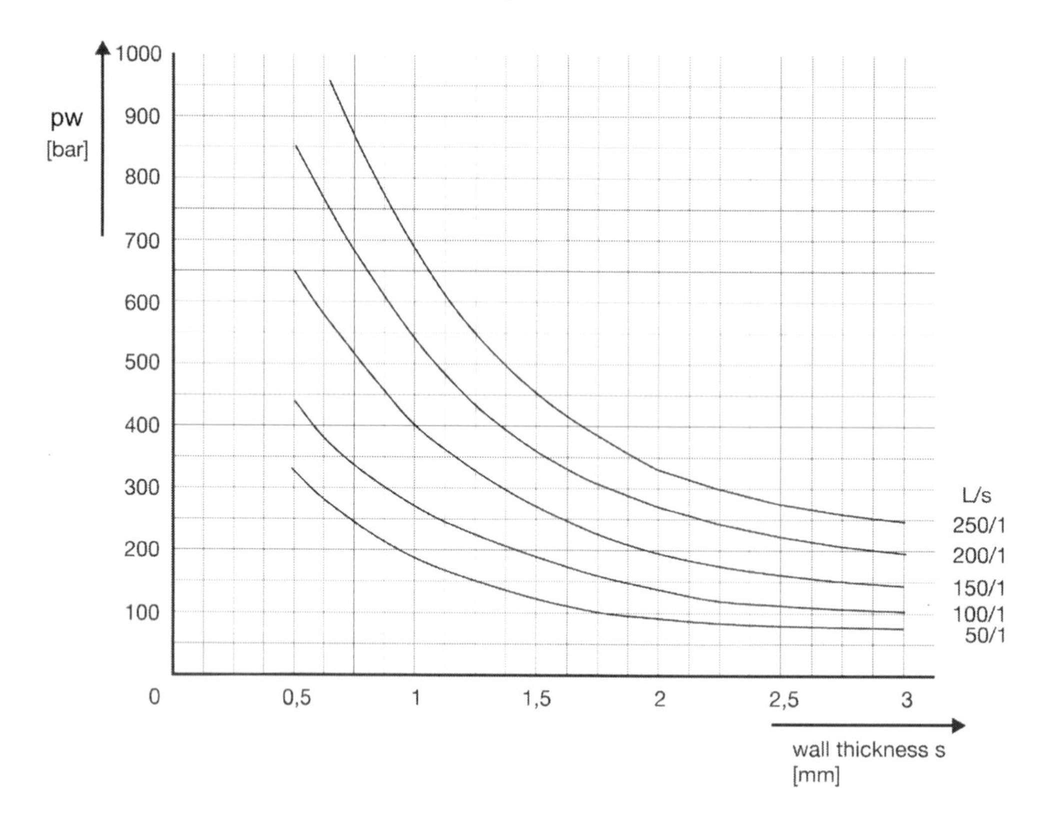

Figure 3.20 Mould cavity pressure and wall thickness

3.13 Other Considerations

The type of clamp and the clamping force are the main specifications of a clamping unit. However, there are other design features which also need consideration. These are:

- Maximum daylight
- Space between tie bars
- Clamp stroke
- Clamp speed
- Knockout stroke.

The first two of these points concern the size of the mould tool that can be physically fitted into the machine. The daylight is the maximum distance between the stationary and moving platen, obviously there must be room in the tool to open and eject the part. The tie-bars restrict the length and width of a tool that can be placed on the platen as they restrict entry. With this in mind, there are tie-bar-less machines on the market which become particularly pertinent with multi-shot moulding (see Chapter 10).

The clamp stroke is the maximum distance the moving platen will move. The clamp speed is the maximum speed at which it will do this. This can affect overall cycle time and therefore efficiency and so can be of great importance.

The maximum knockout stroke determines the movement available for ejection.

3.14 International Standard for the Designation of Injection Moulding Machines

Now that the basic machinery components have been discussed, the way in which injection moulding machines are designated can now be considered. Injection moulding machines are classified by the machinery that they are composed of as set out in the International Standard.

$$\text{1st Designation} = \text{clamping force in kN (1 kN} \approx 0.1\text{ Mp} \approx 0.11\text{ ton)}$$

$$\text{2nd Designation} = \text{shot capacity of inj. unit} = \frac{\text{max. inj. volume [cm}^3] \times \text{max. inj. pressure [bar]}}{1000}$$

Machines with two injection units bear the designation for the second injection unit next to that of the first unit, e.g., 700-210/210

The shot capacity of the injection unit does not depend on the size of the plasticising cylinder used:

Injection volume = screw-piston surface x stroke

Injection pressure = injection force / screw-piston surface

That is:

$$\frac{\text{Inj. volume} \times \text{inj. pressure}}{1000} = \frac{\text{screw - piston surface} \times \text{stroke} \times \text{inj. force}}{1000 \text{ screw - piston surface}} = \frac{\text{stroke} \times \text{inj. force}}{1000}$$

Examples of machine designation

1. ALLROUNDER® 305:
   Clamping force – 700 kN
   max. inj. volume with 35 0 screw – 140 cm^3
   max. inj. pressure with 35 0 screw – 1500 bar
   Shot capacity of injection unit = $\frac{140 \times 1500}{1000} = 210$
   International Standard Designation = 700-210

2. Two-colour ALLROUNDER® 270
   Clamping force – 500 kN
   Shot capacity of first injection unit – 90
   Shot capacity of second Injection unit – 210
   International Designation 500-90/210.

Important!

Until January 1983 the international standard designation was as follows:
1st Designation – shot capacity of injection unit (now 2nd)
2nd Designation – clamping force in Mp (now clamping force in kN as 1st designation)

4 Injection Mould Tooling Basics

4.1 Types of Moulds

An injection mould tool has two major purposes:

- It is the cavity into which the molten plastic is injected
- The surface of the tool acts as a heat exchanger (as the injected material solidifies with contact)

Injection mould designs differ depending on the type of material and component being moulded. Mould tool design and component design are equally important considerations for success. Component design is beyond the scope of this book but the various tooling, gating, temperature control and ejection systems that make up the mould tool will be considered here. After parts are injection moulded they must be ejected. A variety of mechanisms can be employed such as ejector pins, sleeves, plates or rings.

The design standard for injection mould tools is the two-plate design.

4.1.1 Two-Plate Mould

This is the simplest mould design. Mould cavities are formed in one plate only with the stationary half of the mould blank. A central sprue bushing can be placed into the stationary half of the mould or it is possible to have a direct runner system to a multi-impression mould. The moving half of the mould contains the ejection mechanism. This is illustrated in Figure 4.1.

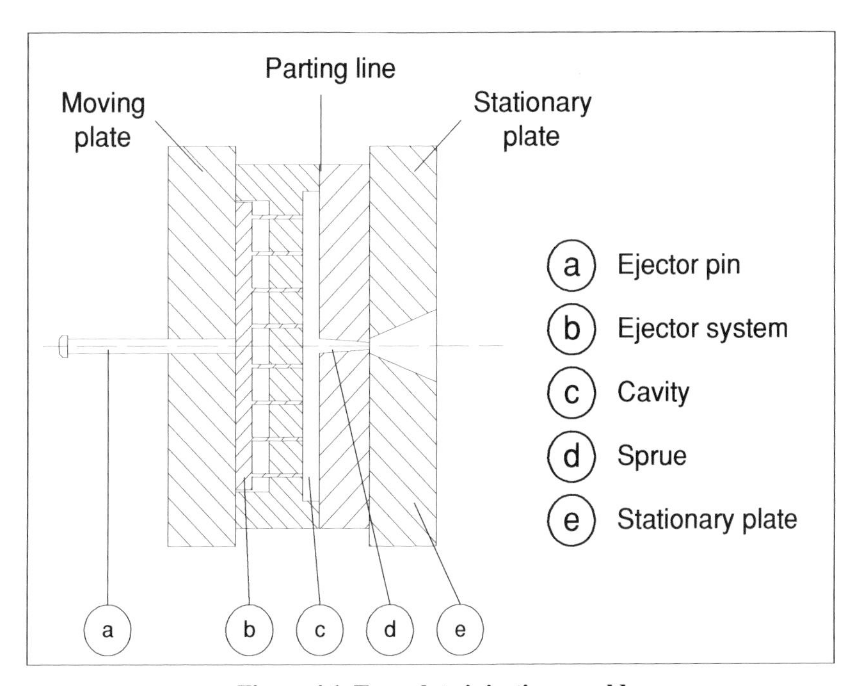

Figure 4.1 Two-plate injection mould

4.1.2 Stripper Mould

A stripper mould is very similar to the standard two-plate mould except for the ejection system. This design has a stripper plate for ejection, whereas the standard one has pins or sleeve as the ejectors. This is illustrated in Figure 4.2. The advantage of a stripper plate is the increased surface area for ejection that it offers.

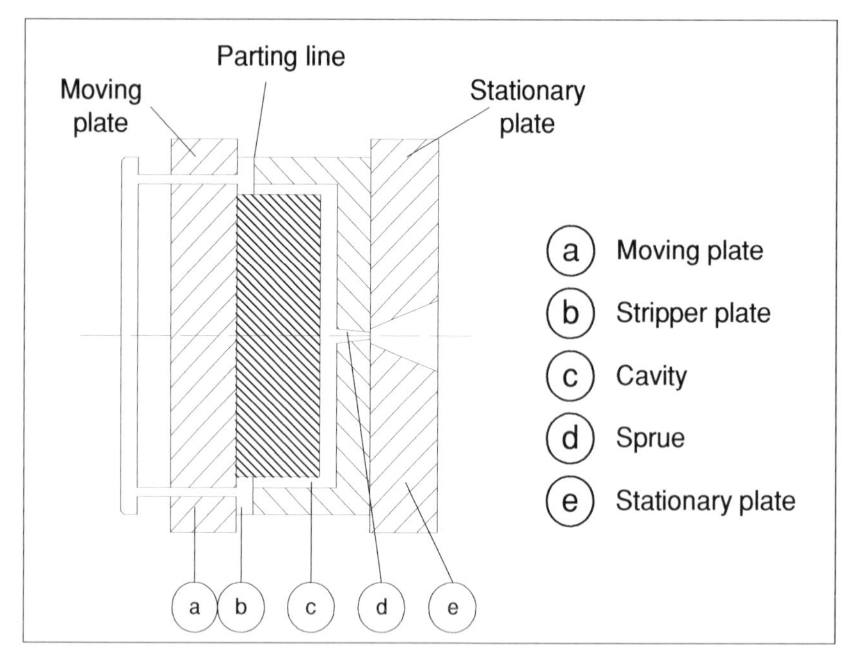

Figure 4.2 Mould with stripper plate

4.1.3 Slide Mould

Like the two previous designs, this is also a two-plate mould. However it has slides and cam pins for additional lateral movement as shown in Figure 4.3. This type of design is suitable for producing parts with undercuts or external threads.

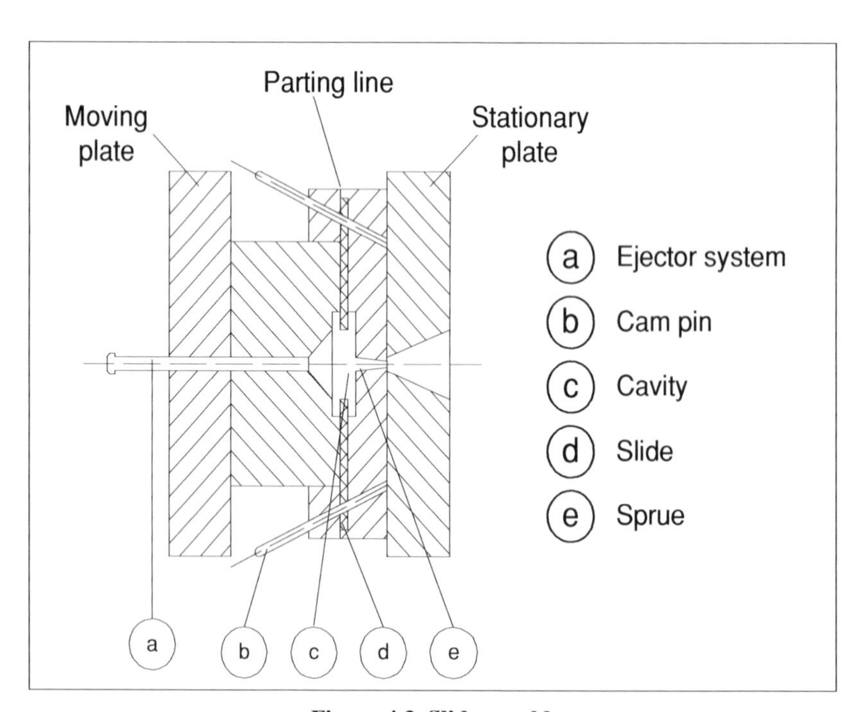

Figure 4.3 Slide mould

4.1.4 Three-Plate Mould

These are normally used when multi-cavities are involved and semi- or fully-automatic working is required. This type of mould, as its name suggests, has an extra plate (see Figure 4.4). This plate usually continues the gate on one of its sides with the complete runner system, preferably trapezoidal. The opposite side of the plate carries part of the mould form (usually the female part).

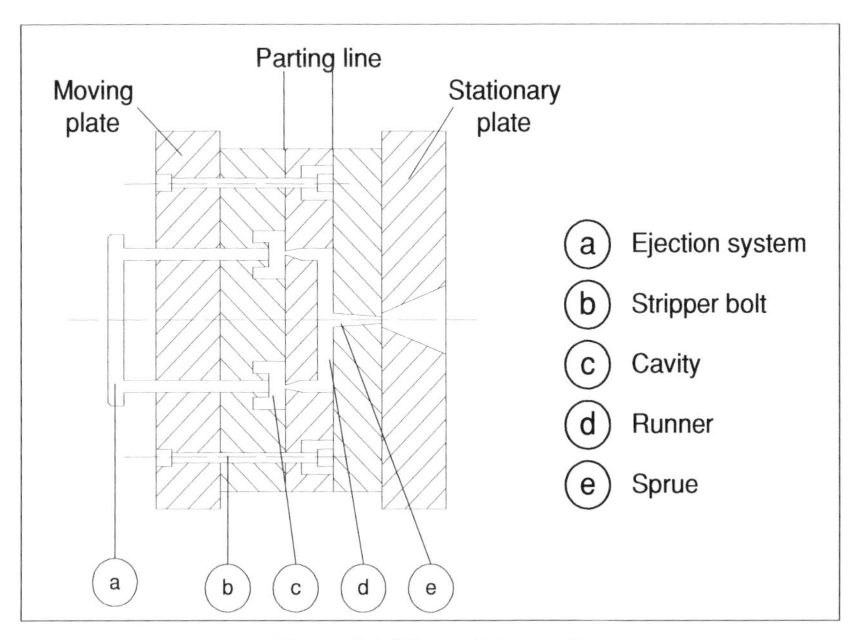

Figure 4.4 Three-plate mould

When the mould opens the plate is separated by means of a delayed action mechanism (e.g., chains or length bolts), so breaking the restricted gate. The mouldings are then ejected from one daylight and the sprue and runner system are ejected from the other.

Successful ejection of mouldings relies on clean separation of the moulding and gate at the parting line. Figure 4.5 illustrates the sequence. With this method of tooling, restricted gates of the correct design must be used.

Multi-plate moulds are usually more expensive than two-plate moulds and can be slower in production if an operator has to remove the sprue and runner system when the mould is open. This can usually be avoided by providing automatic ejection of sprue and runner. The distance travelled by the plates is governed by the length of the chain or the length of the bolts used to separate them.

4.2 The Feed System

4.2.1 Introduction

The feed system accommodates the molten polymer coming from the barrel and guides it into the mould cavity. Its configuration, dimensions and connection with the moulding greatly affect the mould filling process and subsequently, the quality of the product. A design that is based primarily on economic viewpoints, (rapid solidification and short cycles) is mostly incompatible with quality demands. The two main areas that need to be considered are the runner system and the gate.

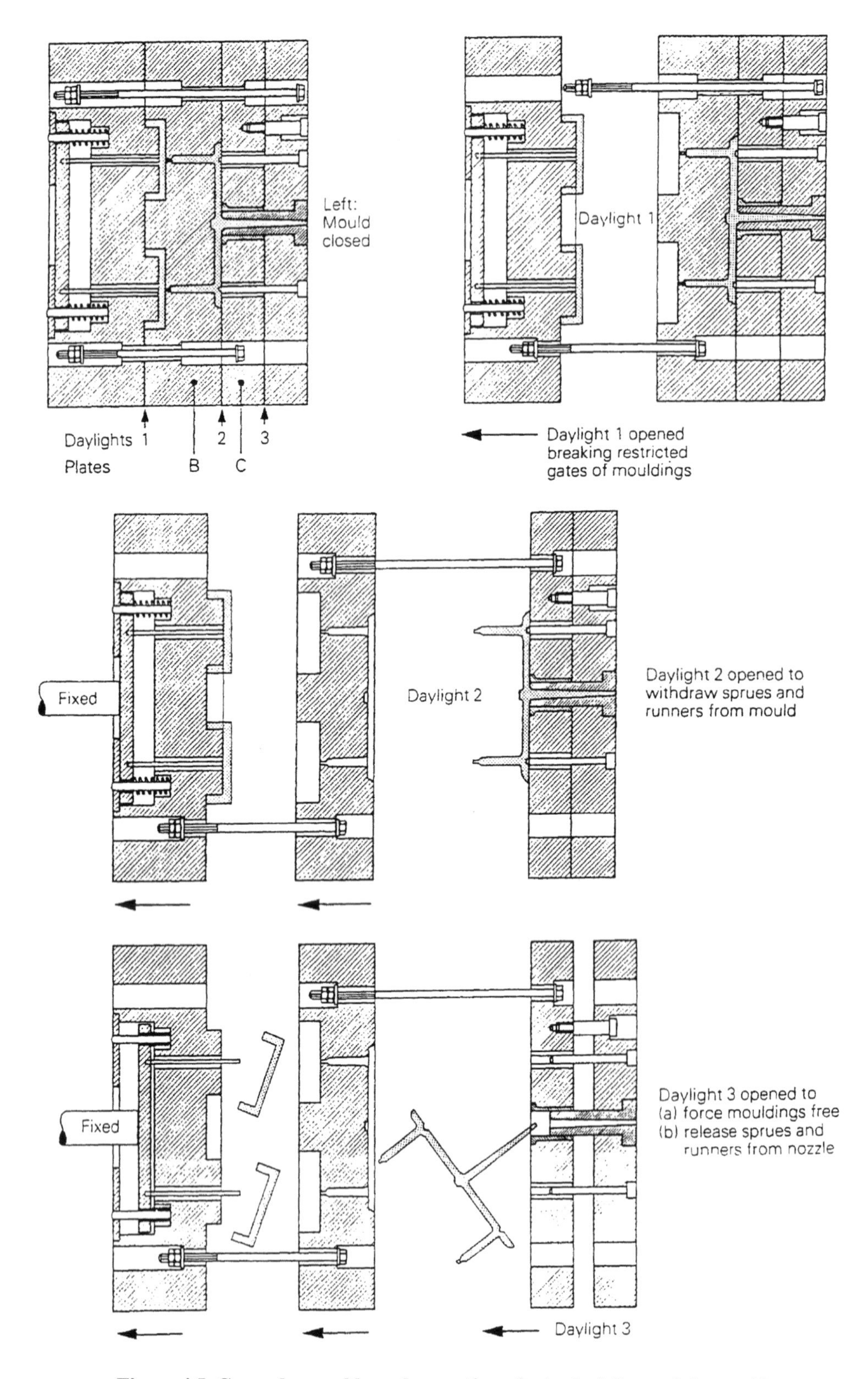

Figure 4.5 General assembly and operation of a typical three-plate mould

4.2.2 Runner System

When designing runner systems the three primary considerations are as follows:

1. The shape of the runner
2. The runner layout
3. The runner dimensions.

More specific demands of the runner design could include the following points:

1. The cavity should fill with a minimum of weld lines
2. The cavities fill at the same time
3. Restrictions to flow should be as low as possible
4. Share of the total shot weight should be as low as possible
5. Should be easily demoulded
6. Appearance of the product should be unaffected
7. Length as short as is technically feasible to reduce losses in temperature and pressure and keep scrap to a minimum.
8. Cross-section as large as required to allow a longer or equal freezing time to that of the component (to allow effective packing of the part).

4.2.3 Runner Shape

The cross-sectional shape of the runner used in a mould is usually one of four forms as shown in Figure 4.6.

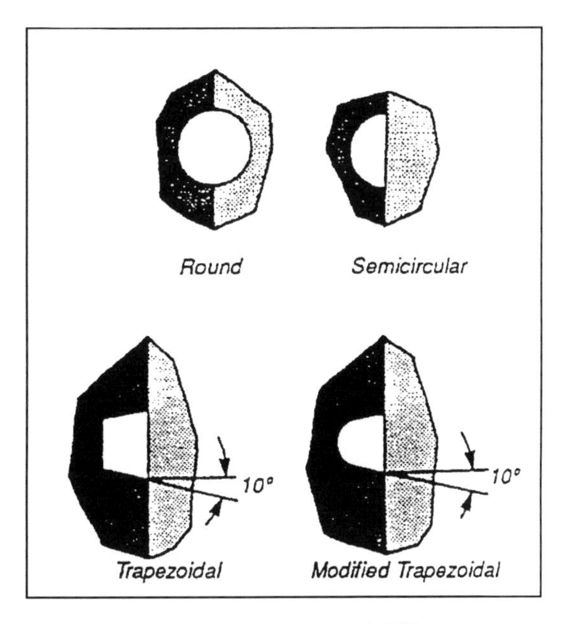

Figure 4.6 Cross-section shapes of different runners

The criteria for efficient runner design is that the runner should provide a maximum cross-sectional area and a minimum contact on the periphery from the point of heat transfer.

As can be seen, the round and modified trapezoidal types of runners are the two most satisfactory designs, whereas the ratios exhibited by the semi-circular and trapezoidal systems make their use less desirable.

As the plastic melt progresses through the runner, the melt touching the cold mould surface will rapidly decrease in temperature and solidify. The material which follows will pass through the centre of this solidified material and, because of the low conductivity that thermoplastics possess, the solidified material acts as an insulation and maintains the temperature in the central melt flow.

Ideally the gate should be positioned in line with the central melt flow. A full round will meet this requirement. The trapezoidal runners are not as satisfactory in this respect since the gate cannot normally be positioned in line with the central flow stream.

The main objection to the fully round runner is that the runner is formed from two semi-circular channels cut into each half of the mould. It is essential that these channels are accurately matched to prevent an undesirable and inefficient runner system being developed. Because of extra machining etc., a fully round runner adds to the cost of the mould, but only marginally. A modified trapezoidal design is preferred if the runner is to be machined in only one half of the mould.

The choice of runner section is also influenced by the question of whether positive ejection of the runner system is possible. In simple two-plate moulds this generally does not pose any problems.

For multi-plate moulds, however, positive ejection of the runner system is not practical. Therefore, the basic trapezoidal runner is always specified, the runner channel being machined into the fixed half from which it is pulled as the mould opens. In this way the runner is free to fall under gravity between the mould plates.

4.2.4 Runner Layout

The layouts of the runner will depend upon certain factors:

1. The number of impressions
2. The shape of the components
3. The type of mould
4. The type and positioning of the gate.

The runner length should be kept as short as possible, to reduce material wastage and to reduce pressure losses. The runners should also be balanced, i.e., the distance the material has to travel from the sprue to the gate should be the same for each cavity. This is illustrated in Figure 4.7.

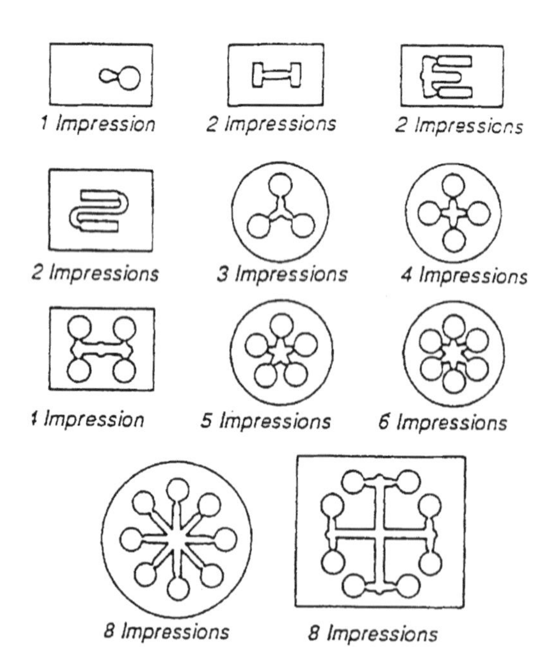

Figure 4.7 Balanced runner layouts: melt has the same length of travel for all impressions

Sometimes, however, it is not always practical to have a balanced runner system (family moulds etc.). In these cases uniform filling of the cavities can be achieved by varying the runner and gate dimensions as shown in Figure 4.8.

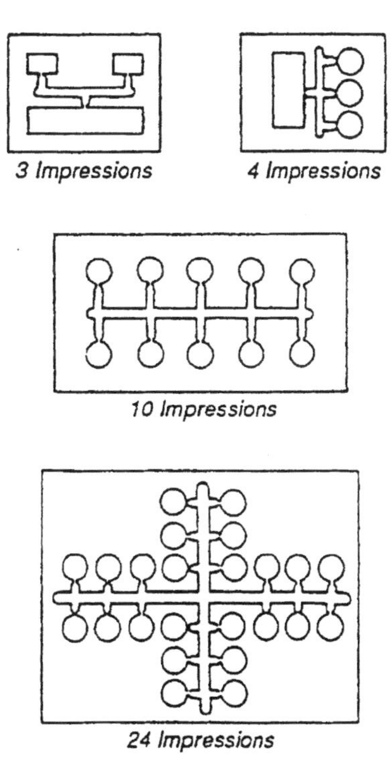

Figure 4.8 Runner layouts which require balanced gating

Runner systems can be divided into three main types: standard runner, cold runner and hot runner.

4.2.4.1 Standard Runner

Standard runner systems are machined straight into the mould plates, their temperature therefore being that of the mould temperature, (i.e., usually 20 °C to 120 °C). The material passes through the runner to the cavities which are filled and packed by holding pressure and then the molten material in the runner freezes with the rest of the component during cooling.

4.2.4.2 Cold Runner

For reactive material such as thermosetting material or rubber, a cold runner is required. This is to prevent a premature reaction of the materials in the runner.

4.2.4.3 Hot Runner

The use of the hot runner technique for feeding multi-impression and large area mouldings is now firmly established. The advantages of hot runner mouldings are as follows:

- Melt enters the cavities in a more controlled condition than with a sprue and runner system, as the temperature control in the hot runner is adjustable to finer limits
- A possible reduction in post-moulding finishing operations to remove large sprue gate witness marks
- The elimination of cold sprues and runners in multi-impression moulds which would normally be scrapped or reworked
- Hot runners enable single impression, large area mouldings to be edge-gated, whilst keeping the moulding in the centre of the machine platen (see Figure 4.9)
- Effective increase in the shot capacity of the machine as, once the hot runner is filled, the injection capacity can be fully concentrated into the cavities.

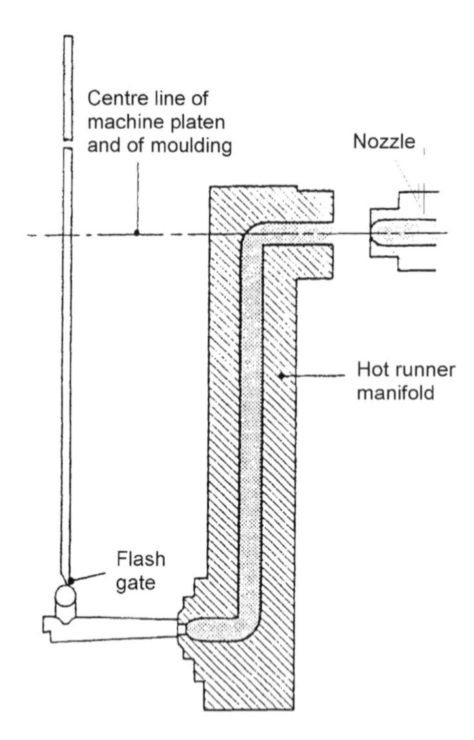

Figure 4.9 Hot runner layout permitting mould to be placed centrally on platen

In designing hot runner moulds the following important points should be observed:

- Provide adequate heating for the hot runner manifold (1.8 watts/cm^3 or 30 watts/in^3) and nozzle (approximately 300 watts)
- Make provision for closely controlling the temperature of the manifold and nozzles with suitable instruments
- Insulate the hot runner manifold and nozzles from the machine platen or mould cavities by air or compressed temperature resistant sheeting
- Provide adequate runner channels in the heated manifold, i.e., minimum 12 mm diameter
- Make the machine nozzle orifice diameter of similar size to the channels in the hot runner manifold
- Ensure that the runner channels are devoid of any sharp corners or blind spots where melt could become trapped and consequently degraded.

Typical hot runner systems for a thermoplastic have a heated manifold. A hot runner system will have a temperature which is in the range of melt temperature of the thermoplastic and therefore significantly higher than the standard runner system. This ensures that the material does not freeze off prematurely in the runner. Less raw material is required because the runner content does not need to be demoulded and is available for the next shot.

However, the disadvantages of the hot runner system can be described as follows:

- More rejects can be expected especially during start-up
- Higher costs of purchase and installation of auxiliary equipment such as heaters, temperature controllers and sensors
- Long flow paths and high shear velocity can increase the likelihood of material degradation
- The uneven melt temperature distribution can result in nonuniform filling.

More designs and components of hot runner systems are illustrated in Figures 4.10-4.13.

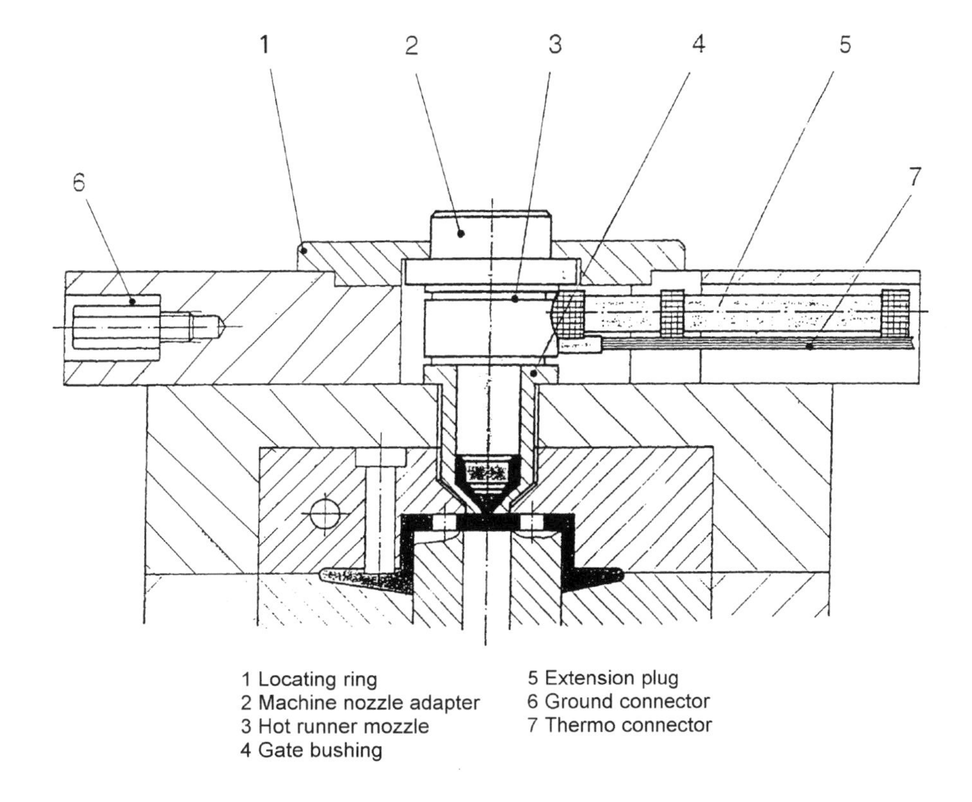

Figure 4.10 Hot runner design 1

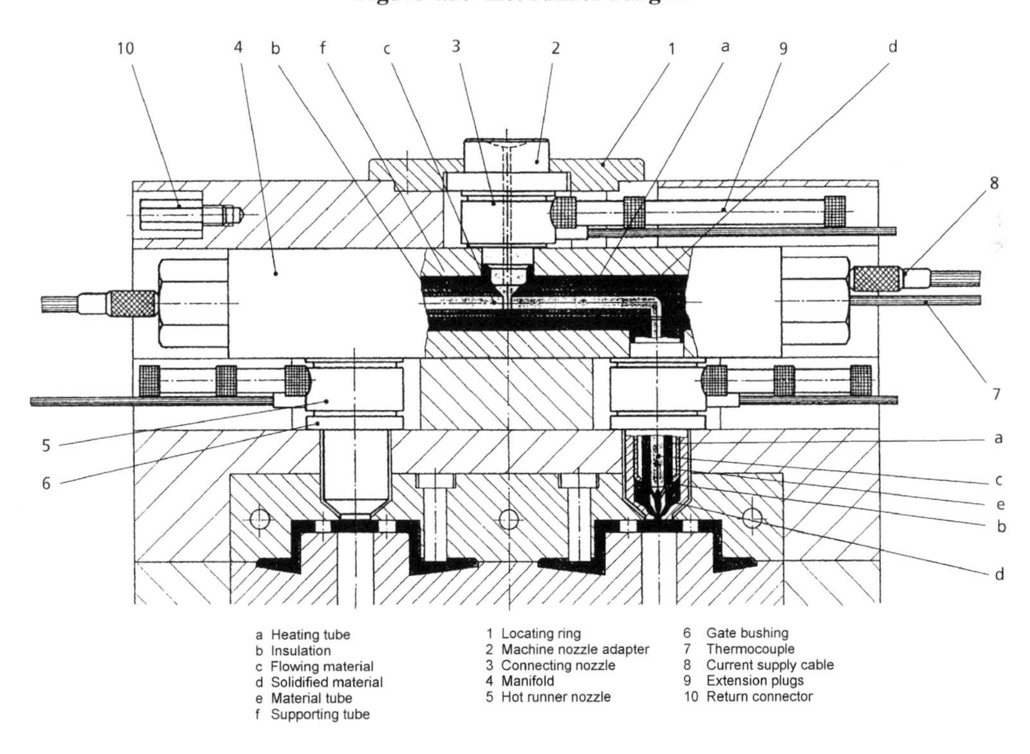

Figure 4.11 Hot runner design 2

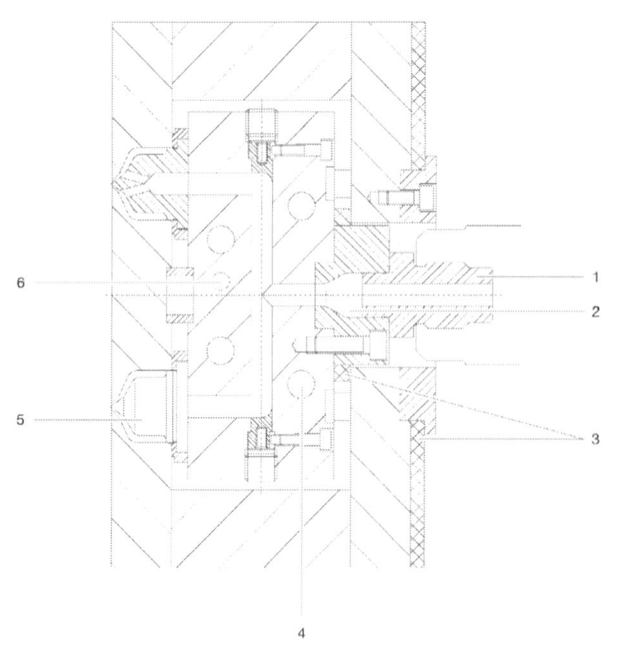

Figure 4.12 Special hot runner (1), Injection through 1, 2, 3 or 4 tips with a multi-tip nozzle

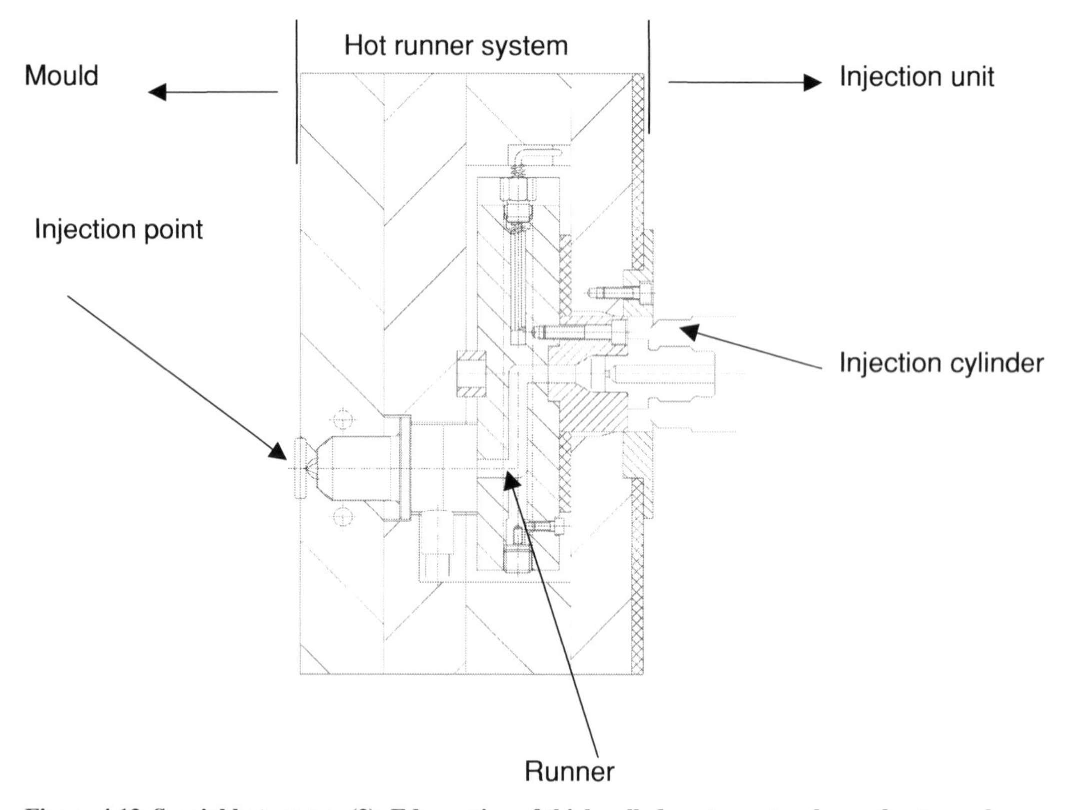

Figure 4.13 Special hot runner (2): Edge gating of thickwalled parts, no touch up of gate mark required

4.3 Gate Design

The gate is the connection between the runner system and the moulded part. The objective of the gate is to allow enough material flow for both mould filling and thermal shrinkage compensation. The moulding process and the properties of the final part are directly affected by the type of gate used, the location within the overall moulding and the size. There are a number of gate designs available, generally they fall into two basic types: large or offering only restricted flow.

The type and position of the gate is often dictated by the design of the component and the number of mouldings to be produced in each cycle. The following sections provide information on different gating methods.

4.3.1 Sprue Gate

This type of gate is the preferred gate and is normally used for single-impression moulds, especially suitable when the component is cup shaped and involves a base. Compared to a side gate, the material flow is more direct, experiencing minimal pressure loss and reduced shearing. However there can be a high stress concentration caused by the polymer at the gate area, as well as problems with material freeze off which can lead to sinking around the gate. There is also the need to remove the gate once the part is finished.This system may be extended to multi-impression moulds in conjunction with a hot runner assembly. It is illustrated in Figure 4.14.

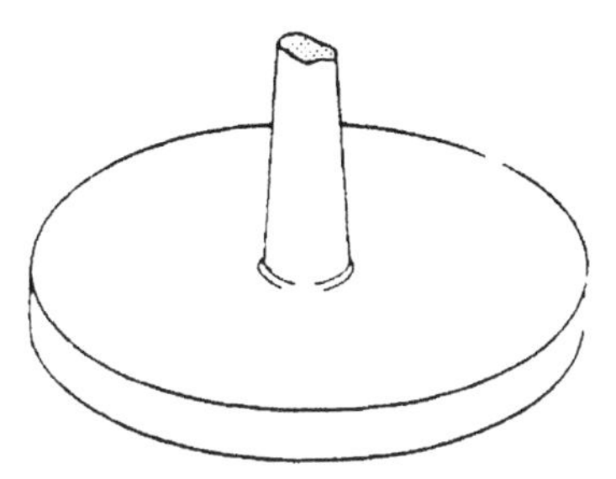

Figure 4.14 Sprue gate

4.3.2 Restricted Gate

This type of gating is used for multi-cavity tools and is shown in Figure 4.15. Finishing operations can often be eliminated because the small gate is broken off during the ejection of the moulding. The gate must not be too small otherwise the filling of the cavity is impaired. Also, under the effect of high injection pressures, frictional heating of the material passing through the gate could lead to splash marking and burning on the finished component.

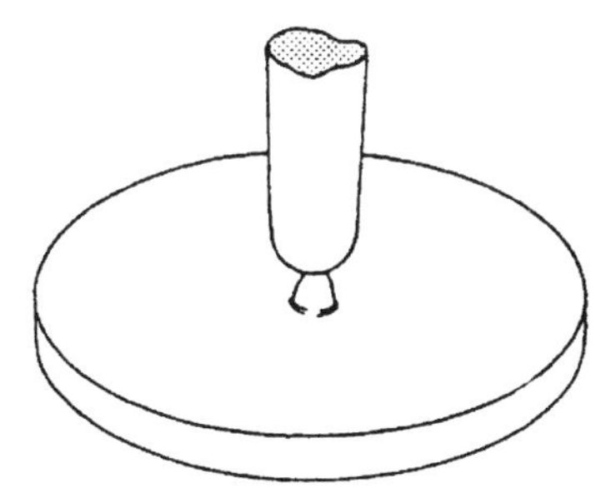

Figure 4.15 Restricted gate

However, the gate must not be made too large otherwise it will not break off satisfactorily during ejection. As a guide restricted gates should not be smaller in diameter than 0.6 mm or greater than 2 mm. It is also essential to have a generous runner system to prevent premature freezing of the melt.

To prevent any cracking around the gate during the ejection of the moulding (particularly where gates of 1.5-2.0 mm are being used) the gate should have a slight back taper so that it breaks off about 1.5 mm from the surface of the moulding.

Owing to the notch effect (the notch is recessed and therefore part thickness and strength are reduced), restricted gates should be located at a point in the moulding subject to low mechanical stresses. Also, where a clean finish is required, the pronounced orientation of the material in the gate area often hinders the removal of the gate mark by milling, due to small cracks occurring along the lines of orientation. Hence care should be taken in the removal of any restricted gates.

4.3.3 Side or Edge Gate

This type of gating is shown in Figure 4.16 and is normally used for multi-impression moulds where components are relatively small and of a flat or shallow nature. The size of the gate is dependent upon the shape and thickness of the moulding. For thick sections the gate thickness should be approximately 75% of the component thickness and as wide as the runner. With multi-cavity moulds where the gates are arranged in series, it is necessary to balance the filling of the cavities. This is not always easy to predict at the design stage of the mould and it may be necessary to complete the balancing operations by trial runs. Generally the gates furthest from the sprue are given the greatest cross-section and those nearest the sprue the smallest.

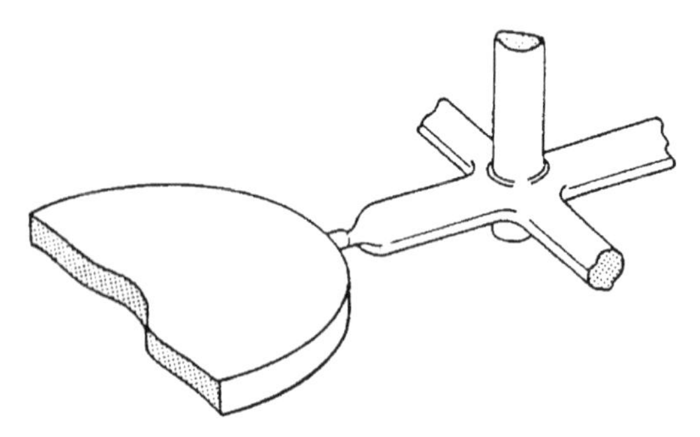

Figure 4.16 Side or edge gate

4.3.4 Flash Gate

This gate type is shown in Figure 4.17 and for long flat components of thin section this type of gate can be used quite successfully. It enables a large cavity to be filled quickly and consistently. The length and width of the article and the flow pattern required dictate the length of the gate. In some instances it is advantageous to have the gate the full length of the article, though usually a gate length which is about 50% of the longer side dimension is sufficient.

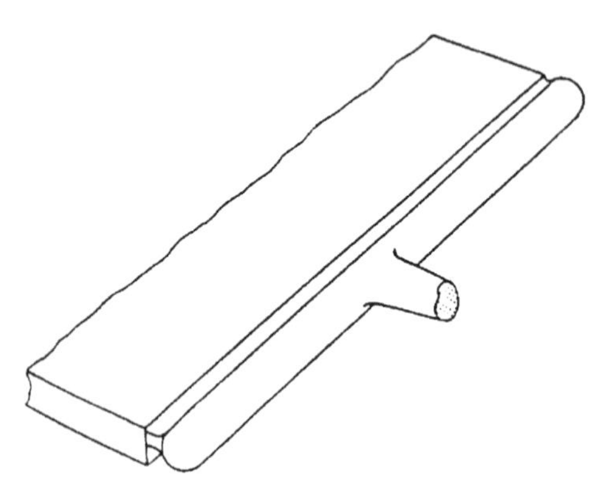

Figure 4.17 Flash gate

4.3.5 Fan Gate

For thick section mouldings such as optical lenses, this type of gating is often used. This is because it enables the runner to be made of an adequate size to help prevent the material from chilling off when it is injected slowly (as is necessary when making these components). It also allows sufficient follow-up pressure into the cavity during the cooling contraction. The design is shown in Figure 4.18.

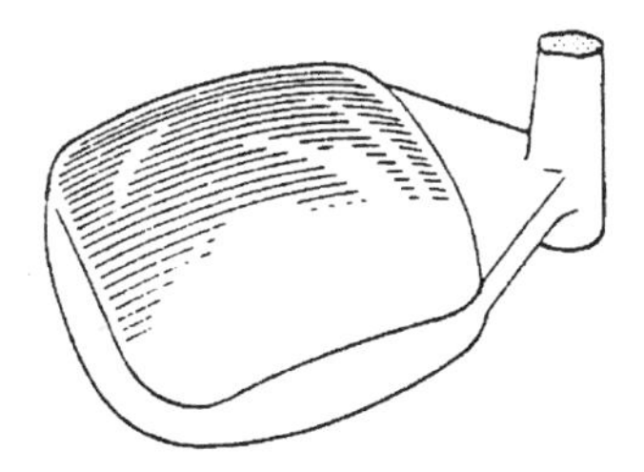

Figure 4.18 Fan gate

4.3.6 Tab Gate

This type of gating can be used as an alternative to side gating, especially in multi-impression moulds, to produce articles of a flat or shallow nature. It has certain advantages over normal side gates in that the design minimises the jetting of material into the mould cavity which leads to weld lines and flow marks. It also creates turbulence which is an aid to dispersion when moulding dry-coloured material.

Tab gates are normally used to produce elongated articles such as radio scales and rules. The tab in these instances is located towards one end so that the mould cavity is filled evenly down the greater part of its length. The longitudinal orientation of the material tends to strengthen the article and, because the gate is remote from the centre point of maximum stress, it avoids the risk of cracks developing at the gate area if the moulding is subsequently flexed.

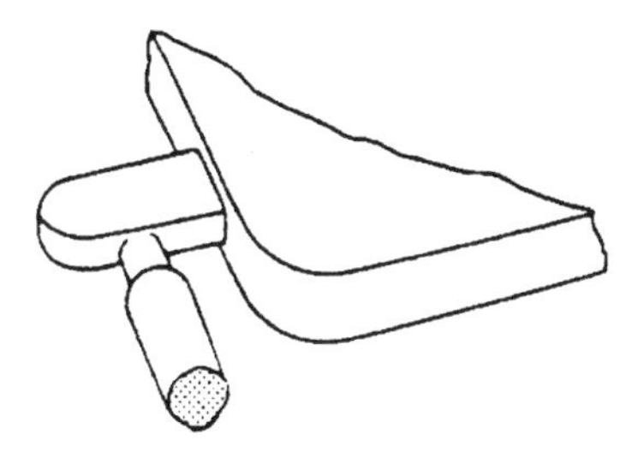

Figure 4.19 Tab gate

4.3.7 Diaphragm Gate

For single-impression moulds which are to be produced with a central orifice, this type of gating can be used to obtain uniform radial mould filling. The diaphragm gate is removed by a subsequent machining operation. It is shown in Figure 4.20.

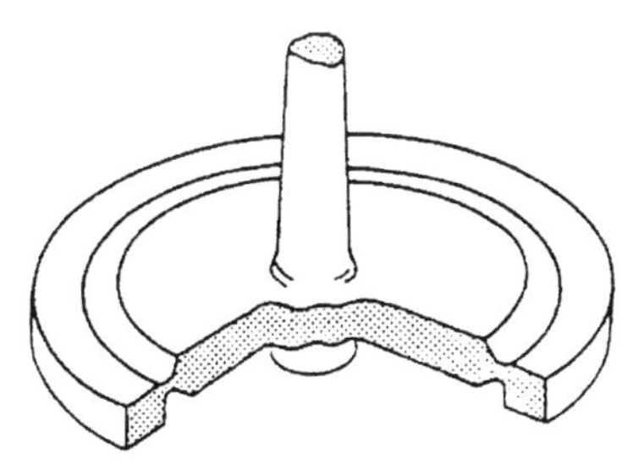

Figure 4.20 Diaphragm gate

4.3.8 Spider Gate

This is a variation of the diaphragm gate and is shown in Figure 4.21. It is normally used for moulding large diameter apertures and helps to reduce material wastage. A disadvantage is that the meeting of the separate flow streams creates weld lines and this factor needs to be considered at the component and mould design stages.

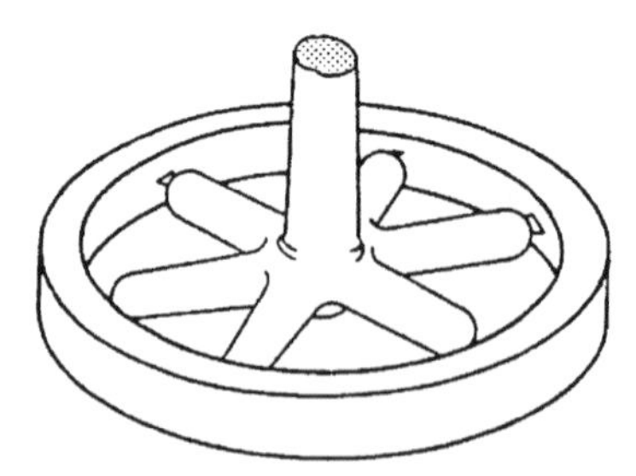

Figure 4.21 Spider gate

4.3.9 Ring Gate

For multi-impression moulds, which are used to produce tubular type articles, this type of gate ensures consistent filling of the moulds. It also helps to ensure that the core pin is central with the cavity, whereas using an ordinary side gate the initial pressure would tend to displace the core pin and so cause the article to have an uneven wall section.

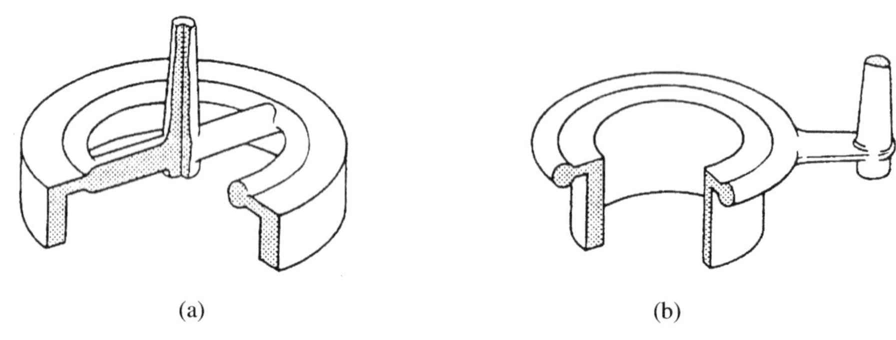

(a) (b)

Figure 4.22 Examples of ring gates

4.3.10 HOT TIP Gating

The HOT TIP gate is the most commonly used gating method. Acceptable for crystalline and amorphous materials, this method relies on gate diameter and gate area cooling to optimise the application. The selection of gate diameter and proximity of gate cooling are therefore critical to final part quality.

HOT TIP gating leaves a small mark on the moulded part surface. The less notch sensitive the thermoplastic or the larger the gate diameter, the more the vestige will protrude. Although the gate mark is only a few tenths of a millimetre high, it is common to recess the gate for larger diameters such that the vestige lies below the part surface. These are referred to as the 'cosmetic' or 'technical' gate as shown in Figure 4.23. Cosmetic gating must leave a sprue mark which is aesthetically acceptable, as opposed to a technical gate which is functional.

Gate diameters range from 0.6 mm to 1.8 mm and should be selected based on material and part considerations. With any gate diameter, the land should be no more than 0.1 mm. A sharp gate opening is essential in producing the required gate separation for cosmetically superior vestiges.

Gate and land tolerances are generally very tight. Adherence to a specific nominal gate diameter and land, although important, is in no way as critical as maintaining uniformity among all gate and lands.

For example, if one gate is 1.05 mm and all other gates in the system are 1.0 mm in diameter, the larger gate has a 10% greater cross-sectional flow area. Assuming a Newtonian laminar flow where flow is proportional to the diameter to the fourth power, the total flow in the larger gate is 22% greater. Therefore, all gates must be opened to the slightly larger gate diameter to prevent unequal part filling.

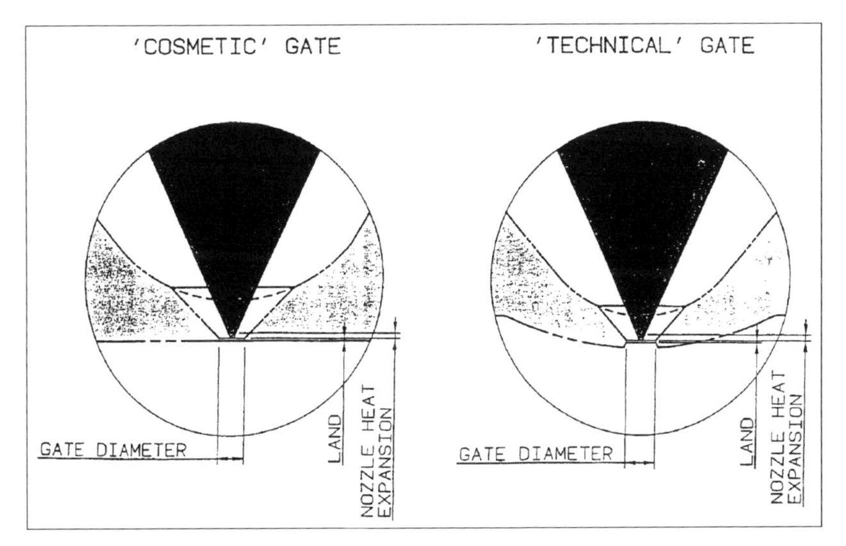

Figure 4.23 HOT TIP gate cross-section

4.4 Mould Temperature Control

With thermoplastics the main purpose of the mould system is to minimise both the cycle time and thermal differences in mould part cooling. Mould cooling is therefore essential for both cost saving and quality control. Uniform cooling improves product quality by preventing differential shrinkage, high residual stress and mould release problems.

Therefore in designing an injection mould tool, the size and layout of the cooling channel is an important part. Water circulation may be used for cooling, or if higher temperatures are required, oil. An oil-base system is suitable for temperatures up to 350 °C, whereas temperatures lower than 25 °C require a water/glycol (antifreeze) mixture system.

In order to get uniform cooling of a moulding, the following points should be considered in conjunction with Figure 4.24.

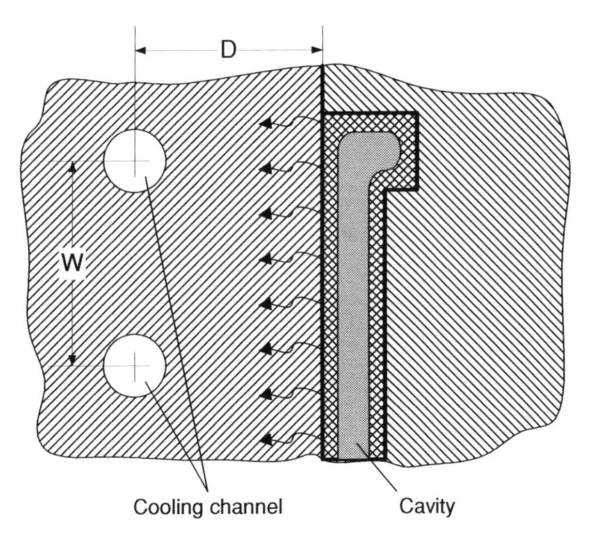

Figure 4.24 Layout of the cooling channel

D lies between d to 5d and W lies between 2d and 5d

where d is the diameter of the cooling channel

D is the depth of the cooling lines from the moulding surface

W is the width of the pitch.

Increases in D lead to a reduction in the heat transfer efficiency, and a large value of W results in a nonuniform tool temperature.

The demoulding temperature requirements of various materials are given in Table 4.1.

Table 4.1 Recommended values for the demoulding temperature of different plastics

Short designation according to DIN 7728	Demoulding temperature (°C)		
	Lower temperature	medium temperature	upper temperature
PS	20-35	35-45	45-60
SB	20-35	35-50	50-65
SAN	35-50	50-70	70-85
ABS	35-55	55-75	75-90
PVC rigid	45-65	65-80	80-100
PVC soft	25-35	35-45	45-55
CA	35-50	50-65	65-80
CAB	30-45	45-60	60-75
CP	30-40	40-55	55-70
PMMA	50-70	70-90	90-110
PPE mod.	65-80	80-95	95-110
PC	60-85	85-110	110-130
PAR	120-140	140-160	160-185
PSU	100-130	130-160	160-190
PES	130-145	145-165	165-185
PEI	135-150	150-170	170-190
PAI	200-220	220-230	230-240
PE soft	30-40	40-50	50-65
PE rigid	40-50	50-60	60-75
PP	45-55	55-65	65-80
PA6	50-70	70-90	90-110
PA 6.6	75-90	90-120	120-150
PA 6.10	40-55	55-70	70-85
PA 11	60-80	80-105	105-130
PA12	40-60	60-80	80-100
PA amorphous	55-70	70-85	85-100
POM	60-80	80-100	100-130
PET	75-95	95-120	120-150
PBT	60-75	75-90	90-120
PPS	120-145	145-170	170-190
FEP	160-180	180-200	200-220
ETFE	140-150	150-160	160-180
PAEK	120-145	145-160	160-180
LCP	60-100	100-140	140-180
TPE-E	25-35	35-50	50-65
PF	mould temperature		
UF	mould temperature		
MF	mould temperature		
UP	mould temperature		
EP	mould temperature		
LSR	mould temperature		

Notes:
PF = phenol-formaldehyde, UF = urea-formaldehyde, MF = melamine-formaldehyde,
PAR = polyacrylate, PSU = polysulfone, PAI = polyamidimide, PAEK = polyaryl ether ketone,
TPE-E = thermoplastic elastomer (type E ethylene), UP = unsaturated polyester

Thermosets unlike thermoplastics require heating in the mould and not cooling. The demoulding temperature is therefore the temperature the mould is set at for crosslinking. More details on controlling the cooling or heating systems can be found in Chapter 8.

4.5 Ejection Systems

After a component has solidified and cooled down, it needs to be removed from the mould cavity. Ideally, this is done by gravity and the part falls to the floor as shown in Figure 4.25. However, some components with design features such as undercuts, adhesion or internal stresses may have to be removed from the mould manually or by robots.

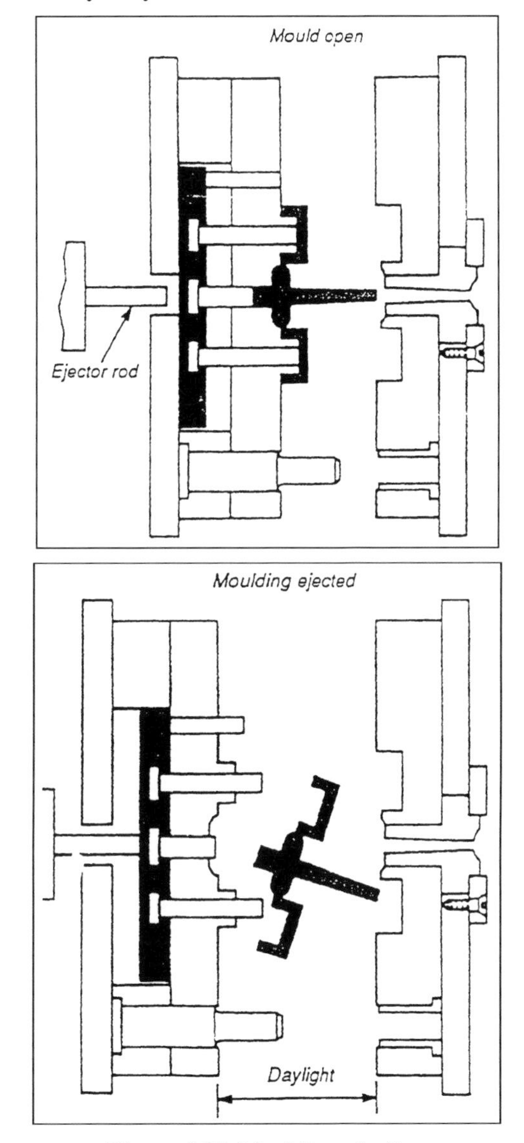

Figure 4.25 Moulding ejection

Ejection equipment is usually actuated mechanically by the opening stroke of the moulding machine. If this simple arrangement is insufficient, ejection can be performed pneumatically or hydraulically.

The ejector system is normally housed in the movable mould half. Mould opening causes the mechanically actuated ejector system to move towards the parting line and to eject the moulding. The result of this procedure is that the moulding stays on or in the movable mould half. This can be achieved by undercuts or by letting the moulding shrink onto a core. Taper and surface treatment should prevent too much adhesion. Methods of ejection for various components are shown in Table 4.2.

Table 4.2 Types of ejection systems

Type	Ejection method	Components of operation	Applications
Standard system for small parts	During opening stroke in direction of demoulding. Ejection with pins, sleeves or stripper plate	Various: mechanical, hydraulic, pneumatic, manual, machine stop, lifting cylinder, cam, pivot, inclined plane, thrust plate	Mouldings without undercuts
Direction of ejection towards movable side, stripping is used but usually for circular parts only	During opening stroke pull in direction of demoulding, ejection with stripper plate	Mechanical, hydraulic, pneumatic, stripper bolt, lifting cylinder, pin-link chain	Cup-like mouldings with internal gate
Demoulding at two parting lines for automatic operation including separation of gate	During opening stroke thrust in direction of demoulding, ejection with pins, sleeves or stripper plate	Mechanical, stripper bolt	Mouldings with automatic gate separation
Demoulding of parts with local undercuts (slide mould)	During opening stroke thrust in direction of demoulding, ejection with pins, sleeves or stripper plate after release of undercut	Mechanical, cam pins, slide mechanism	Flat parts with external undercuts (threads)
Demoulding of large, full-side undercut (split-cavity mould)	During opening stroke thrust in direction of demoulding, ejection with pins	Mechanical, hydraulic, springs, links, pins, cams	Parts with external undercuts (ribs) or opening in side wall
Air ejectors usually provide support Breaking is done mechanically	Thrust in direction of demoulding causes a first air shot followed by ejection with compressed air	Mechanical-pneumatic in stages	Cup-like, deep parts

4.6 Venting

Another design aspect of tooling is the need to provide vents for compressed air and gases to escape during moulding. Trapped air and gases can cause a variety of moulding defects which are more fully described in Chapter 9, such as:

- short shots (incomplete filling of the mould)
- scorching or burning
- shrinkage (often seen as ripples or depressions in finished parts)
- in extreme cases volatile gases may cause etching on the mould surface.

Common venting methods are to provide parting-line vents, vent plugs and pins. More recent developments include the use of porous metals that allow gas to escape but not the polymer. These materials also often allow for the venting area to be increased.

4.7 Conclusion

This chapter has introduced the basics of tooling for injection moulding. It can now be appreciated how the machine, material and tool all play a part in the production of a successful injection moulding.

Further Reading

G. Menges, W. Michaeli, P. Mohren., *How to Make Injection Molds*, Third Edition, Hanser, 2001.

D.V. Rosato and D.V. Rosato, *Plastics Engineered Product Design*, Elsevier Advanced Technology, 2003.

5 Process Control Systems

5.1 Introduction

The control system is there to ensure repeatability during moulding operation. It monitors both the hydraulic system and the process parameters such as temperature, injection speed, screw retraction speed and injection and back pressure. The ability to control the process has a direct impact on final part quality, part to part consistency and economy. The nature of the control system may vary from a simple relay switch to a complex microprocessor system with closed-loop control. Some of the components of the machine control system will now be introduced.

5.2 Explanation of the Different Concepts in Control and Regulation Technology

5.2.1 Pump

The hydraulic pump generally draws the hydraulic fluid from the supply reservoir and delivers it to the pump outlet. From here, it is conveyed through valves to the consumer and then returned. Figure 5.1 illustrates the symbol used to show a pump.

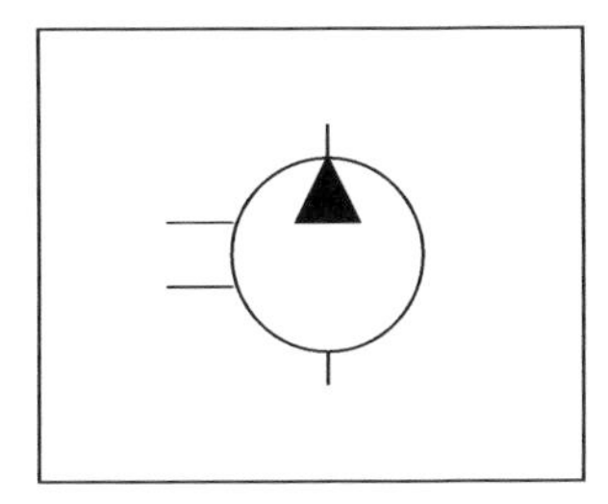

Figure 5.1 Hydraulic pump with constant displacement volume

5.2.2 Motor

Hydraulic motors transform the hydraulic energy supplied by the pumps back into a mechanically-utilised working force with a rotary motion. Figure 5.2 illustrates the symbol used to represent a motor.

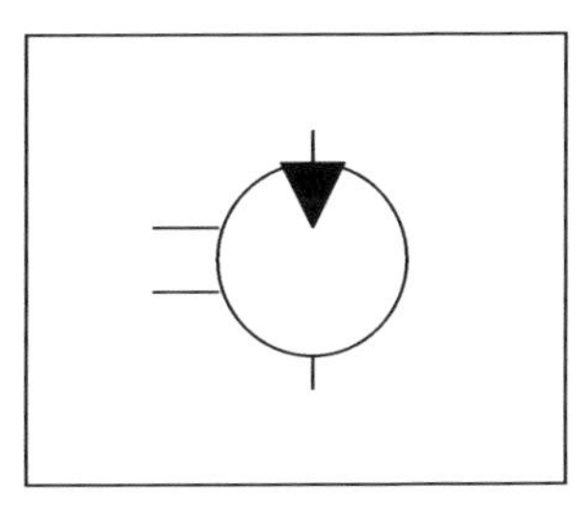

Figure 5.2 Hydraulic motor with constant displacement volume

5.2.3 Cylinder

The cylinder (located behind the injection unit) is charged with hydraulic fluid through valves in the base and the head. Through this, a motion is transferred through the piston surface of the working cylinder to the piston's connecting rod. Single and double-acting cylinder are shown in Figure 5.3 and Figure 5.4.

Figure 5.3 Single-acting cylinder with single-side connecting rod (force applied in only one direction)

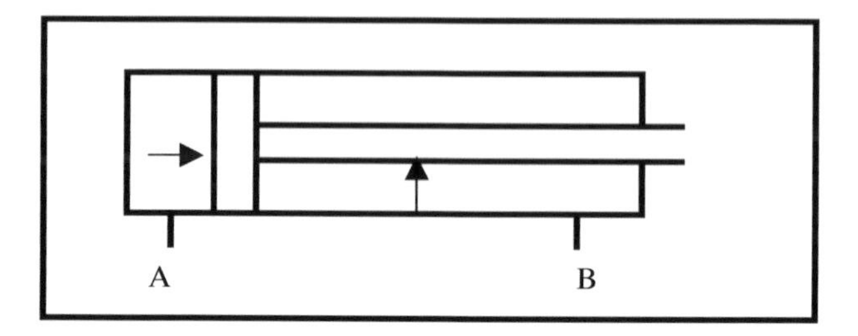

Figure 5.4 Double-acting cylinder (A and B are separate cylinders) with single-side connecting rod (force applied in both directions, uneven surfaces)

5.2.4 Directional Valves

The function of directional valves is to block different hydraulic lines from one another or to open them, and to continually create alternating line connections. In this manner, the effective direction of pressures and volume flows is influenced, and the starting, stopping and the direction of motion of the consumer (cylinder or hydraulic) motor are thus controlled. The number of connections and operational positions of a directional valve is of great significance. These are indicated as a prefix before every designation. Every operational position is represented by a square. Arrows and dashes within the square identify the connections between the lines. The simplest form of a directional valve has two connections and two operational positions as shown in Figure 5.5 and Figure 5.6. A more complex 4/3 directional valve is shown in Figure 5.7 (see also connection designations in Table 5.1).

The result of the operational position is clear as soon as the total set of operating symbols has been displaced into the fixed section of the service lines.

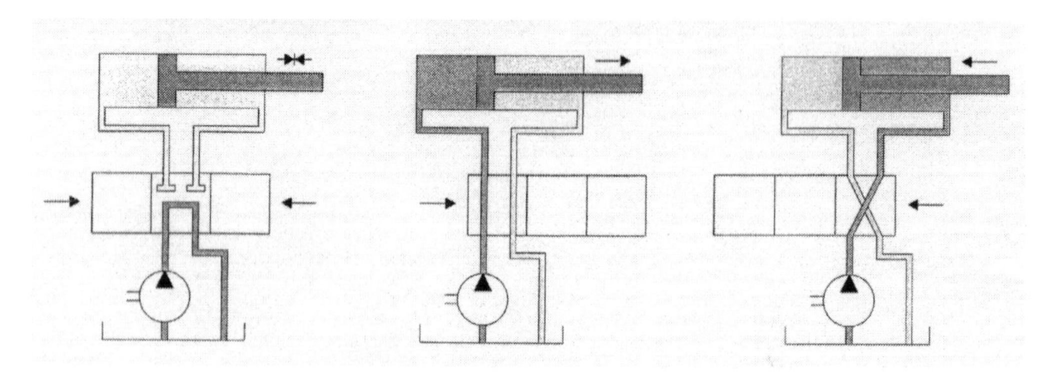

Figure 5.5 Directional valves

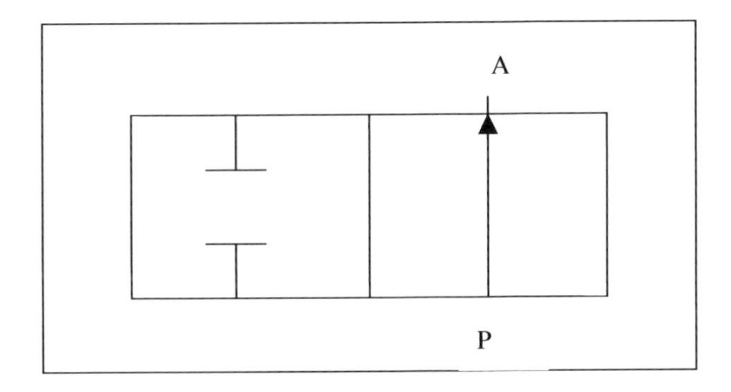

Figure 5.6 2/2-Port directional valve

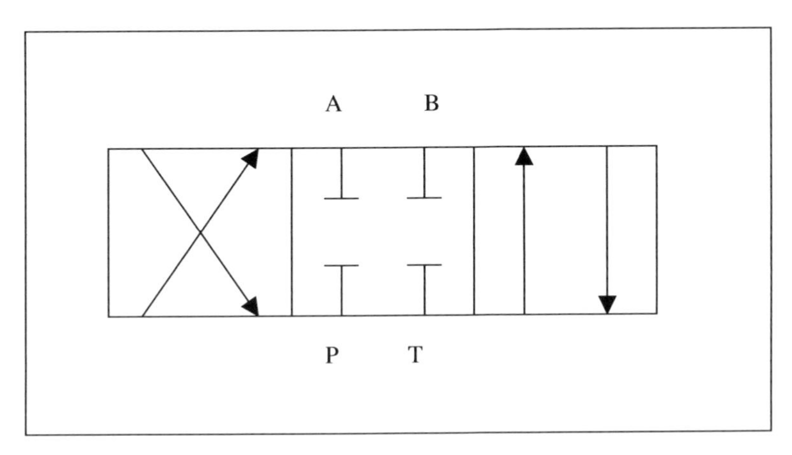

Figure 5.7 4/3-Port directional valve
(4-number of connection numbers, 3-number of operational positions)

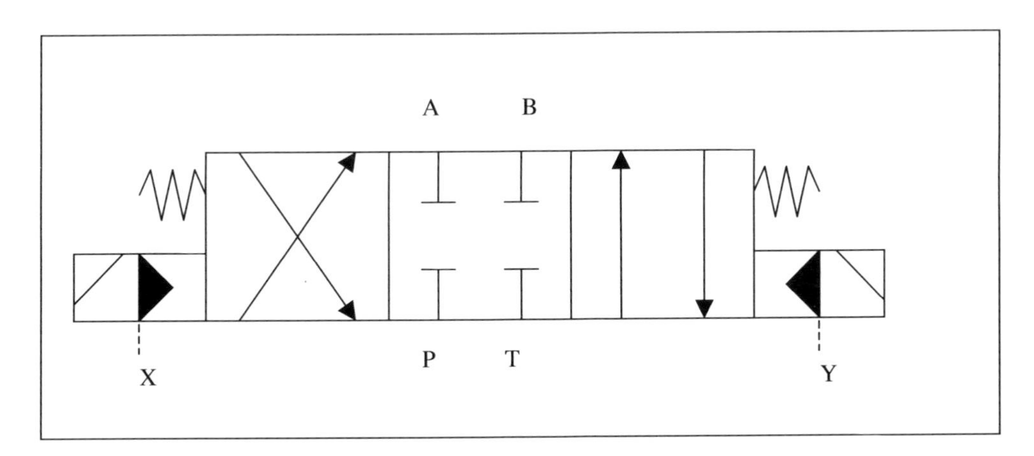

Figure 5.8 Piloted directional valves

Larger directional valves are hydraulically actuated by a pilot valve. The pilot valve is controlled either electrically or pneumatically. A piloted directional valve is shown in Figure 5.8. The connection designations are shown in Table 5.1.

Table 5.1 Connection designations	
P	Pressure connection
T	Tank connection
A, B	Working connections
L	Leakage return connection
X,Y	Control connections

5.2.5 Pressure Valves

Pressure control valves have the primary task of limiting pressure in the system and thus protecting individual components and lines from rupturing or overloading. The valve opens when a predetermined pressure is reached and conveys the pump's excess delivery flow back into the tank. A pressure control valve is shown in Figure 5.9.

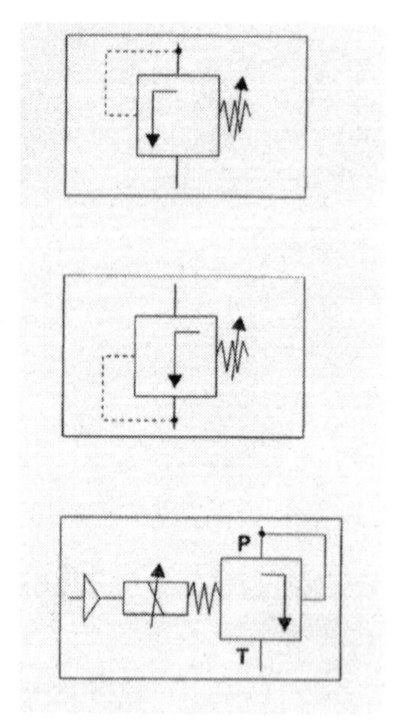

Figure 5.9 Pressure control valve (top), pressure limiter valve (middle), proportional pressure valve (bottom)

Pressure limiter valves limit the pressure in a specific circuit for a specific consumer. They close an open connection directly when the applied pressure exceeds a pre-set value. This is shown in the middle figure in Figure 5.9.

Proportional pressure valves convert an electrical input signal in the form of a voltage from 0...10V proportional to a hydraulic pressure. In principle, they are electrically controlled pressure valves in which the manual setting device is replaced by an electrical setting actuator called the proportional solenoid. This is shown in the bottom drawing in Figure 5.9.

5.2.6 Flow-Regulator Valves

The task of the flow-regulator valve is to influence volume flow by changing the diameter of the valve governor, and thus to control the speeds of cylinders and hydraulic motors.

The 2-port directional flow-regulator valve is installed when a constant speed is required for a consumer, regardless of load. Through-flow is independent of the pressure differential imposed upon it.

Proportional control valves or flow-regulator valves directly change through-flow in proportion to an electrical input current (U=0...10V). The valves consist essentially of a sliding valve with precise control stages whose opening diameter is changed by a proportional solenoid. A proportional flow-regulator valve is shown in Figure 5.10.

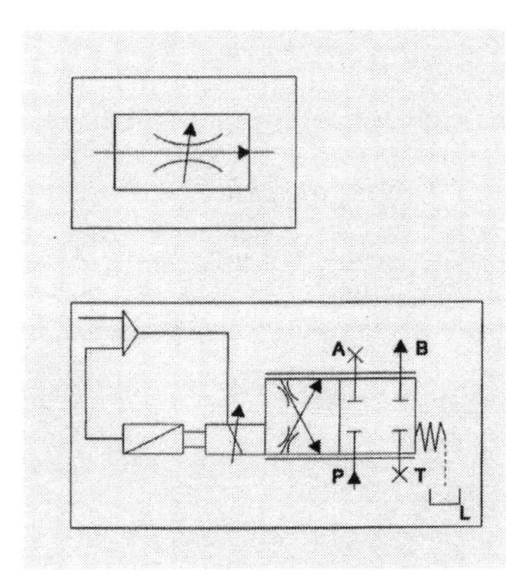

Figure 5.10 2-Port directional flow-regulator valve (top), proportional flow-regulator valve (bottom)

5.2.7 Location and Setting of Proportional Valves

The proportional valves for combined pressure and flow-regulation are set directly on the radial piston pump. A further proportional pressure valve (mounted on the outer right hand side of the hydraulic manifold) is responsible for the back pressure regulation.

It is therefore possible to digitally set the pressure and volume with a position controlled proportional valve.

- With proportional valves, a proportional offset of the magnetic armature is obtained and is dependent on the strength of the current applied to the magnets
- The proportional magnet is supplied with 24V DC. The regulation is pulse-width modulated
- The stroke of the magnetic armature is detected by an inductive displacement transducer and is transmitted to the electronic regulation on the PVS board where the difference between the command signal (input voltage) and the feedback signal (actual voltage proportional to the displacement of the transducer) is regulated
- We are dealing with position control of the magnetic armature.

5.2.7.1 Valve Bodies

Depending on the type of valve the valve body is designed as follows:

- as a throttle valve with the volume proportional valve
- as a cylinder seat valve with the pressure proportional valve

5.2.7.2 Function of the Volume Proportional Valve

The volume proportional valve is located on the pump output channel and is double supplied from the pump from P and T to A and B as shown in Figure 5.11.

The armature of the position controlled proportional-magnet acts directly on the valve piston and pushes it directly against a centering spring. The position of the valve piston dictates the cross-section of the opening and consequently the flow. When the main switch is turned on the proportional magnet is supplied immediately with DC voltable and the valve piston is set in the (P and T to A and B is closed) pre-adjusted zero position.

The position regulation of the proportional amplifier in conjunction with the displacement transducer see to it that the magnet and valve piston go to and remain in the pre-adjusted position.

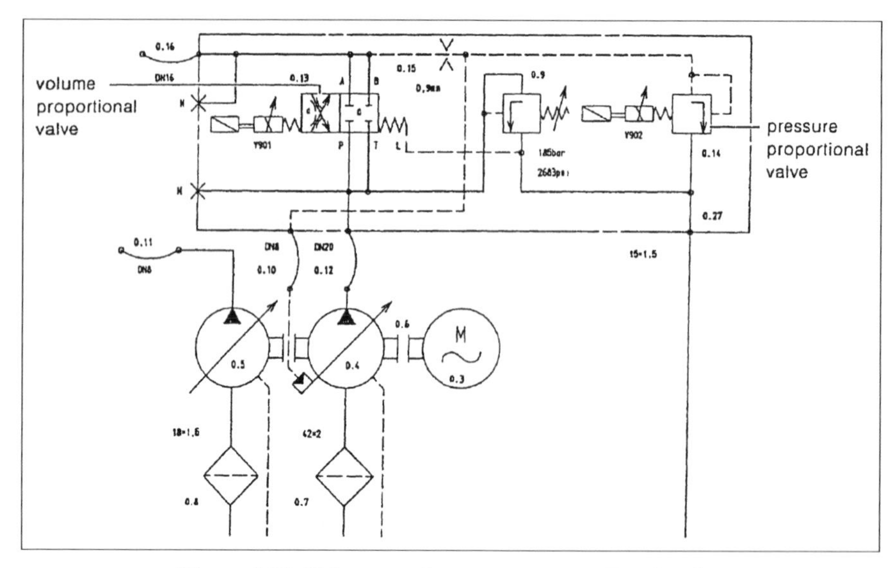

Figure 5.11 Volume and pressure proportional valve

5.2.7.3 Function of the Pressure Proportional Valve

Should the operating pressure of the pump rise above the pre-set value, this valve signals a reduction of the feed flow pressure to the pump regulator.

The flow supplied by the pump automatically reduces when the demand from the users is zero, i.e., the pump idles.

The 0.9 mm orifice plate acts as a flow control limit as the pressure proportional valve so demands (volume reduction because the valve Y902 in Figure 5.11 can only drain off 3 l/min).

When the main switch is turned on the proportional magnet is immediately supplied with DC voltage and it pushes the valve spring to the pre-set zero position.

The position regulation of the proportional amplifier in conjunction with the position transducer see to it that the magnetic armature (pre-tensioned valve spring) goes to and remains in the pre-selected position.

A block diagram of a typical system is shown in Figure 5.12.

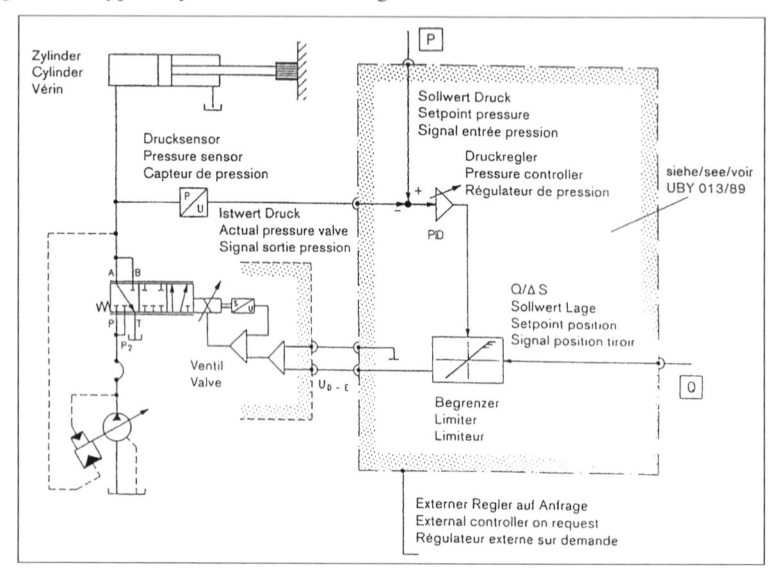

Figure 5.12 Hydraulic system block diagram

5.2.8 Check Valves

Non-return valves have the task of blocking the volume flow in one direction and allowing free-flow in the opposite direction. The blockage should provide completely leakproof sealing. Balls or cones are used primarily as sealing elements.

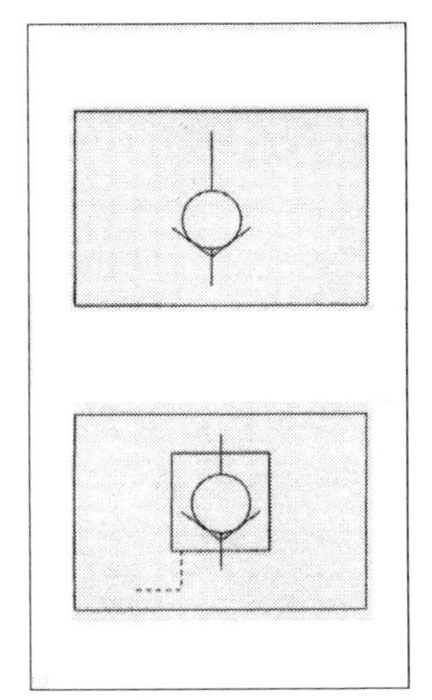

Figure 5.13 Non-return valve (top) and unlocking non-return valve (bottom)

In order to hold a cylinder upon which external forces are being applied firmly in position, it is not enough just to block the working connections of the directional valve. Leakage in this operational position leads to a decrease in the load. In order to avoid this, unlocking non-return valves are used. With these valves, the closed position can be overridden by the control of the valve cone. Throughflow which was blocked in one direction is thus permitted. Both types of non-return valves are illustrated in Figure 5.13.

5.2.9 Receivers

Hydropneumatic receivers have the task of collecting and storing hydraulic energy, and then releasing it on demand. This type of receiver is used in conjunction with injection moulding machines with very rapid injection (with accumulator). Here, a high volume flow which can be partially accessed from the receiver is required periodically for brief intervals. The benefits in the application of a hydropneumatic receiver are in the use of relatively small pumps, drive motors and oil reservoirs.

The working principle of a receiver is that it is virtually impossible to compress the hydraulic fluid. If it is nevertheless to be stored under pressure, a gas is utilised, in this case nitrogen. The gas is compressed in a pressure reservoir by the hydraulic fluid and decompresses as needed through the release of fluid. In order to ensure that the gas does not mix with the hydraulic fluid, the pressure reservoir is divided into two chambers by an elastic separation wall (membrane). A hydropneumatic receiver symbol is shown in Figure 5.14.

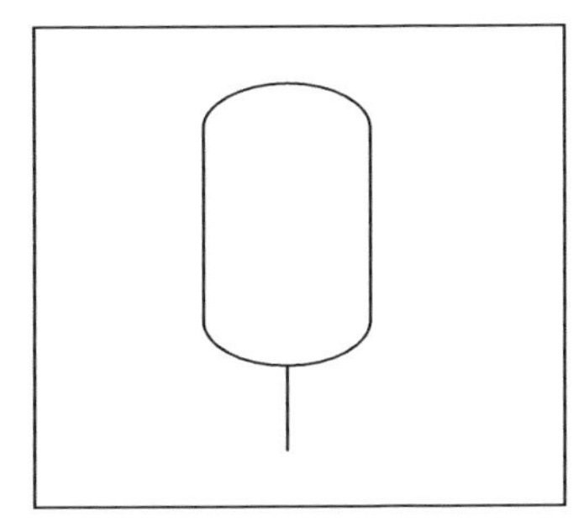

Figure 5.14 Symbol for hydropneumatic receiver

5.3 General Information Concerning Control Units, Regulators, Position Regulation and Injection Process Regulation

5.3.1 Control Unit

In the control unit, a desired valve – the nominal value – is adjusted, for instance the injection speed in the SELOGICA control unit. This adjustment valve is transmitted as a command through the electronic control of a hydraulic valve. The valve opens to a specified position, thus allowing the hydraulic fluid to flow through at a desired volume rate per second, and so powering the stroke of the screw. The electronic controls and the control valve must be matched here so that there is no through-flow of the fluid at an input of 0% and for the 100% setting a maximum fluid volume is allowed to flow. The amount of fluid which actually flows at injection, or the speed with which the screw is moved are not monitored here. Hence, temperature changes in the hydraulic fluid, viscosity variations in the plastic and other negative factors can create deviations between the adjusted or desired speed (the nominal value) and the true speed (the actual value).

Differences between the actual value and the nominal value are also possible when the characteristics curve of the control valve is not completely linear.

Finally, the most important disadvantage of the control unit should be mentioned: differences between the nominal value and the actual value are not independently recognised or adjusted.

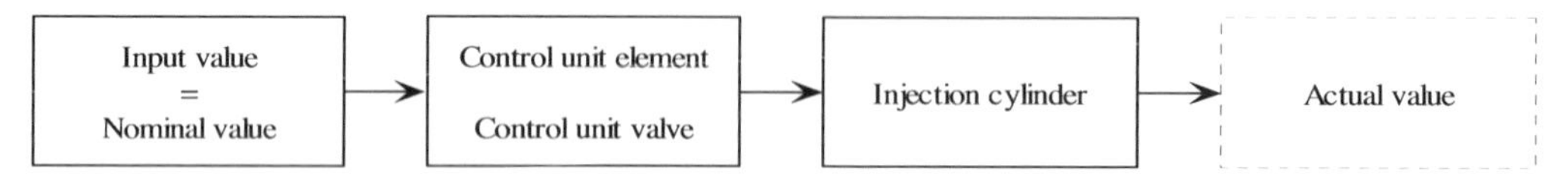

5.3.2 Regulator

With a regulator, the actual value is measured and compared by the regulator electronics to the adjustment value. When there are deviations of the actual value from the nominal value, the hydraulic valve is reset until the true speed (actual value) agrees with the desired speed (nominal value).

With the regulator, there is hence no relationship between a precisely fixed setting for the hydraulic valve and an adjustment value. The hydraulic valve is reset for as long or as often as is necessary for the measured actual speed to be equal to the desired adjustment.

For this reason, the regulator requires greater expenditure from a technological standpoint than a control unit. Test data gauging devices for the measurement of the actual valves are additionally required (speed and pressure gauges). Also, faster hydraulic valves are necessary than with the control unit so that deviations can be corrected immediately.

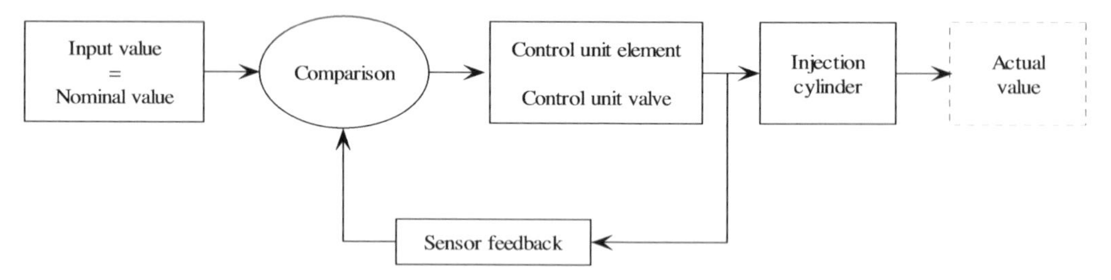

5.3.3 Screw with Position Regulator

Position regulation of the screw is made possible by a double-action cylinder. In contrast to a single-action cylinder, this is referred to as a 'contained system'.

With a single-action system, pressure is applied on only one side of the injection piston. Using a position-sensing system and time factor, a highly precise speed profile can be realised here. However, in contrast to position regulation the piston can be accelerated only: it cannot be braked. The way this type of system works is comparable to an automobile without brakes. The defined speed cannot be maintained as quickly and precisely, because the only control options here are 'apply pressure' and 'withdraw pressure.'

With the 'contained system', pressure may be applied to both sides of the injection piston. The actual pressure value of both piston chambers is measured by pressure gauges, the screw position is read by the position-sensing system. With the contained system, it is possible to define and regulate the position of the screw precisely. With this system, precise injection speed or holding pressure profiles may be followed, since there are now control commands for 'apply pressure injection side,' 'withdraw pressure injection side,' and also 'apply pressure retraction side' and 'withdraw pressure retraction side.'

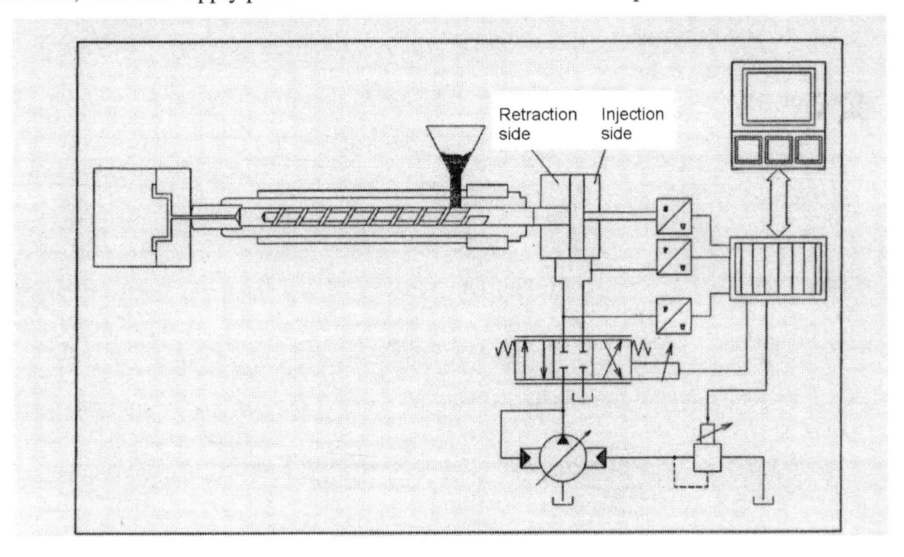

Figure 5.15 Screw with position regulation

The screw with position regulation as shown in Figure 5.15, ensures especially high motive power and reproducibility of the injection process. Here, the hydraulic valve sits directly on the injection unit (for injection governing with a screw which does not have position regulation, the valve sits on the control block of the injection side) and thus possesses significantly higher switching force through the short fluid column from valve to cylinder. Furthermore, inaccuracies in the volume and pressure of the hydraulic fluid, such as losses, which occur due to the friction, and the elasticity of the pressure lines, can be avoided. Two pressure sensors are positioned directly on the injection apparatus for determining the pressure differential between the injection side and the return side. Position regulation of the screw has enormous benefits during injection, especially in injection profiles, as well as in the maintenance of a position which has been achieved. Position regulation balances outside forces such as gravity in vertical devices.

At injection with a receiver, the receiver energy is proportioned through the regulator valve. Because of this, position regulation and pressure regulation are ensured even with high injection performance demands.

5.3.4 Injection Process Regulation

The properties of an injection moulded part depend upon the working material and upon the processing conditions. In the production of a series of parts, a certain deviation in quality features such as weight, dimensional consistency and surface characteristics must always be considered. These deviations vary from machine to machine and from material to material. Furthermore, external influences or negative factors have an effect on the quality of an injection moulded part. Examples of such negative factors are changes in the viscosity of the melt, temperature changes in the mould, viscosity changes of the hydraulic fluid and changes in the characteristics of the plastic.

The causes through which these negative factors may arise are, for example, machine start-up after a long period of non-operation, changes in material properties in the processing of a new lot or a different colour, and environmental influences such as the ambient temperature at the time of processing.

The design purpose of injection process regulation is to make these negative influences ineffective, and thus to attain an even higher reproducibility of the parts.

The decisive factor for all quality features that are concerned with dimension and weight is the internal pressure of the mould. Constant maintenance of this pressure curve in every cycle guarantees uniformity of the quality of injection moulded parts. If the mould internal pressure curve is maintained at a constant, all of the negative factors mentioned above are compensated.

During injection moulding without injection process regulation, a specified pressure curve is established for injection and holding pressure, which can also be maintained with assurance with a regulated machine. However, the mould internal pressure curve that arises can only be assumed. Pressure losses through the runner manifold as well as the mould-specific filling behaviour cannot be identified.

With the application of injection-process regulation, the mould internal pressure is first measured and compared with a nominal value. If there is a deviation, a hydraulic valve that applies pressure to the injection cylinder is actuated. It is thus possible to follow the nominal value precisely and independently of negative factors. The switch over from injection to holding pressure also occurs as a function of internal pressure. Thus, no pressure spikes can occur since the switch over takes place when a specified threshold value is reached.

The following benefits are achieved through the application of injection process regulation:

- Significant reduction in start-up cycles. The required consistency in quality characteristics is achieved after just a few cycles.
- Better reproducibility of the parts. The deviation spread of the dimensions lies significantly below that of a non-regulated machine.
- Cycle-time reduction. By the ability to visualise the internal pressure signal, the sealing point can be determined much more easily and accurately.
- Re-starts. If the same internal pressure curve is applied at a re-start, the resulting parts are exactly alike.
- Improved quality of the parts through effective speed and pressure profiles. Internal pressure profiles without spikes make possible the production of parts with low residual stresses. Switch over as a function of internal pressure prevents over-injection of the part, regardless of the selected dosage stroke.

The enormous significance of a mould internal pressure curve is characterised by the large number of parameters that can influence the appearance of the curve. An example of such a curve is shown in Figure 5.16.

The most important influencing factors on mould internal pressure are:

- *In the injection phase*: the injection speed, the flow resistance as a function of the type of plastic, the material temperature and the mould wall temperature.
- *In the pressure holding phase*: the material temperature, the mould temperature, the level of the holding pressure and the duration of the holding pressure.
- *In relation to the maximum mould internal pressure*: the injection speed, the material temperatures, the switch over point and the material flow.

The appearance of the internal pressure curve additionally influences the following quality data:

- *In the injection phase*: the appearance, the surface characteristics, the orientation and the degree of crystallinity of the moulded part.
- *In the pressure holding phase*: the formation of ridges, the weight, dimensions, shrinkage, shrink holes and sink marks, and the orientation.

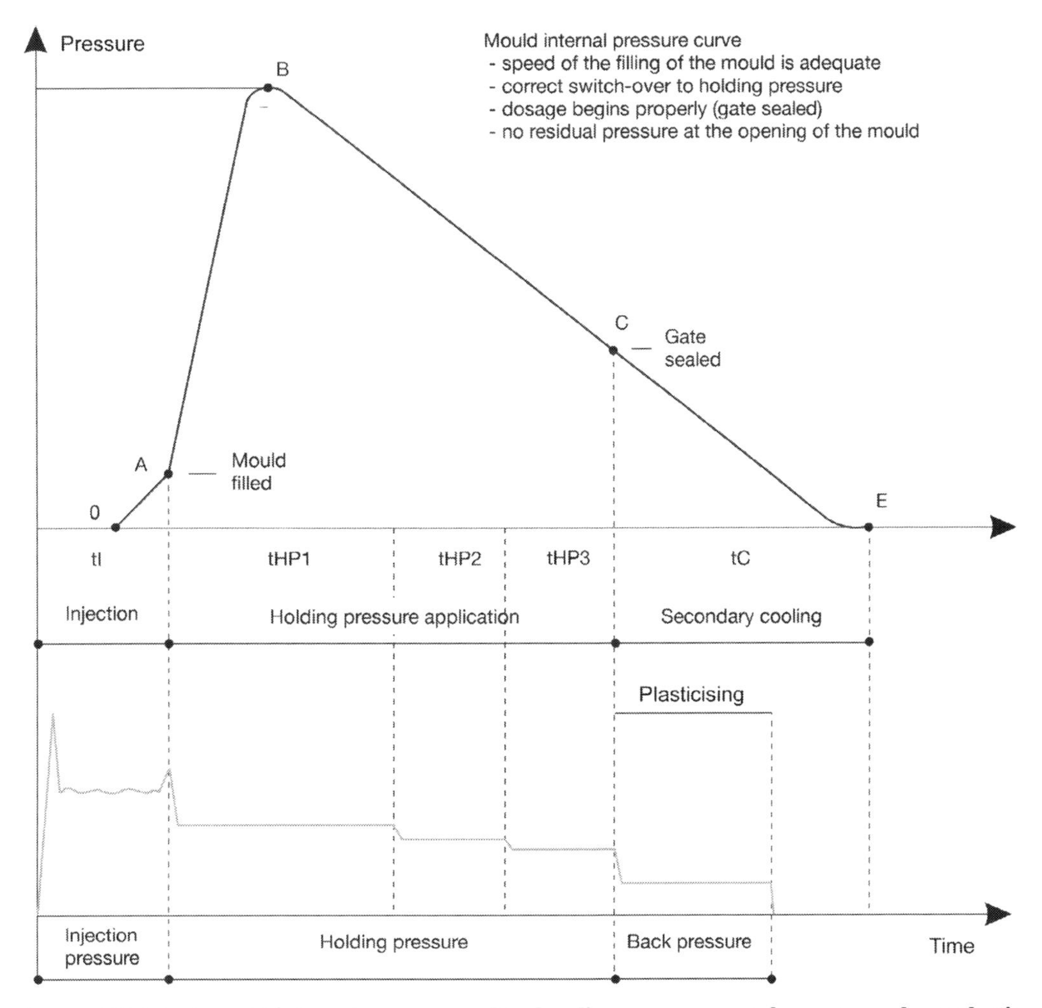

Figure 5.16 Ideal mould internal pressure and hydraulic pressure curve for an amorphous plastic

5.4 The User Interface

The injection machine operator will set the injection moulding machine by inputting settings to a controller unit. The basic specifications of three common units now follow.

5.4.1 Multitronica Control Unit – Standard Equipment

- Multronica controller (computer programmed microprocessor system)
- Digital data input via keypad and display on the monitor
- Data input as absolute values for pressures, forces, speeds, times and temperatures
- Data input in % for pressures, forces and speeds
- Diskette storage for 60 mould data sets
- Display of the operating condition of the monitoring switches, adjustment elements, etc., on the monitor
- Operator instructions and malfunction display in plain language on the monitor
- Optical malfunction display with selectable malfunction display duration
- Parts counter, cycle counter, counter pre-set, run time counter
- Mould blow programme, variable start and duration.

5.4.2 Dialogica Control Unit – Standard Equipment

- Dialogica controller (computer programmable microprocessor system)
- Digital data input via keyboard and visual display on monitor
- Remote control
- Data input in absolute values for pressures, forces, strokes, speeds, times and temperatures
- Extensive tolerance band inputs and monitoring
- Data diskette stores 100 complete moulding set-ups
- V24 interface for printer
- Automatic switch on'/off with weekly programme
- Operating modes:
  - automatic switch on
  - dry cycle without screw movement
  - automatic programmes
- Display of the operating condition monitored by switches, diagnostic display, etc., on the monitor
- Operator hints and fault diagnosis in clear text on the monitor
- Fault evaluation programme (alarm after selected number of errors)
- Switch off automatic with selected alarm display
- Various alarm follow-on functions (purging etc.), selectable before switching off the machine
- Optical alarm display with selectable duration
- Part counter, cycle counter, run time counter
- Mould blow programme, variable blow and start time

5.4.3 Selogica Control Unit – Standard Equipment

- Selogica controller (modular multi-microprocessor system)
- Regulation and monitoring of the control cabinet
- Application specific controller architecture with self recognising Bus system
- High definition colour graphic monitor in desk top format
- Data set administration via diskette
- Actionica manual control panel in mould proximity
- V24 interface for printer
- Control cabinet directly connected to machine on operator side
- Function panel selection via control keys
- Cycle sequence programming with representative symbols
- Cycle stage display as a flow diagram
- Graphical representation of various process signals
- 2 useable text pages
- User authorisation via key
- Text input via keyboard
- Automatic switch on/off and automatic shut down
- Quality monitoring with varying error evaluation programmes and tolerance monitoring
- Optical alarms
- Mould blow programme, variable start and blow time.

All these systems offer user-friendly systems and allow the operator to monitor and control the process. Sample Selogica screen shots are shown in Figures 5.17 and 5.18. For ease of use, they can be set for the language of the user. Figure 5.19 outlines the five function areas.

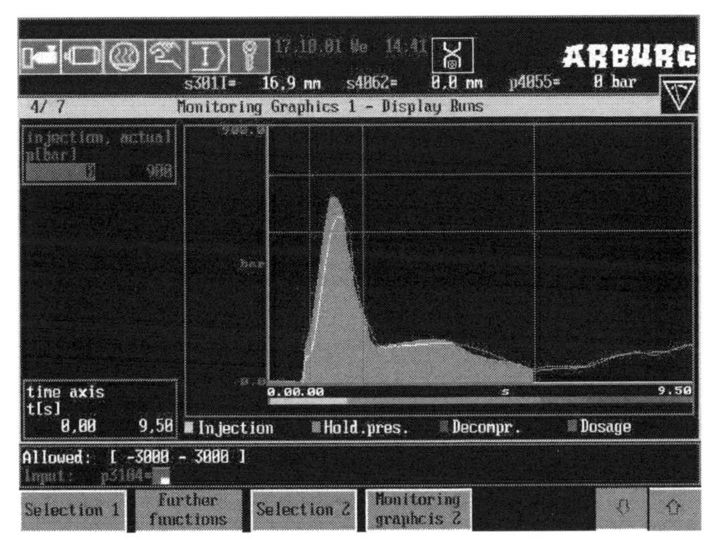

Figure 5.17 Sample screen shot – integral monitoring

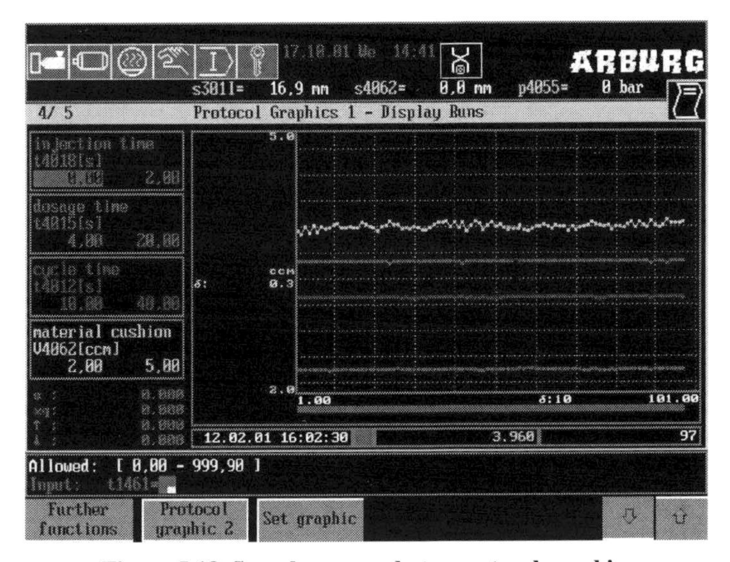

Figure 5.18 Sample screen shot – protocol graphics

The system is controlled via a keypad, through which the user can access all the various machine functions. Selogica control units split pages into five function areas:

1. Alarm line
2. Status display
3. Display area
4. Input area
5. Function keys display

Through these the operator can programme and control the entire processing sequence. Many systems also have production control, quality assurance and documentation packages to ensure the machine operates at as high a quality and performance as possible.

6 Processing Amorphous and Semi-Crystalline Thermoplastics

6.1 Introduction

Thermoplastics can be subdivided into two distinct classes based upon differences in molecular structure. These differences can have a bearing on the performance of mouldings in service, and have a most significant effect on the behaviour of the material during processing.

Materials such as polystyrene (PS), polycarbonate (PC), acrylics (PMMA), acrylonitrile-butadiene-styrene (ABS) and polyvinyl chloride (PVC) are said to be amorphous thermoplastics. This signifies that in the solid state their molecular structure is random and disordered, the long chain molecules being all entangled rather like solidified spaghetti.

Materials such as most of the nylons (PA), polyacetal (POM), polypropylene (PP), polyethylene (PE) and the thermoplastic polyesters (PET) have a much more ordered structure in the solid state, a considerable proportion of the long chain molecules being closely packed in regular alignment. These materials are known as semi-crystalline thermoplastics. It should be noted however, that with both the semi-crystalline and amorphous materials at sufficiently high temperature (this is when the material is in its melt state) the molecular structure is amorphous. Table 6.1 classifies some common materials into these two groups.

Most amorphous thermoplastics are transparent in their natural, unpigmented form, although ABS for example is an exception. Most semi-crystalline thermoplastics in their solid unpigmented form are translucent or an opaque white colour. It is interesting to observe (for example when purging injection moulding machines) that fully molten natural polypropylene or acetal are initially transparent, but as the melt cools it clouds over becoming translucent in the case of polypropylene, and opaque white in the case of acetal. This clouding is due to the material's molecular structure gradually rearranging itself from the tangled amorphous state in the melt to the more ordered semi-crystalline state in the solid.

Table 6.1 Classifying plastics

Amorphous	**Semi-crystalline**
Acrylic	Acetal
PVC	Nylon
SAN	Polyester
Polystyrene	Polyethylene
Polycarbonate	Polypropylene
ABS	PTFE

The main differences in behaviour between the amorphous and semi-crystalline materials observed during injection moulding are:

a) *Melting and solidification*

Amorphous thermoplastics exhibit a progressive softening over a wide temperature span, whereas the semi-crystalline materials rapidly change from the solid melt condition over a quite narrow temperature band.

Conversely, when amorphous materials are cooled they slowly solidify over a wide range of temperature, as against the semi-crystalline plastics, which change from melt to solid over a narrow range of temperature.

b) *Shrinkage*

Amorphous thermoplastics display very low shrinkage when they solidify, typically between 0.5% and 1%. Semi-crystalline materials shrink very much more, usually between 1.5% and 5% depending upon the particular material.

The higher shrinkage with the semi-crystalline materials is due to the repeat units along the molecular chains being of such a form that they can pack very closely together in an ordered manner. By use of appropriate moulding conditions it is possible to vary the extent of the crystalline areas. For example, when semi-crystalline thermoplastics are moulded in hot moulds, cooling rates are slow allowing more time for the molecular chains to disentangle themselves and take up their crystalline formation. This results in a greater proportion of the material being in its crystalline state (higher crystallinity) giving a product with superior mechanical strength and dimensional stability, but with relatively high shrinkage. If the same material is moulded in a cold mould, the more rapid cooling will inhibit the formation of crystalline areas. The resulting lower level of crystallinity will give the product inferior mechanical properties, and a lower shrinkage. This is accompanied by a tendency for dimensional instability and distortion during later service due to aftershrinkage.

The next section will describe the properties of some typical amorphous materials, semi-crystalline materials are discussed in Section 6.3.

6.2 Amorphous Plastics

This section gives details of properties and applications for the following eleven amorphous materials.

- Standard polystyrene (PS)
- Styrene copolymers (SB, SAN, ABS)
- Polyphenylene oxide (PPO), mod. (e.g., Noryl)
- Polyvinyl chloride (PVC)
- Cellulose acetates (CA, CAB, etc.)
- Polymethyl methacrylate (PMMA)
- Polycarbonate (PC)
- Polyacrylates (PAR, PAE, APE, PEC)
- Polysulfones (PSU, PES)
- Polyetherimide (PEI)
- Polyamide-imide (PAI)

6.2.1 Overview and Common Properties

Amorphous plastics are basically transparent (exceptions: styrene copolymers containing butadiene (B), such as SB and ABS). Amorphous plastics have lower shrinkage values than semi-crystalline compounds, so with amorphous compounds higher levels of accuracy can be obtained at lower cost than with semi-crystalline materials.

Table 6.2 Amorphous plastics: examples and properties

Section	Abbreviation as per DIN 7728	Tensile modulus of elasticity[2] (N/mm^2)	Tensile impact strength (Nmm/mm^2)[3]	Maximum operation temperature in air without load: Permanent[4] (°C)	Maximum operation temperature in air without load: Temporary[5] (°C)	Plasticising (freezing) temperature (°C)	Processing temperature (°C)	Flow behaviour[6]
6.2.2	PS	3200	1.5-3	50-70	60-80	90	160-250	I
6.2.3	SB	1800-2500	5-20	50-70	60-80	90	160-250	I
6.2.3	SAN	3600	2-3	85	95	100	180-260	M
6.2.3	ABS	1900-2700	7-25	75-85	85-100	105	180-260	MS
6.2.4	PPO mod.[1]	2500	8-15	80	150		280-320	S
6.2.5	PVC hard	1000-3500	2-5 or 50[7]	55-65	70-80	90	160-180	S
6.2.5	PVC soft		3 WB[8]	50-55	55-65	65	150-170	IM
6.2.6	CA	2200	2-40	50-70	80-90	100	160-230	I
6.2.6	CAB, CP	1000-1600	8-15	60-115	80-120	125	160-230	I
6.2.7	PMMA	2700-3200	1.5-3	65-90	85-100	105	200-250	M
6.2.8	PC[1]	2100-2400	20-35	135	145	150	250-300	S
6.2.10	PSU[1]	2600-2750	3-4	150	200	195	320-400	S
6.2.10	PES[1]	2450	4-6	200	260	230	320-400	S

1. These compounds also exist in reinforced forms (with fibre glass, etc.), which will exhibit differing properties from those shown
2. The ranges given are indicative of different grades and different processing conditions
3. 1 Nmm/mm^2 = 1 KJ/m^2 = 1 $kpcm/cm^2$
4. Months to years
5. Up to a few hours
6. I = easily flowing, M = average flow characteristics, S = flows with difficulty
7. Impact-resistant types
8. WB = without break

Amorphous plastics are subject to strong elastic deformation between melting point and freezing point. They must therefore be processed with a holding pressure low enough to avoid overloading in the moulds, which results in ejection difficulties and internal stresses. There is therefore a high risk with amorphous plastics as opposed to semi-crystalline plastics, that the injection moulded components will be encumbered with stress. If a higher injection pressure is required for filling the mould, two pressure stages (holding pressure control) must be used in processing. General properties of selected materials are given in Table 6.2.

6.2.2 Standard Polystyrene (PS) Homopolymer

Typical Characteristics and Applications

- Compound plastic
- Unplasticised, brittle
- Transparent
- Very low dielectric losses
- Used in economy price mass articles (housings, packs)
- Used for toys, office equipment

Processing data is given in Table 6.3.

Table 6.3 Processing data for polystyrene

Pre-drying	Generally not required; only if high surface quality is desired: forced-air drying cabinet, 2-3 h, at 70-80 °C. High surface moisture content leads to formation of waviness.
Compound temperature	Inside the cylinder: 160-280 °C depending on type. Excessively low compound temperature causes internal stresses. If compound temperature is too high, a sweet smell, a yellow colour and the formation of waviness on components indicate the onset of decomposition.
Mould temperatures	20-50 °C
Flow behaviour	Good to very good.
Holding pressure	Must be sufficiently low to avoid overloading, especially because PS is susceptible to stress cracking, so it is usually advisable to inject with two pressure stages (holding pressure control). Excessive holding pressure causes a tendency to rupture during ejection, as well as high internal stresses.
Processing shrinkage	0.2-0.6% Practically no aftershrinkage. If lifting tapers are too slight on external contours: difficulties in ejection possible.
Special features	PS is suitable for antechamber through injection process.

6.2.3 Styrene Copolymers

Typical Characteristics and Applications

Styrene-butadiene (SB)

- Somewhat more resistant to impact, but less hard and rigid than PS
- Not transparent (opaque) due to butadiene (B)
- More susceptible to ageing than PS due to B
- Used in housings and components for record players, electrical equipment, household equipment, refrigerator parts, camping equipment, shockproof packs.

Styrene-acrylonitrile (SAN)

- More rigid and harder than PS, impact resistance higher than PS, but lower than SB
- Can be transparent
- Used in technical components as per SB, especially when transparency required; light covers, scales, precision equipment housings.

Acrylonitrile-butadiene-styrene (ABS)

- Significantly higher impact resistance than PS and SB
- Not transparent due to butadiene (B)
- More susceptible to ageing than PS due to B
- Galvanisable types generally relatively problem-free
- Used in technical components, as per SB, however, with higher susceptibility to stress

Processing data is given in Table 6.4.

Table 6.4 Processing data for styrene copolymers

Pre-drying	SB: As for PS, usually not required SAN: Pre-drying (moisture leads to formation of waviness in components) ABS: Advisable for surface moisture: Forced-air drying cabinet, 2-3 h at 70-80 °C
Compound temperature inside the cylinder	SB: 160-250 °C SAN: 200-260 °C ABS: 180-240-(260) °C (Onset of decomposition indicated by yellowing and formation of waviness)
Mould temperatures	SB: 50-70 °C (80 °C if high surface gloss is desired) SAN: 40-80 °C ABS: 50-85 °C (higher values for high surface gloss)
Flow behaviour	SB: Somewhat worse than PS SAN: Worse than PS and SB ABS: Comparable to SAN
Holding pressure	Sensitive to overloading as with PS, for all three
Processing shrinkage	SB: Up to 7% SAN: 0.5-0.6% ABS: 0.4-0.7%

6.2.4 Modified Polyphenylene Oxide (PPO mod.)

Typical Characteristics and Applications

- The properties here represent a polyblend (mixture) of PPO and PS or PAN (polyacrylonitrile), about 50/50
- Rigid and impact resistant, similar to ABS, however slightly less tendency to creep and higher temperature resistance
- Generally self-extinguishing
- Not transparent
- High value mechanical and electrical components, similar to ABS only with susceptibility to stress
- Often used to replace metallic materials.

Processing data is given in Table 6.5.

Table 6.5 Processing data for modified polyphenylene oxide

Pre-drying	In general only if high demands on surface quality is required, forced-air drying cabinet, approx. 2 h at about 85-100 °C. Drying too abruptly can lead to material feeding difficulties.
Compound temperature inside the cylinder	260-300 °C depending on type. Higher temperatures only for short dwell times. Overheating the melt reduces mechanical properties, and in particular, causes brittleness.
Mould temperatures	80-150 °C depending on type. Higher temperatures lead to better flow behaviour, fewer internal stresses, higher surface finish at higher temperatures.
Flow behaviour	Lower than for ABS, rapid filling of mould is required, as solidification begins as early as 240 °C; the moulding system should be designed accordingly.
Injection pressure	1000-1400 bar, in order to fill mould faster.
Processing shrinkage	0.5-0.9%, practically no aftershrinkage.
Special features	Only fibre glass reinforced types should be used for inserts.

6.2.5 Polyvinyl Chloride (PVC)

Rigid PVC = PVC without plasticiser
Soft PVC = PVC with plasticiser

Typical Characteristics and Applications

- Compound plastic similar to PS, however it is more often processed into half-finished products by extrusion moulding, as there is less play in the processing temperatures with injection moulding and thermal decomposition releases corrosive hydrochloric acid (HCl)
- Used in components which must be flame-resistant and self-extinguishing, or for requirements for chemical resistance (salt solutions, many acids, alkalis, petrol, mineral oils, fats and alcohol)
- Unplasticised PVC: Hard and rigid like PS, slightly less brittle
- Plasticised PVC: Rigidity depends on plasticiser type and content, often brittle at low temperatures
- Even unplasticised PVC cannot be used at above 60-70 °C
- Slight tendency to stress crack formation
- Can be transparent.

Processing data for PVC is given in Table 6.6.

Table 6.6 Processing data for PVC

Pre-drying	Generally not required
Compound temperature inside the cylinder	Unplasticised PVC: 160-180 (200) °C. Decomposition can be recognised by brown or black colouration of moulded components and pungent smell of HCl. Plasticised PVC: 150-170 °C. Depends on type and addition of stabilisers. Short temperature loading, i.e., short dwell time in cylinder reduces danger of decomposition.
Mould temperatures	30-70 °C, higher values advisable to improve flow behaviour.
Flow behaviour	High melt viscosity for unplasticised PVC. Injection speed not too high to avoid overheating through excessive shear rate. Sprue runners and gates as wide and short as possible. Avoid pin-point gates and sharp turns.
Injection pressure	From 400 bar (plasticised PVC) and 1000 bar (unplasticised PVC) up to 1500 bar
Screw circumferential speed, back pressure	To reduce overheating due to friction, work with a lower screw speed (circumferential speed max. 0.08-0.1 m/s) and low back pressure (40-80 bar).
Processing shrinkage	Unplasticised PVC: 0.2-0.5% Plasticised PVC: 1-2.5% Practically no aftershrinkage with unplasticised PVC
Special features	Use only suitable material grades for injection moulding. Only screw injection moulding machines are suitable for PVC injection moulding. Only in exceptional cases can a few types of unplasticised PVC be processed on plunger-type machines. To keep the dwell times in the cylinder short, the shot volume should be not less than 20% of the maximum cylinder volume. Dead spots in the compound flow are to be avoided so work only with open nozzles and also if possible without a non-return valve. Even in normal processing, decomposition phenomena cannot entirely be avoided. Since the HCl released is very corrosive, cylinder fittings and moulds must be protected against corrosion (mould hard-chrome plated or made from corrosion-resistant steel).

6.2.6 Cellulose Materials

Cellulose acetate, CA
Cellulose acetobutyrate, CAB
Cellulose acetopropionate, CAP
Cellulose propionate, CP

Typical Characteristics and Applications

- Brilliant and scratch-resistant surfaces due to self-polishing effect
- High viscosity
- Rigidity depends on type and quantity of plasticiser (always present)
- High water absorption
- Low tendency to becoming dusty
- Can be transparent
- Used in handling elements of all kinds, which must have high surface finish, such as control keyboards, tool handles, filling container housings, and the like, and specially those with metal inserts.

Processing data is given in Table 6.7.

Table 6.7 Processing data for cellulose materials

Pre-drying	If water absorption above 0.2%: Forced-air drying cabinet, 3 h at approx. 80 °C. Excessively high drying temperatures lead to migration of plasticiser (in some circumstances leading to corrosion).
Compound temperature inside the cylinder	180-230 °C depending on type Higher compound temperatures increase strength When running in, start at lower limiting values, in view of danger of decomposition and separating off of plasticiser
Mould temperatures	40-70 °C
Flow behaviour	Good: aim for highest possible injection speed
Injection pressure	800-1200 bar
Processing shrinkage	0.2-0.5-0.7% Lower values with lower wall thickness. Practically no aftershrinkage, but dimensions change due to absorption of moisture

6.2.7 Polymethyl Methacrylate (PMMA)

Copolymers:
Acrylonitrile-methyl methacrylate (AMMA) has a higher chemical resistance than PMMA.
Methyl methacrylate-butadiene-styrene (MBS) has a higher impact strength than PMMA, without any significant impairment of its transparency.

Typical Characteristics and Applications

- High transparency and high resistance to ageing
- Brittle, hard, rigid
- Breaks without plastic deformation, but does not splinter
- High static charge
- Used in high class optical components (lenses, prisms), weather resistant elements such as light covers, reflectors, light windows, advertising panels, visual education models.

Processing data is given in Table 6.8.

Table 6.8 Processing data for PMMA	
Pre-drying	Efficient pre-drying essential. Forced-air drying cabinet, 2-3 h. (without air circulation, 4-6 h) at 70-110 °C, depending on type. Moisture leads to formation of blisters and waviness of components.
Compound temperature inside the cylinder	200-250 °C To avoid sink marks on very thick-walled components, it may be necessary to work at 170 °C. This gives a very high melt viscosity which requires a high screw torque (risk of overload).
Mould temperatures	50-90 °C High temperature improves the flow behaviour and reduces the internal stresses.
Flow behaviour	Average to high melt viscosity, depending on type Filling speed should not be too high, excessively strong shearing action leads to overheating of compound and causes formation of waviness. Large runner and gate cross-sections are especially necessary where high demands are made on optical quality.
Injection pressure	400-1500 bar (2000) Select sufficiently low holding pressure because of overload risk. Injection with two pressure stages may be necessary.
Screw circumferential speed and back pressure	Excessively high screw speeds lead to overheating through frictional heat.
Special features	Where the component has to fulfil high optical requirements, care should be taken to ensure that the screw, cylinder, etc., are as clean as possible, the granules being free from dust (risk of attracting dust through electrostatic charge). Subsequent post cure of the component is advisable, if it can come into contact with media which may generate stress cracks, or it is to be glued or lacquered. 5 °C below the temperature at which deformation begins to occur (approx. 60-90 °C, depending on type), cool slowly.

6.2.8 Polycarbonate (PC)

Typical Characteristics and Applications

- Hard and rigid
- Extremely impact resistant from –100 °C up to +135 °C (for short periods)
- Low creep tendencies
- Low water absorption (max. 0.5%) and low heat expansion, giving components with stable dimensions
- Susceptible to stress cracks with specific chemicals; susceptible to long-term effects of hot water and steam
- Can be transparent (often with light yellow or blue tinge)
- Very impact resistant (transparent) construction elements, housings and protective covers, especially for electrical units and installations, spools, plug boards, precision components, requirements, eating utensils.

Processing data for PC is given in Table 6.9.

Table 6.9 Processing data for polycarbonate	
Pre-drying	Absolutely essential. Forced air drying cabinet, 8-14 h at 120-130 °C. Important to keep dry until fed into the screw. Injection using pre-warmed granules improves the surface quality, and using heated granule hoppers or dry conveyor units is recommended. The moisture absorption of PC is relatively low (max. 0.5%), but absorption takes place very rapidly, however, for processing it may only reach 0.02% at most, any moisture in excess of this leads to a reduction in the mould-filling capacity and to the formation of waviness on the component.
Compound temperature inside the cylinder	250-330 °C Thermal damage (discolouration) begins above 340 °C and results in a reduction in impact resistance, etc.
Mould temperatures	80-120 °C High temperatures to improve flow behaviour and reduce risk of internal stresses.
Flow behaviour	Melt very viscous. Mould filling more strongly influenced by rise in temperature than by increase in pressure. Select injection speed as high as possible. Sprue runners and gates with suitably large cross-sections and as short as possible.
Processing shrinkage	0.7-0.8% in direction of flow and almost the same transversely (low tendency to orientation); practically no aftershrinkage.
Injection pressure	800-1600 bar
Special features	If production is interrupted, lower cylinder temperatures to 160-170 °C only, as cool PC adheres strongly to the cylinder wall. Cylinder, screws and nozzles must usually be thoroughly cleaned if the injection moulding compound is changed. Otherwise, dark particles may become detached from the heat-damaged surface layer upon re-starting and influence the component quality.

6.2.9 Polyacrylates (PAR, PAE, APE, PEC)

Polyacrylate (PAR)
Polyacrylic ester (PAE)
Aromatic polyester (APE)
Polyester carbonate (PEC)

The construction and properties of polyacrylates are similar to those of polycarbonates (PC). The processing temperatures range is between PC and polysulfone (PSU).

Typical Properties

Similar to polycarbonates, but:

- Higher rigidity
- Higher processing temperature
- Lower viscosity
- Electrical properties similar to PC, durability similar to PC (chemicals, ambient influences)

Typical values for modulus, transition temperature and tensile impact strength are shown in Figure 6.1.

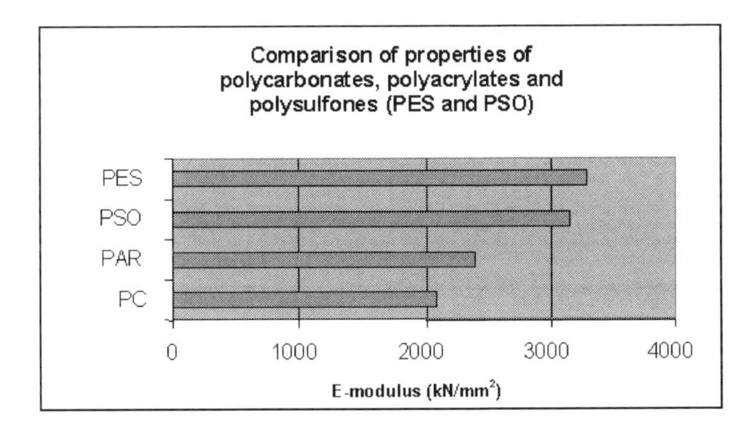

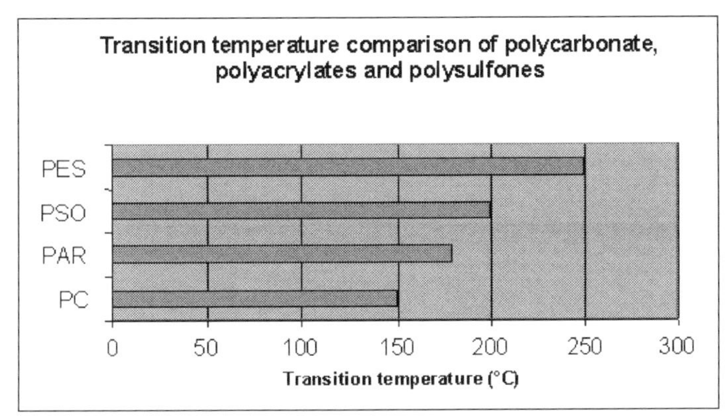

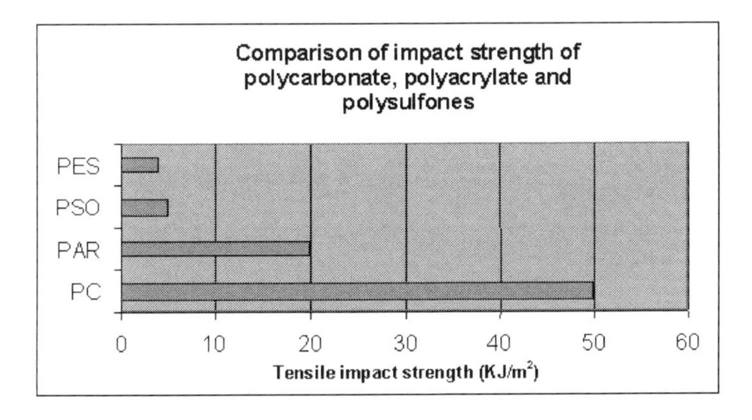

Figure 6.1 Typical properties of polyacrylates

Typical applications as PC, but also for higher temperatures.

Typical processing data is given in Table 6.10. Polyacrylate processing is similar to PC but with higher material temperatures (330-370 °C) and higher melt viscosity.

Table 6.10 Processing data for polyacrylates

Pre-drying	Absolutely necessary: Circulating air dryer 5-8 h, 110-120 °C dehumidifier 3-5 h, 120-140 °C. PAR only absorbs little humidity (max. 0.5%), but it absorbs it very quickly; for processing the humidity should not exceed 0.02%; more humidity leads to a reduction of the mould fill behaviour and leads to part surface waviness. To avoid the absorption of humidity inside the feed hopper of the injection moulding machine the hopper should be heated or a dry feeder used.
Compound temperature inside the cylinder	(320)-330-370 °C Longer dwelling with more than 350 °C (e.g., in the case of faults) should be avoided. Thermal damage begins from 370-380 °C, perceivable by discolouration. Consequence: reduction of the impact resistance, etc.
Mould temperatures	40-100 °C Upper range temperatures improve the flow behaviour of the mass and reduce the formation of internal stresses. Higher temperatures can make demoulding more difficult.
Flow behaviour	The melt is more viscous than PC, similar to PSU. The viscosity of the melt does not diminish with increasing shear speed. The sprue channels and the feed orifices should have corresponding cross-sections. The diameters of the feed orifices should not be smaller than 1.2 mm.
Injection speed	Should not be too high to keep the internal stress as small as possible
Injection pressure	1000-1600 bar
Holding pressure	600-1300 bar to keep internal stress as small as possible
Maximum internal pressure	350-650 bar
Back pressure	80-120 bar
Shinkage	Processing shrinkage: 0.7-0.9%, independent of the flow direction Aftershrinkage: almost 0
Special features	Reduce the cylinder temperatures to 160-170 °C when production is interrupted as frozen PAR adheres strongly to the cylinder wall. When you change the material type it is usually necessary to clean cylinder, screw and nozzle carefully. Otherwise dark particles of the thermal damaged material from the wall layer will reduce the quality. With processing of transparent PAR, one special nozzle should be reserved exclusively for this purpose. When you change the material to PAR you should inject PC first, as PAR has to be processed with high temperatures. For cleaning, PC is also recommended.

6.2.10 Polysulfones

Polysulfone (PSU)
Polyether sulfone (PES)

Typical Characteristics and Applications

- Hard, rigid, and at the same time extremely impact-resistant
- Similar to PC, however low creep tendency over even greater temperature range (–100 to +170 °C)
- Stress crack formation can occur in some media
- May be transparent, but with yellowish-brown colouring
- Used in construction elements subjected to high mechanical, heat and electrical stress, especially if transparency is required.

Processing data for polysulfones is given in Table 6.11.

Table 6.11 Processing data for polysulfones

Pre-drying	Required; forced-air drying cabinet, 120-140 °C, 4-6 h vacuum cabinet: 120-140 °C, 2.5-3 h
Compound temperature inside the cylinder	330-400 °C No processing difficulties worth mentioning, in spite of this high processing temperature.
Mould temperatures	100-160 °C (180 °C)
Injection pressure	700-1400 bar
Processing shrinkage	0.7-0.8%
Aftershrinkage	Practically no aftershrinkage
Special features	Do not use silicone based mould release agents with PSU. Post cure of the component is recommended to reduce the internal stresses and improve the mechanical properties: up to 5 min. in oil or glycerine, up to 5 hours in air at approx. 165 °C.

6.2.11 Polyetherimide (PEI)

E.g., Ultem (GEP)

Typical Properties and Applications

- Very high solidity and rigidity, also not reinforced, up to +210 °C low tenacity
- Low creep tendency
- High endurance under completely reversed stress
- Low thermal expansion coefficient
- Very low dielectric losses, high dielectric rigidity, very resistant to chemicals, very resistant to hydrolysis, very resistant to ambient influences, very resistant to radiation
- High non-flammability, low development of smoke
- Used for parts which are put under a lot of mechanical, thermal and electrical stress, e.g., bearings, ball bearing guides, cogs, gear pans, carburettor housings, valve covers, connector pans, flow solderable flat connector blocks, integrated circuit housings, parts of microwave ovens
- Natural condition: transparent amber.

Processing data is given in Table 6.12.

Table 6.12 Processing data for polyetherimide (PEI)	
Processing	Very high flow capability Frozen melt tends to adhere strongly to the cylinder
Pre-drying	Preliminary drying absolutely necessary. Circulating air dryer 4-6 h, 150 °C. Dehumidifier 3-5 h, 150 °C For processing the humidity should be lower than 0.05% (absorption of humidity up to 0.25% within 24 h)
Compound temperature inside the cylinder	340-425 °C In many cases a temperature of 360 °C is optimal (Ultem)
Mould temperatures	60-175 °C In many cases a mould temperature of 95 °C is optimal. A high mould temperature improves the mould fill and the rigidity of the joint line. It reduces internal stresses (can stand higher temperatures and has higher resistance to chemicals)
Flow behaviour	Very high flow capability Minimal wall thickness of 0.25 mm possible Frozen melt tends to adhere strongly to the cylinder
Injection speed	Select mid-range values
Injection pressure	750-1500 bar
Holding pressure	400-750 bar
Maximum internal pressure	350-650 bar
Back pressure	40-80 bar
Processing shrinkage	0.7-0.5-0.1% reinforced with fibre glass
Special features	Be careful that the melt does not solidify in the cylinder (extremely adhesive). Reduce the cylinder temperature to 180-200 °C when the production is interrupted. Clean the cylinder carefully after having finished the production.

6.2.12 Polyamide-Imide (PAI)

E.g., Torlon (Amoco)

Typical Properties

- Very high strength and rigidity between –200 and +260 °C
- High impact resistance
- Low creep tendency
- High endurance under completely reversed stress
- Very resistant against abrasion
- Low thermal expansion coefficient (9-30 x 10^{-8} °C)
- Very low dielectric losses
- Low stress cracking susceptibility
- Very resistant against chemicals
- Very stable against UV and high energy radiation
- Not resistant against hot water
- Low exhalation under high-vacuum
- High non-flammability
- High resistance against oxidation
- Typical applications: for pans which are put under a lot of mechanical, thermal and electrical stress up to +260 °C and under abrasion stress such as cams, bearings, sliding rings, vanes for hydro and pneumatic motors, connector parts, spark suppression covers
- Natural condition: brown.

Typical processing data is given in Table 6.13.

Table 6.13 Processing data for polyamide-imide (PAI)	
Pre-drying	Necessary: l6 h at 150 °C or 8 h at 180 °C
Compound temperature inside the cylinder	340-360 °C
Mould temperatures	ca. 230 °C
Flow behaviour	Very low Very high melt viscosity after injection, heat treatment is recommended
Injection speed	As high as possible It is recommended to use an accumulator
Injection pressure	750-1550 bar
Holding pressure	550-1050 bar
Maximum internal pressure	450-750 bar
Back pressure	40-80 bar Some types need a special screw with a compression ratio of 1:1
Special features	To reach optimal properties a heat treatment is necessary; up to 3 mm wall thickness: 48 h (2 days), for thicker walls: 3-4 days, for some types: 7-17 days

6.3 Semi-Crystalline Plastics

This section gives details of properties and applications for the following eleven material types:

Polyethylene (PE)
Polypropylene (PP)
Polyamide (PA)
Polyacetal (POM)
Polyterephthalate (PET, PBT)
Polyphenylene sulfide (PPS)
Polyfluoroolefins (PFA, FEP, ETFE, PVDF)
Polyacrylic acid (PM)
Polyphthalamide (PPA)
Polyaryl ether ketone (PAEK) (PEEK, polyetheretherketoneketone (PEEKK), polyether ketone (PEK))
Liquid crystal polymer (LCP)

6.3.1 Overview and Common Properties

Semi-crystalline plastics are naturally of a milky opaque colour (thus non-transparent) as a result of light diffusion at the crystallite borders. Semi-crystalline thermoplastics as a whole have higher shrinkage values than amorphous plastics between solidifying and transition temperatures. The risk of overloading is therefore considerably less; thus ejection difficulties due to this reason scarcely occur in semi-crystalline components. At the same time increased shrinkage is equalised by favourable anti-friction behaviour to give a positive effect. An overview of the general properties is shown in Table 6.14.

Table 6.14 Semi-crystalline plastics: examples and properties

Section	Abbreviation as per DIN 7728	Tensile modulus of elasticity[3] (N/mm^2)	Tensile impact strength as per DIN 53453[4] (Nmm/mm^2)	Maximum operation temperature in air without load		Plasticising (freezing) temperature approx. (°C)	Processing temperature (°C)	Flow behaviour[7]
				Permanent[5] °C	Temperature[5] °C			
6.3.3	PE soft	200-500	18-WB	60-75	80-90	105-115	160-240	I M
	PE rigid	700-1400	4-WB	70-80	90-120	125-140	190-280	M S
6.3.4	PP [1]	1100-1300	4-18	100-110	130-140	158-168	200-270	M S
6.3.5	PA 6.6 [1, 2]	2000-2900	5-WB	80-120	170-200	250-265	260-300	
	PA 6 [1,2]	1400-3200	3-WB	80-100	140-180	215-225	230-260	I
	PA 6.10 [1,2]	2000	7-10	80-100	150	210-225	220-260	I
	PA 11[1]	1000	30-40	70-80	140-150	180-190	200-250	I
	PA12 [1]	1600-1700	10-20	70-80	140-150	175-185	190-250	I M
	PA amorphous [1]	2000	13	80-100	130-140	150-160	260-300	M
6.3.6	POM [1]	2800-3200	5-10	90-110	110-140	165-175	200-220	I
6.3.7	PET [1]	3100	3-6	100	200	255-258	260-280	I
	PBT [1]	2000	3-6	100	165	220-225	230-270	I
6.3.8	PPS [1]	3400	3-8	200	300	280-288	315-360	I
6.3.9	FEP	350-650	WB	205	250	285-295	340-360	S
	ETFE	1100	WB	150	220	270	315-365	S
6.3.10	PAA [1]	11300-17700	7.5-27.5	115-145	-230	235-240	250-290	I
6.3.11	PPA [1]	6300-14500[10]		to 185	-280	310	320-350	I
6.3.12	PAEK [1]	4000	WB	250	-300	335-370	350-420	I
6.3.13	LCP [1]	10000-35000[9]	20-70	120-240		270-380	300-450	I M

1. These compounds also exist in reinforced forms (with fibre glass, asbestos fibres, etc.), then some of the values given in the table are replaced by others
2. The mechanised values given are valid for the conditioned state, i.e., with a sufficiently high water content
3. The ranges given are conditioned by differences of type and different processing conditions
4. 1 Nmm/mm^2 = 1 KJ/m^2 = 1 $kpcm/cm^2$, WB = without break
5. Months to years
6. Up to a few hours
7. I = easily flowing, m = average flow characteristics, 5 = flows with difficulty
8. Impact-resistant types
9. With optimal orientation
10. Filled or reinforced material

6.3.2 Polyethylene (PE)

Polyethylene can be purchased in a variety of densities and properties dependent upon the formation of the polymer chains. The most common materials encountered in injection moulding are low density polyethylene (LDPE) and high density polyethylene (HDPE).

The density range for LDPE is 0.91-0.94 g/cm^3. The low density being due to strongly branched molecules.

For HDPE the density range is 0.94-0.96 g/cm^3. The high density is due to molecules with few branches.

Operating and processing properties of polyethylenes are strongly dependent on the density (molecule shape) and the degree of polymerisation (molecule length or molecular weight).

It can be seen that as the density and the degree of polymerisation increase and melt flow index (MFI) falls there are:

- Increases in, for example, hardness, rigidity, upper operating temperature (from 80 to 105 °C, 120 °C for brief periods)
- Decreases in, for example, brittle temperature, tendency to stress crack formation
- It becomes more difficult to process by injection moulding because flowability is reduced.

Typical Characteristics and Applications

- Flexible (LD) to tough and hard (HD)
- Cold resistant down to –50 °C and below
- Practically indestructible, however, becomes brittle under the effect of UV (can be stabilised against UV)
- No water absorption
- Very low dielectric losses
- High chemical resistance (against nearly all common solvents)
- Gas permeability higher than for many other plastics, though steam permeability very low
- Very transparent (semi-crystalline)
- Used in drain lugs, covers, containers, bottle crates, household articles, toys. Not used for precision components because of high shrinkage and aftershrinkage.

Processing data is given in Table 6.15.

Table 6.15 Processing data for polyethylenes

Pre-drying	Generally not required (no water absorption). Short pre-drying period only needed if surface moisture could have built up after long storage in unsealed container.
Compound temperature	160-280 °C depending on flow behaviour. At lower temperatures, dull points appear on component.
Mould temperatures	20-70 °C Higher surface gloss and higher degree of crystallisation at higher temperatures thus producing less aftershrinkage. At higher temperatures above 70 °C, ejection difficulties due to compound adhering to mould.
Flow behaviour	Good to medium, depending on density and molecular weights.
Shrinkage	Processing shrinkage 1.5-3.5% for easy flowing types. 2-4% for difficult flowing types. Postshrinkage unavoidable, as even at 70 °C mould temperature the cooling speed is still so great that the crystallisation in the mould cannot be completed.
Injection pressure, holding pressure time	500 bar (LD)-1200 bar (HD). If holding pressure time ends before gate freezes, this leads to formation of sink marks or cavities, because of the high shrinkage.
Special features	PE is suitable for pre-chamber through injection process. Gluing, printing, lacquering, and other processes only possible after pre-treatment (because of high chemical resistance).

6.3.3 Polypropylene (PP)

The way the atoms are arranged in the polymer molecule has a great influence on the properties of a polymer component. This order is named the tacticity and with polypropylene there are two different arrangements commonly available isotactic and atactic. The degree of tacticity determines the level of crystallisation possible. Atactic polymers are those with side groups placed in a random order. Atactic PP with nonuniformly arranged methyl groups has a density of 0.94 g/cm^3. Isotactic polymers are those whose side groups are all on the same side. Isotactic polypropylene has a density of 0.90 g/cm^3. There are also copolymers with ethylene, to increase impact resistance below 0 °C.

Typical Characteristics and Applications

- Higher rigidity and hardness than PE
- Impact resistance falls off sharply
- Upper operating temperature 110 °C and possibly higher in stabilised types
- Good electrical properties similar to PE
- No tendency to stress crack formation
- Lower chemical resistance than PE
- Colourless, more translucent (less opaque) than PE
- Used in applications such as fan wheels, heating ducts in vehicles, internal components for washing machines and dish washers, transport cases, containers with film hinges.

Processing data for PP is given in Table 6.16.

Table 6.16 Processing data for PP

Pre-drying	Usually not required, as with PE
Compound temperatures	(170 °C)-200-270 °C-(300 °C)
Mould temperatures	50-100 °C Upper temperatures give better surface gloss, dull points at lower temperatures
Flow behaviour	Flowability better than PE, specially in lower processing temperature range. Shut-off nozzles are usually required, if working without material decompression.
Shrinkage	Processing shrinkage in direction of flow 1.3-2%, crosswise to direction of flow 0.8-1.8%.
Injection pressures, holding pressure time	1000-1500 bar Select sufficiently long holding pressure time (similar to PE)
Special features	PP is suitable for pre-chamber through injection process. The plasticising efficiency of the machine is lower, due to the lower density (only approx. 70% compared to polystyrene).

6.3.4 Polyamide (PA)

The most important types of PA are shown in Table 6.17.

Table 6.17 Common types of PA

Type	Density	Water absorption
PA 6	1.12-1.15 g/cm^3	2.8-3.2%
PA 6.6	1.12-1.15 g/cm^3	2.5-2.7%
PA 6.10	1.06-1.08 g/cm^3	1.2-1.4%
PA 11	1.04 g/cm^3	0.8-0.9%
PA 12	1.01-1.02 g/cm^3	0.7-0.8%
PA amorphous types	1.06-1.08 g/cm^3	0.3-1.1%

The PA types with numbers are semi-crystalline. They are distinguished by the number of C atoms of the monomers (building blocks) from which the thread shaped molecules have been formed thus, for example, PA 6 is made from a kind of monomer with 6 C atoms. PA 6.10 is made from two different kinds of monomer, one of which has 6 C atoms and the other 10 C atoms.

The properties of polyamides are vitally affected by the degree of crystallinity and the water content. Water absorption leads to dimensional changes.

As the number of C atoms increases (higher index numbers), the water absorption capability falls rapidly (increased dimensional stability). Strength and rigidity show a slight tendency to fall. The price rises. (PA 12 is twice as expensive as PA 6).

Typical Characteristics and Applications

- High rigidity and toughness, good friction and wear properties
- PA 6.6 has the greatest strength and wear resistance
- As water absorption increases, the rigidity decreases and toughness increases
- Cold resistant down to around –40 °C. Upper operating temperature 80-120 °C, depending on type, 140-210 °C for brief periods (cf. melt temperature)
- Partly self extinguishing
- High dielectric losses (not suitable for high frequency (HF) use)
- Low tendency to become dusty (water absorption reduces surface resistance)
- Good creep resistance
- Resistant to petrol and oil, and numerous solvents, weather resistant (use stable types for thin walled components)
- Low tendency to stress crack formation
- Cannot be transparent apart from amorphous types
- Used in construction elements subjected to friction stress and wear stress, such as slide elements, bearings and similar coupling elements operating on a ball or roller bearing principle, tread rollers, cams, etc., fan wheels, fittings, housings for electrical tools, electric motors, pumps, etc. (usually with fibre glass reinforcement)
- PA 6 and PA 6.6 are not suitable for precision components on account of high water absorption (low dimensional stability).

Processing data for polyamides is shown in Table 6.18.

Table 6.18 Processing data for PA	
Pre-drying	Absolutely necessary, vacuum drying cabinet, a few hours at 80 °C. Exception: granules packed in tin canisters in same condition as when delivered. Moisture causes surface waviness and impairment of mechanical properties.
Compound temperatures	10-40 °C (50 °C) above melt temperature: PA 6 215 °C, PA 6.6 250 °C, PA 6.10 200 °C, PA 11 187 °C, PA 12 177 °C If material feed difficulties arise, select cylinder temperatures rising toward throat side. Dwell time in cylinder to be as short as possible. Melt tends to oxidise if air enters (yellowing). Semi-crystalline polyamides have a sharp melt temperature, so take care that the gate does not freeze too early, so that the holding pressure can be effective as long as possible (gate dimensions, compound temperature, mould temperature).
Mould temperature	Generally 50-90 °C. If higher crystallinity desired: 100-120 °C. The higher the degree of crystallisation (crystallinity) the lower the aftershrinkage due to after crystallisation. The degree of crystallisation (if low) can also be increased by subsequent re-heating treatment. If crystallinity is uniform and rising, the wear resistance, in particular, also rises.
Flow behaviour	Very good, melt very thin. Shut-off nozzles are required if it is not possible to work with compound decompression. Fluidity causes tendency to formation of scratches on moulds not closing well.
Shrinkage	Processing shrinkage 0.2-2.5%, depending on type and component. Post shrinkage: decreases as processing shrinkage increases.
Injection pressures	Injection pressure 450-1550 bar Holding pressure 350-1050 bar
Special features	Because of the high heat of fusion, usually only around 60% of melting efficiency is achieved in comparison with PS. The typical properties of PA such as, for example, good toughness, only come into play if the water absorption is sufficient. In injection fresh (dry) components, this means that the moisture content to be expected under operation conditions is deliberately introduced by storage in water or by atmosphere humidity. This is also important for checking the operating dimensions.

6.3.5 Polyacetals (POM)

Homopolymers or copolymers made from formaldehyde.

Typical Characteristics and Applications

- High levels of hardness and rigidity
- Good elastic properties
- Low creep tendency
- Good friction and wear properties (slightly lower wear resistance than PA 6.6, but lower friction coefficient)
- Cold resistant down to –50 °C, upper operating temperature 100 °C, 150 °C for brief periods
- Low moisture absorption
- Favourable dielectric behaviour
- Resistant to petrol, oil and numerous solvents, weak acids; copolymers also resistant to alkali solutions
- Practically no tendency to stress crack formation in air
- Natural colour strongly whitish-opaque
- Applications: precision components, parts subjected to friction stress and wear stress (bearings, gears, etc), similar to PA, however with better dimensional stability. All kinds of parts with high permanent load, frequently replacing metals. Components with flexible parts, especially for tear off closures.

Typical processing data is given in Table 6.19.

Table 6.19 Processing data for polyacetals

Pre-drying	Generally not required, only necessary for very moist material, very high humidity (associated) reduces heat resistance during injection moulding On microcomponents, to increase flowability. Treatment in vacuum drying cabinet recommended to remove volatile constituents.
Compound temperature	Homopolymer: 215±5 °C Copolymer: 205±5° C Copolymers are somewhat less susceptible to overheating than homopolymers. Dwell time in cylinder as brief as possible, decomposition starts at temperatures above 230 °C. If possible, work with open nozzles.
Mould temperatures	50-120 °C Upper limits on precision components, to keep aftershrinkage very low.
Flow behaviour	Very good
Shrinkage	Processing shrinkage 1-3.5%, depending on wall thickness (upper values for large wall thickness) and processing conditions. Post shrinkage decreases as processing shrinkage increases. Aftershrinkage can be forestalled by re-heating.
Injection pressures	800-1700 bar Holding pressure must be sufficiently high and of sufficient duration. No pressure jumps should occur during the holding pressure phase, so as not to interfere with the uniform crystallisation cycle. For precision components, gates of around 2/3 of maximum wall thickness are required (so that the holding pressure can operate as long as possible)

6.3.6 Polyterephthalates (Linear Polyesters)

Polyethylene terephthalate (PETP or PET)

Polybutylene terephthalate (PBTP or PBT) is identical to polytetramethylene terephthalate (PTMP)

PET has a very low crystallisation rate. Certain types remain amorphous (transparent), if cooling is rapid (low mould temperatures).

Typical Characteristics and Applications

- High levels of hardness and rigidity
- Very low creep tendency
- Friction and wear properties better for PET than for POM at correspondingly high levels of crystallinity, about the same wear resistance as PA 6.6 but a lower friction coefficient
- Cold resistance: PET cold resistant down to –40 °C, PBT down to –60 °C. Upper operating temperatures for (crystalline) PET, 100 °C, higher for short periods; for PBT 110 °C, 170 °C for short periods
- Very low water absorption
- Favourable dielectric behaviour
- Resistant to oil, petrol, dilute acids and alkalis, salt solutions, not resistant to hot water and hot steam (hydrolysis)
- No tendency to stress crack formation in air
- With, amorphous PET, cloudiness can arise at high temperatures owing to crystallisation
- Used in components subject to friction stress and wear stress and components under high permanent load, similar to POM. High crystallinity required for semi-crystalline precision components (with PET obtainable only through high mould temperatures).

Processing data is given in Table 6.20.

Table 6.20 Processing data for polyterephthalates

Pre-drying	Moistened granules 3-4 h at 100-120 °C Moisture decreases thermal stability during processing
Compound temperature	PET: 260-280 °C PBT: 230-270 °C Only brief dwell times at upper temperatures, because of risk of heat damage.
Mould temperatures	PET: 30-60 °C (amorphous) up to 140 °C (semi-crystalline) PBT: 30-60 °C For precision components made from PET: 140 °C With semi-crystalline PET, excessively low temperatures lead to nonuniform crystallisation (mottling on non pigmented compounds)
Flow behaviour	For PBT favourable, similar to PA, for PET somewhat less good
Shrinkage	Processing shrinkage 1-2%. Postshrinkage with semi-crystalline PET only becomes negligibly small at mould temperature of 140 °C.
Injection pressures	1000-1700 bar Secondary pressure should be sufficiently high and of sufficient duration Gate cross-sections should be made sufficiently large, so that sealing does not occur too early (see POM)

6.3.7 Polyphenylene sulfide (PPS)

Typical Characteristics and Applications

- Very high strength and rigidity even at relatively high temperatures (up to 230 °C)
- Very low tendency to creep, high wear resistance, low deformation capability
- Very low moisture absorption
- High electrical resistance, very low dielectric losses
- Resistant to solvents up to 200 °C and to many acids and alkalis, good resistance to hydrolysis
- Flame-resistance
- Non-transparent (semi-crystalline)
- Used for components subjected to mechanical, thermal, electrical and chemical stress.

Processing properties are given in Table 6.21.

Table 6.21 Processing data for polyphenylene sulfide

Pre-drying	Neither necessary nor advisable
Compound temperature	315-360 °C Because of the high temperature, cylinder and nozzle should first be thoroughly cleaned to remove traces of other compounds.
Mould temperature	20-200 °C Low mould temperatures give higher impact resistance values and less processing shrinkage (but, at the same time, they produce higher post-shrinkage at relatively high operating temperatures). With mould temperatures above 130 °C, components have better surface finish values and higher dimensional stability at high operating temperatures (lower aftershrinkage as a result of lower aftercrystallisation).
Shrinkage	Processing shrinkage 0.5-1.5% Higher value for higher mould temperature and large wall thicknesses.
Injection pressures	Injection pressure: 750-1500 bar Holding pressure: 350-750 bar
Special features	PPS low molecular weight polymers can be thermally crosslinked, heat treated under an oxygen atmosphere. Linear type PPS is also available where a high molecular weight polymer is used. These materials are less brittle as well as having increased tensile strength and flexural modulus properties.

6.3.8 Polyfluoroolefins

Polytetrafluoroethylene (PTFE), density = 2.1-2.2 g/cm^3, has such a high viscosity above the crystallite melting point (327 °C) that injection moulding is not possible.

Tetrafluoroethylene-perfluoropropylene copolymer (FEP), density = 2.1-2.2 g/cm^3

Ethylene-tetrafluoroethylene copolymer (ETFE), density = 1.7 g/cm^3

Typical Characteristics and Applications

PTFE

- Flexible to tough and hard (depending on the crystallinity)
- Relatively high creep tendency
- Very low friction coefficient (same at rest and in motion, so no slip-stick effect), however wear resistance relatively low
- Operating range: (–270 °C) –200 to +260 °C (280 °C)
- Non-flammable
- No water absorption
- Very high electrical resistance, even at high atmospheric humidity, very low dielectric losses
- Very high chemical resistance to almost all corrosive media, no stress crack formation
- Non transparent (semi-crystalline).

FEP and ETFE copolymers

- Properties similar to PTFE, but rigidity significantly higher, anti-friction properties and electrical insulating properties not quite as good. Chemical resistance of ETFE not quite as good.
- Application ranges: FEP from –100 °C to +205 °C; ETFE from –100 °C to +155 °C-180 °C
- Flame resistant and self-extinguishing
- Used in low stressed sliding mountings, seals, (rod-packing glands, piston rings, electrical components (resistant to cold and heat) including HF technology, high resistance components in chemical equipment construction, for pumps and laboratory equipment and similar applications.

Processing data for FEP and ETFE is given in Table 6.22.

Table 6.22 Processing data for FEP and ETFE

Pre-drying	**Not required**
Compound temperatures	FEP: 340-370 °C ETFE: 315-365 °C
Mould temperatures	FEP: 150-180 °C ETFE: 80-120 °C
Flowability	Work with low injection rate only, because of high susceptibility to shear breaks. For the same reason, sharp bends and abrupt cross-section alterations in the mould are to be avoided.
Shrinkage	ETFE (at 3-5 mm wall thickness) 1.5 to 2% in direction of flow 3.5 to 4.5% in transverse direction Re-heating treatment for preform recommended, to forestall possible aftershrinkage.
Special features in processing	If melt comes into contact with iron (or steel), fluorine is split off (corrosive). Plasticising unit and mould must therefore be protected against corrosion. Good suction (ventilation) in working areas vital.

6.3.9 Polyacrylic Acid (PAA)

E.g., IXEF (Solvay) – PA MXD6; available only as fibre glass reinforced compound

Typical Properties and Applications

- Excellent mechanical properties
- Apparent limit of elasticity at 255 N/mm^2 with 2% breaking elongation
- Modulus of elasticity up to 20,000 N/mm^2, modulus of bending up to 17,500 N/mm^2
- Stiffness 10 times higher than non-reinforced polyamides
- Low creep tendency
- High resistance to oscillation (higher than with other glass reinforced PA)
- High continuous usage temperature
- Crystalline melt temperature 235-240 °C
- Low thermal expansion coefficient (11-21 x 10^{-6}/°C), i.e., within the same range as metals
- Low water absorption
- Natural condition: non transparent (partly crystalline and filled) light colours possible
- Typical applications: high requirement mechanical parts used in machine engineering, automotive industry, precision mechanics, electrical industry, etc.

Processing data is given in Table 6.23.

Table 6.23 Processing data for PAA	
Processing	Easy to inject due to high flow capability. Injected parts are practically torsion free due to low shrinkage and small differences in shrinkage as well as minute pressure differences in the mould as a consequence of the low viscosity. The joint lines are weakened mechanically (as with all reinforced compounds), especially when they are far from the sprue (solidity of the joint line approx 80 N/mm^2).
Pre-drying	3-5 h at 100 °C with warm air, 12 h with max. 80 °C, or in vacuum cabinet 5 h with max. 120 °C The remaining humidity has to be less than 0.3%
Compound temperature inside the cylinder	250-290 °C For types in flame protected execution max. 270 °C With longer dwell time above 300 °C the thermal decomposition starts.
Mould temperatures	120-150 °C The effects of lower mould temperatures are insufficient and inhomogeneous crystallisation and after-crystallisation with torsion tendency. Mould temperatures in the upper range improve the surface quality.
Flow behaviour	Very high flow path/wall thickness ratio at least 200:1 with 2 mm wall thickness; we recommend using a shut-off nozzle; but you can also work with an open nozzle.
Injection speed	High The filling time should be as short as possible.
Injection pressure	1000-1500 bar
Holding pressure	350-800 bar Higher dimensional stability in the upper range
Maximum internal pressure	300-700 bar
Back pressure	40-80 bar
Screw circumferential speed	Up to 10 m/min Higher back pressures and higher circumferential speeds can lead to damaging of the fibre glass.
Processing shrinkage	0.2-0.6%
Aftershrinkage	Low The higher the mould temperature and therefore the processing shrinkage, the lower the aftershrinkage.
Special features	If you have to interrupt production for more than one hour you should empty the cylinder and clean it with a cleaning material.

6.3.10 Polyphthalamide (PPA)

E.g., Amodel (Amoco) only available in reinforced fibre glass and/or mineral filled varieties.

Typical Properties and Applications

- Excellent mechanical properties
- Modulus of elasticity at 225 N/mm^2 with 2% breaking elongation
- Flexural strength up to 305 N/mm^2
- Modulus of bending up to 14.500 N/mm^2
- High tenacity
- Low creep tendency
- High resistance to oscillation
- High continuous usage temperature, up to 165 °C for 20,000 h
- Crystalline melt temperature 310 °C
- Low thermal expansion coefficient
- Low absorption of humidity; practically no influence on the mechanical properties
- High tracking resistance

- Very resistant against chemicals
- High non-flammability
- Natural condition: non transparent (partly crystalline and filled)
- Typical applications: high requirement mechanical parts used in car industry, aerospace industry, chemical industry, sanitary industry, etc.

Processing data is given in Table 6.24.

Table 6.24 Processing data for PPA

Processing	Easy to inject, melt is not aggressive. Injected parts almost do not contort due to small differences in shrinkage as well as low absorption of humidity. Contrary to PC and other materials pre-drying is not critical.
Pre-drying	Not necessary when the material is processed directly from the rolls; otherwise 16 h at 80 °C
Compound temperature inside the cylinder	320-350 °C
Mould temperatures	135-165 °C
Flow behaviour	High, similar to PA
Injection speed	Middle to high
Injection pressure	700-1500 bar
Holding pressure	350-800 bar
Maximum internal pressure	300-700 bar
Back pressure	40-80 bar
Processing shrinkage	0.2-1.1% according to the type

6.3.11 Polyaryl Ether Ketones (PAEK)

Polyaryl ether ketones are composed of aryl ether groups and aryl ketone groups.

Generally they are partly crystalline thermoplastics with high mechanical, thermal and chemical strength. Examples include:

Polyether ketone (PEK)
Polyether etherketone (PEEK)
Polyether ketone ketone (PEKK)
Polyether etherketone ketone (PEEKK)
Polyether ketone etherketone ketone (PEKEKK)

A high share of ketone groups increases the processing temperature due to higher transition temperatures and higher crystallisation melt temperatures. The ether groups have positive effects on the processability of the material.

Typical Properties and Applications

- Very high strength and rigidity, almost constant up to 140 °C
- Processable up to 250 °C, in the short run up to 300 °C
- Very low creep tendency
- High endurance under completely reversed stress
- High abrasion resistance up to 250 °C, favourable sliding behaviour
- Low thermal expansion coefficient
- Low absorption of water
- Up to 220 °C constant good electrical properties, especially small dielectric losses
- Very resistant against chemicals (except against acetones), resistant against hydrolysis up to 280 °C
- Very resistant against stresses
- Very resistant against radiation of high energy; low resistance to UV radiation when not coloured
- High non-flammability
- Least development of smoke among all thermoplastics

- Natural condition: non transparent (partly crystalline)
- Typical applications: For parts which are put under a lot of mechanical, thermal and electrical stress, e.g., bearing, ball bearing guides, cogs, valves, seals (car and aerospace industry). Parts for hot water flow meter, pump turbines, connector parts, printed circuit boards.

Processing data is given in Table 6.25.

Table 6.25 Processing data for PAEK	
Pre-drying	3 h at 150 °C
Compound temperature inside the cylinder	350-420 °C 440 °C should not be exceeded.
Mould temperatures	150-180 °C
Flow behaviour	Flows easily
Injection speed	Middle to high
Injection pressure	800-1500 bar
Holding pressure	450-800 bar
Maximum internal pressure	400-700 bar
Back pressure	60-90 bar
Processing shrinkage	1% non-reinforced, 0.1-0.4% reinforced

6.3.12 Liquid Crystal Polymers (LCP)

Liquid crystal polymers are block copolymers. The molecules consist of rod shaped hard segments (mesogenes), which are connected to each other by short and more flexible segments. Therefore there is already an ordered structure in the liquid condition caused by the parallel order of the little rods (semi-crystallisation).

The above described LCPs, called thermotropic LCPs (as they show a liquid crystalline order in the melt), are suitable for injection. Lyotropic LCPs are liquid crystalline in solvents and are used to produce fibres (e.g., aramid). There are also side chain LCPs with little rods which are connected at the side with the main chain (which is flexible in the melt).

During injection, the rods orientate into the flow direction. Therefore LCPs have exceptional mechanical and thermal properties. This so-called self-reinforcing effect only appears along the orientation direction. In the orientation direction, LCPs have a fibril structure that is reminiscent of wood. Along the orientation direction (privileged direction) the LCPs do much better than conventional thermoset plastics: vertical to the orientation direction they are in the range of technical thermoset plastics.

In most LCPs there are aromatic components. There are different types of LCPs with specially developed properties. If the material has very high strength, rigidity, resistance to abrasion and high thermal resistance, there are generally weaknesses in impact resistance and processability, (high temperatures, high viscosity of the melt). Materials with the highest impact resistance and good processability surrender the extreme strength and rigidity. LCPs are available which are very resistant against chemicals. Materials with the best processability with high strength and rigidity, lose some of the high thermal resistance.

Typical Properties and Applications

- Very high strength and rigidity in orientation direction: strength = 150-300N/mm^2, rigidity = 10000-35000 N/mm^2
- The processing temperature is generally above 200 °C
- Very little creep tendency
- Danger of splitting parallel to the orientation direction, high to modest impact resistance. (depends on the type)
- High endurance under completely reversed stress
- Very small thermal expansion coefficient (values as with steel or smaller), adjustable by processing conditions; therefore excellent dimensional stability

- Low dielectric losses
- Very resistant against chemicals, no tendency to cold crack formation, very resistant against ambient influences, very resistant against radiation of high energy, high non-flammability
- Low smoke generation
- Natural condition: in general non-transparent (already in the melt) as they are semi-crystalline, certain types can be transparent.
- Typical applications: for parts which are put under a lot of mechanical, thermal, electrical and chemical stress, possibly to replace metallic materials, ceramic or thermoset plastics, for the car industry, aerospace industry, precision mechanics (electronics, optical industry), electrotechnical industry, also for motor parts, gearing, safety systems, etc.

Processing data can be found in Table 6.26.

Table 6.26 Processing data for LCP

Processing	Pre-drying is necessary. The flow behaviour is modest to excellent, depending on the type. Due to the molecule orientation, the joint lines represent weak points. The cycle periods are shorter than with conventional partly crystalline thermoplastics.
Pre-drying	2-4 h at 150 °C Permitted remaining humidity = 0.015%
Compound temperature inside the cylinder	280-450 °C, depending on the type
Mould temperatures	30-160 °C, depending on the material type
Flow behaviour	Flows easily
Injection speed	High
Injection pressure	400-1500 bar
Holding pressure	350-1000 bar
Maximum internal pressure	300-800 bar
Back pressure	40-60 bar
Screw circumferential speed	0.15 m/s
Processing shrinkage	–0.1 – +0.6% Lower values in orientation direction, upper values vertical to it
Aftershrinkage	Practically nil
Special features	Parts behave very anisotropically, e.g., they are very solid in the orientation direction. Position and form of the sprue therefore are of decisive importance

6.4 Conclusion

This chapter has detailed common properties for a number of amorphous and semi-crystalline materials. It has also highlighted any major processing issues attached to using these materials for injection moulding. In conjunction with the processing chapters (8 and 9), this should contain all the necessary information to successfully mould these materials. The next chapter will consider other types of injection mouldable materials, for example, thermosets and liquid silicone rubber.

7 Processing Thermoset, LSR and Ceramic Materials

This chapter is split into three sections covering processing of thermosets, elastomers (specifically LSR) and briefly ceramic and metal powder injection mouldable materials.

7.1 Thermosets

7.1.1 Overview and Common Properties

Thermoset plastics are chemically setting synthetic resins. Thermoset plastics set in the hot mould by crosslinking (chemical process), as opposed to thermoplastic freezing in a sufficiently cold mould (physical process). Synthetic resins tend to form shrinkage cracks, since a plastic deformation is practically impossible. The various resin types are shown in Table 7.1. Phenolic resins may also be referred to as Bakelite (after the chemist Baekeland who invented phenolic resin).

Table 7.1 Resin types

<table>
<tr><td>PF</td><td>Phenol-formaldehyde resins</td><td>Phenoplasts</td><td rowspan="3">Polycondensate, i.e., crosslinking by elimination of H_2O</td></tr>
<tr><td>UF</td><td>Urea-formaldehyde resin</td><td rowspan="2">Aminoplasts</td></tr>
<tr><td>MF</td><td>Melamine-formaldehyde resin</td></tr>
<tr><td>UP</td><td>Unsaturated polyester resins</td><td colspan="2" rowspan="3">Crosslinking by polymerisation or addition, without formation of fission products</td></tr>
<tr><td>DAP</td><td>Diallyl phthalate</td></tr>
<tr><td>EP</td><td>Epoxide (epoxi-, ethoxyl) resins</td></tr>
</table>

Thermoset plastics are almost always processed with fillers and reinforcing materials. This is done to extend and reinforce synthetic resin and especially to minimise shrinkage. Examples of common fillers and reinforcing materials are shown in Table 7.2.

Table 7.2 Filling/reinforcing compounds

	Form	Powder flour	Fibre	Pellets	Chips	Mats Webs	Weave
Inorganic	Stone	X					
	Asbestos		X	X	X	X	
	Glass		X	X		X	X
Organic	Wood	X					
	Cellulose		X		X	X	
	Textiles		X		X		X

Traditionally thermoset plastics are processed by compression moulding. But today injection moulding is one of the most economical ways of processing thermosets. If it is to be cost effective, the technology used must be appropriate to the shape and the properties of the material for all of the components. Thermosets can be supplied in a variety of forms, from granulates to rods to doughy polyesters (dough moulding compound (DMC), bulk moulding compound (BMC), sheet moulding compound (SMC)).

7.1.2 Materials for Screw Injection Moulding Machines

The decisive factor for processing thermosets is the very structure of the material.

The right material must meet the following requirements:

- Free flowing
- Fast plastification possible
- Extended stay in cylinder possible
- Fast and high quality setting in the mould.

Therefore, a range of thermosets from soft to medium is recommended for processing on injection moulding machines. A temperature guide for continuous processing is shown in Table 7.3.

Table 7.3 Temperatures for continuous processing of thermosets

	Temperatures (°C)	60		80		100		120		140		160		180		200		220		240	
PF	Phenol-formaldehyde resins						░	░				▓									
UF	Urea-formaldehyde resin			░																	
MF	Melamine-formaldehyde resin					░		▓													
MP	Melamine-phenol-formaldehyde					░		▓													
UP	Unsaturated polyester							▓	▓	▓	▓	▓	▓	▓							
DAP	Diallyl phthalate											▓	▓	▓	▓	▓	▓	▓	▓		
EP	Epoxide (epoxi-, ethoxyl)				▓	▓	▓	▓	▓	▓											
		░	Organic filled																		
		▓	Inorganic filled																		

7.1.3 Processing Procedures for Thermoset Plastics

7.1.3.1 Compression Moulding

For the classic compression moulding process, a blank, prepared in shape and weight, is put into the opened mould and pressed under high pressure as shown in Figure 7.1.

The disadvantages of this production method are:

- The long cycle times
- The high material consumption
- A considerable additional treatment is necessary
- No accurate parts are obtained
- Only semi-automatic operation is possible.

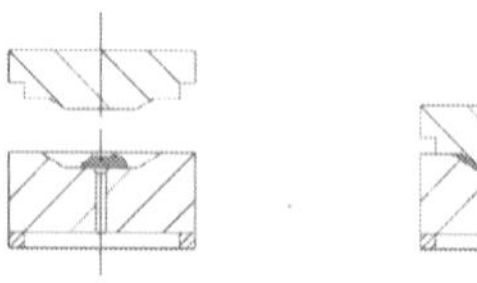
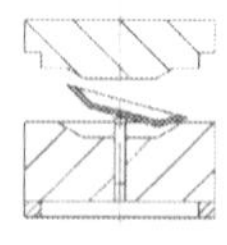

Figure 7.1 Compression moulding

7.1.3.2 Injection Moulding

The more economical process is the injection moulding of thermosets on screw injection moulding machines. The material is plasticised with a special screw. The optimum preparation of the melt means a plasticising unit with a corresponding feed unit and screw for material specific dosage.

Dosing and plasticising for a new shot takes place, while the material is still forming crosslinks in the mould. The whole process can run automatically.

Advantages of processing thermosets on injection moulding machines:

- Considerably shorter cycle times compared to alternative processes
- The material is heated and plasticised homogeneously
- High quality, high precision parts
- Material saved
- Parts with minimum flash, because of injection into closed mould
- Less wear on the mould.

7.1.4 Screw Injection Machine for Thermosets

For injection machines to be used for processing thermosets, the following equipment is required:

- Thermoset cylinder assembly
- Mould blow unit
- Dosage delay
- Venting control
- Special control for cleaning or brushing the mould.

Screws and cylinders for thermosets are treated, for example Arbid. The screws have a special geometry, a suitable tip and no back-flow prevention valve. Other versions with back-flow prevention valves are also available.

There will be wear on the screw and cylinder especially when processing materials with abrasive fillers, such as asbestos, rock meal and glass fibre. The main reason for this is the high injection pressure. The back-flow of material will put additional strain on the screw and the front part of the cylinder.

To increase the wear resistance and durability the following measures can be taken:

- Use hard alloy bushings for the cylinder
- Use a screw with back-flow prevention valve
- Use a hard alloy screw
- Use a screw with double thread in the metering section.

For cylinder heating two temperature control devices are required. Inside the cylinder a temperature between 50 °C and 80 °C is needed in the feed zone and between 75 °C and 115 °C in the metering zone. Generally a temperature control device operated with water is used for the feed zone and an oil operated device is used for the metering zone.

7.1.5 Mould

The mould is heated electrically. It is recommended that control of both mould halves is separate. The heating power for small moulds should be between 1,600 and 2,200 Watt per mould half, varying according to the mould size (e.g., 215 x 300 mm). The mould temperatures vary according to the thermoset material. Standard values range from 150 to 240 °C. The mould should have exchangeable inserts and insulating platens of 10 mm thickness.

7.1.6 Guide Values for Cylinder and Mould Temperature Settings

Always observe specific instructions by the material manufacturers.

When adjusting cylinder temperatures please note:

- The melt temperature is not only influenced by the cylinder temperature, but also by the frictional heat created by the plasticising screw
- The higher the screw speed and the higher the back pressure the higher the melt temperature will be (at the same cylinder temperature)
- The upper temperature limits should be selected only when the melt stays for a very short time in the cylinder (high cycle speed).

A guide to the temperature effects of the various stages of injection moulding of thermosets is given in Figure 7.2.

As a result of the injection operation, the heat of friction is increased through the nozzle and cross-sections, while the material is compressed at holding pressure. The residual heat of reaction, which is necessary for crosslinking of the injected part, is conveyed via the mould-tempering unit. Since the material is cured while heat is supplied, temperature management during plastication and injection operations is of greatest significance for cost-effective production of high quality injected parts. To achieve this, special temperature control units are installed. The injection cylinder is divided into several temperature zones in order to adapt the heating equilibrium and shot volume ideally to one another. This principle is illustrated in Figure 7.2. The heat introduced into the melt during plastication can therefore be maintained within strict limits.

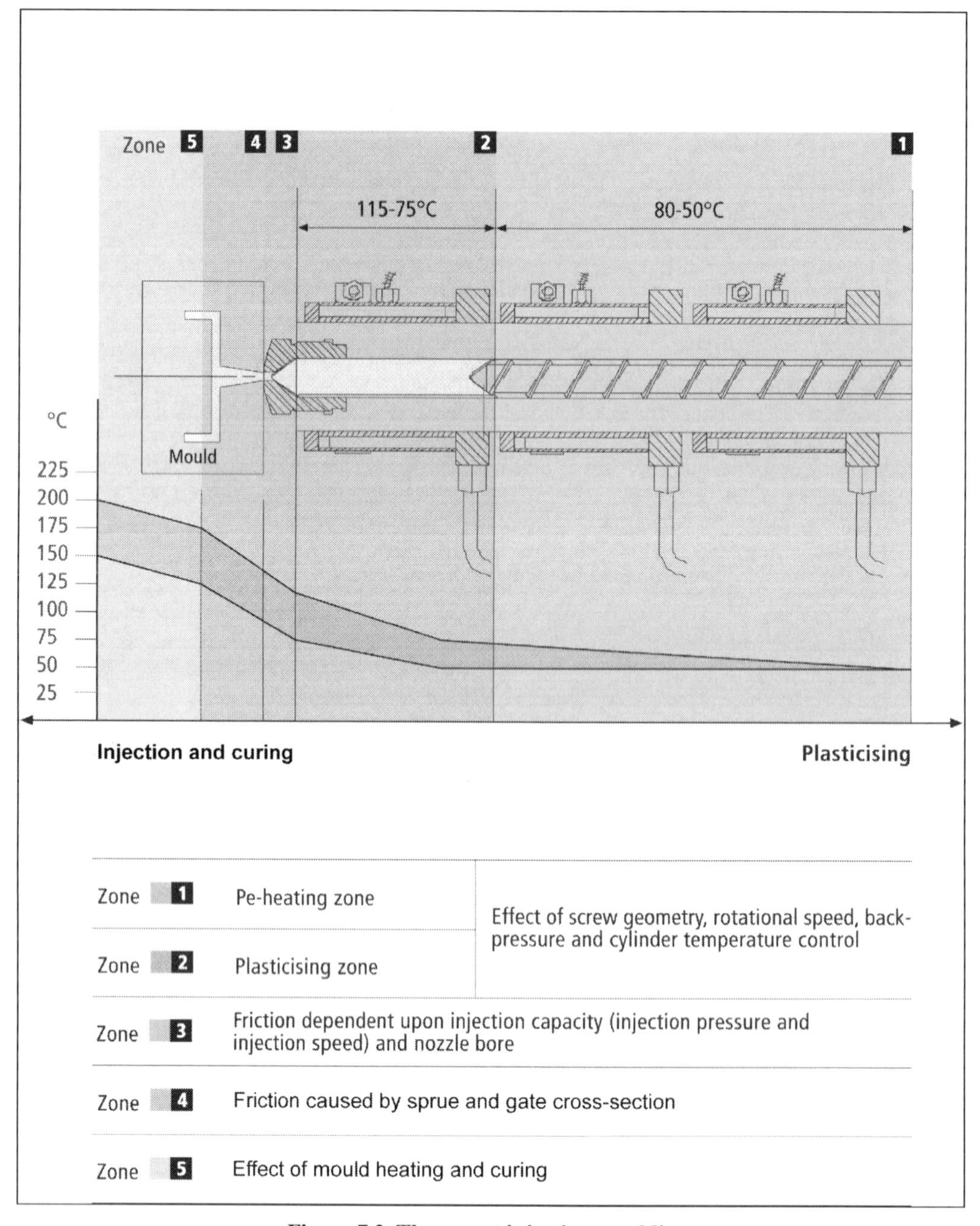

Zone		
Zone 1	Pe-heating zone	Effect of screw geometry, rotational speed, back-pressure and cylinder temperature control
Zone 2	Plasticising zone	
Zone 3	Friction dependent upon injection capacity (injection pressure and injection speed) and nozzle bore	
Zone 4	Friction caused by sprue and gate cross-section	
Zone 5	Effect of mould heating and curing	

Figure 7.2 Thermoset injection moulding

In the next phase the friction of the nozzle significantly increases the melt temperature. As a result, the viscosity of the material is reduced, the melt becomes thin and can be introduced into the cavity in the optimum manner. Machines with high injection capacities (injection pressure and speed) are required to handle this.

The mould heating and the curing time influence the last phase of heat supply. Only a small amount of heat is fed to the melt in order to form the part and to complete the forming process.

The typical processing sequence for the production of thermoset parts follows. The crosslinking of the prepared material is delayed in the relatively cool injection cylinder for a period until the melt reaches the relatively hot mould where it cures quickly. In order to prevent premature curing of the melt in the cylinder, back pressure and screw rotation speeds can be adjusted in stages on thermoset machines. This allows travel through the critical melt range with low back pressure and slow screw rotational speed. The temperature of the melt remains high and shortened curing times is also possible. In addition, the more uniform temperature structure has positive effects on the stress, distortion and shrinkage characteristic of the moulded part.

When compared to thermoplastics, thermosets have several unique and clear benefits which make them an ideal material for the electrical industry:

- No softening temperature region
- Rigidity, even at higher temperatures
- High heat resistance
- Greater hardness
- Greater rigidity
- Excellent dimensional stability
- Lower expansion coefficient
- Lower cold flow
- More favourable combustion behaviour
- Excellent price/performance ratio.

7.1.7 The Injection Unit for Granulated Thermosets

The screw in the injection unit is not compressed and is characterised by its relatively low L/D ratio of 15:1 (compare this to 20:1 for thermoplastics). This is brought about by having the feed zone shifted forward. A surface-cured screw is used as standard. A fully cured version is advised for use with extremely abrasive materials. Bimetal construction of the cylinder makes for extremely long service life. Generally, the heating control systems can be adjusted across three or four independent temperature zones. An open, cured nozzle is screwed directly into the nozzle socket and can therefore be tempered precisely.

7.1.8 The Injection Unit for Processing Moist Polyester

The screw diameter in this case is greater, since lower pressures and a shorter L/D ratio of around 12:1, are sufficient for processing purposes. In comparison with units for granulated thermosets, there are also differences in the following areas:

The screw has geometry specifically designed for moist polyester melts and has a plain check valve, to reduce fibre destruction. This was illustrated in Figure 3.6 in Chapter 3.

The feed aperture is optimised for non-destructive introduction of material.

7.1.9 Nozzle Temperature (if nozzle heating is applied)

The nozzle temperature should be adjusted to 5-15 °C above the temperature suggested for the metering zone of the plasticising cylinder. Suggested values are given in Table 7.4.

Table 7.4 Temperature guide for thermosets

Type of material		Temperatures (guide values)		
		Mould (°C)	Plasticising cylinder	
			Metering zone (°C)	Feed zone (°C)
Phenol-formaldehyde	PF	145-175	75-90	45-60
Urea-formaldehyde	UF	140-165	80-95	45-55
Melamine-formaldehyde	MF	150-180	75-100	50-60
Melamine-phenolic	MP	150-180	70-100	45-55
Polyester	UP	140-165	80-105	40-65
Allyl	DAP	150-190	85-110	50-60
Epoxy	EP	155-200	85-115	50-60
Silicone (LSR)	Si	180-250	5-20	5-20

7.1.10 Common Thermoset Materials: Properties and Applications

This section will detail properties and processing data for the following five classes of thermoset materials:

- Phenol-formaldehyde (PF)
- Urea-formaldehyde (UF)
- Melamine-formaldehyde (MF)
- Unsaturated polyester (UP, DAP)
- Epoxide compounds (EP)

7.1.10.1 Phenol-Formaldehyde Compounds (PF)

This includes grade such as:

- FS 11 – with inorganic (e.g., mineral) fillers
- FS 31 – with wood (e.g., wood flour)
- FS 51 – with paper (e.g., paper fabric)
- FS 71 – with fabric (e.g., cotton)

NB: The higher the first digit in the designation the higher is the percentage of fillers in the resin. The specific properties depend largely on the kind and the percentage of the filling material in the resin.

Typical Properties and Applications

- Hard and stiff
- Good heat resistance, even in long-term application (up to 150 °C for resins with inorganic fillers)
- Strong water absorption of resins with organic fillers
- Low rate of creep
- No light colours
- Often physiologically problematic (contact with foodstuffs forbidden)
- Applications: mass products, for the electrotechnical industry, such as housings, sockets for switches and relays, connectors etc.

Processing data for PF compounds is given in Table 7.5.

Table 7.5 Processing data for PF

Melt temperature inside the plasticising cylinder	45 °C to 90 °C
Back pressure	10 to 15% of injection pressure
Mould temperatures	145 °C to 175 °C
Flow	Depending on the material medium to low flow
Shrinkage	Immediate mould shrinkage: 0.5 to 0.9% in flow direction 0.7 to 1.1% normal to flow direction Post shrinkage: Up to 0.4%
Injection pressure	Injection: 800 to 1500 bar Holding pressure: 40 to 60% of injection pressure Processing practically without material cushion
Storage	Limited storage (6-24 months), material forms crosslinks

7.1.10.2 Urea-Formaldehyde Compounds (UF)

E.g., materials such as FS 131 – with organic fillers (e.g., cellulose)

Typical Properties and Applications

- Hard and stiff
- Maximum temperature stability for long-term use
- Application at 80 °C
- Stress cracking due to strong shrinkage

- Age at high temperatures
- Low rate of creep
- Light colours possible
- Often physiologically problematic (contact with foodstuffs forbidden)
- Applications: light coloured mass products for the electrotechnical industry, especially installation materials.

Processing data is given in Table 7.6.

Table 7.6 Processing data for UF	
Melt temperature inside the plasticising cylinder	45 °C to 90 °C
Back pressure	10 to 15% of injection pressure
Injection pressure	Injection: 800 to 1500 bar Holding pressure: 40 to 60% of injection pressure Processing practically without material cushion
Mould temperatures	140 °C to 165 °C
Flow	Depending on the material medium to low flow
Immediate mould shrinkage	0.2 to 0.6% in flow direction 0.6 to 1.3% normal to flow direction
Post shrinkage	Up to 1.7% depending on the filler
Storage	Limited storage (generally no longer than 6 months), material forms crosslinks.

7.1.10.3 Melamine-Formaldehyde Compounds (MF, MP)

FS 150 – Melamine-formaldehyde-resins (MF)
FS 180 – Melamine/phenolic-formaldehyde-resins (MP)

Both with organic and/or inorganic fillers, e.g., wood flour, cotton wool and/or rock meal, asbestos.

Typical Properties and Applications

- Hard, stiff, low rate of creep
- Good heat resistance, even in long-term application (up to 130/150 °C for resins with inorganic fillers)
- MF tends toward stress cracking, due to strong shrinkage at high temperatures
- Higher creep resistance compared to PF and UF
- MF hardly flammable, self-extinguishing
- Light colours possible (especially with MF)
- Some MF types physiologically harmless (contact with foodstuffs allowed)
- Applications; light coloured parts for the electrotechnical industry (housings, sockets for switches and relays, connectors etc.), even for higher creep resistance requirements. Household appliances, tableware.

Table 7.7 Processing data for MF and MP	
Melt temperature inside the plasticising cylinder	55 to 100 °C
Flow	Depending on the material medium to low flow
Injection pressure	800 to 2000 bar
Back pressure	10 to 15% of injection pressure
Holding pressure	40 to 60% of injection pressure Processing practically without material cushion
Mould temperatures	150 to 180 °C
Immediate mould shrinkage	0.2 to 0.6% in flow direction 0.6 to 1.3% normal to flow direction
Post shrinkage	Up to 1.7% depending on the filler
Storage	Limited storage (generally not longer than 6 months), material forms crosslinks.

7.1.10.4 Unsaturated Polyester (UP, DAP)

UP – General abbreviation for unsaturated polyester
DAP – Diallyl phthalate, special unsaturated polyester (allyl resin)

Both with mostly inorganic fillers (glass fibre, minerals)

Typical Properties and Applications

- Hard and stiff
- Good heat resistance, even in long-term application (up to 180 °C according to type, DAP up to 230 °C)
- Even at low temperatures relatively insensitive to impact
- Low water absorption
- Weathering resistance dependent on coupling agent between resin and glass fibre
- Very high creep resistance
- Light colours possible
- There are some physiologically harmless types (contact with foodstuffs allowed for these)
- Applications: electrotechnical and precision mechanical pans for high mechanical, thermal and electrical requirements (coil cores, connectors, sockets, housings, covers, etc.)
- Due to the very low shrinkage varying material concentrations and sharp edges are less critical than with other materials.

Processing conditions for UP and DAP are given in Table 7.8.

Table 7.8 Processing guide for UP and DAP

Melt temperature inside the plasticising cylinder	UP (generally): 65 to 105 °C DAP: 55 to 110 °C
Back pressure	10 to 15% of injection pressure
Injection pressure	Injection: 800 to 2500 bar Holding pressure: 40 to 60% of injection pressure Processing practically without material cushion
Mould temperatures	UP (generally): 140 to 165 °C DAP: 150 to 190 °C
Flow	Depending on the material medium to low flow
Shrinkage	Immediate mould shrinkage: up to 0.3%
Post shrinkage	Practically zero
Storage	Limited storage (generally weeks to a few months). Material forms crosslinks. In addition to the free flowing materials there are some pasty materials, mostly filled with long glass fibres. These also can be processed in connection with an INJESTER type feeding unit.

7.1.10.5 Epoxy Compounds (EP)

Only with inorganic fillers (glass fibre, minerals)

Typical Properties and Applications

- Hard and stiff
- Tougher than UP
- Strong heat resistance, even in long-term application (up to 130 °C special types even higher)
- Even at low temperatures relatively insensitive to impact
- Very low water absorption (less than for UP)
- High weathering resistance, since there is no coupling agent necessary between resin and glass fibre (EP resins are excellent adhesives)
- High creep resistance
- Light colours may be possible (mostly not fast to light)

- EP resins are physiologically harmless, once they have set (are crosslinked). Avoid contact with the raw material, however. Skin rashes may occur.
- Applications: electrotechnical and precision mechanical parts for high mechanical thermal and electrical requirements (coil cores, connectors, sockets, housings, covers, etc.)
- Due to the very low shrinkage, varying material concentrations and sharp edges are less critical than with other materials.
- Generally high weathering resistance allows for outdoor use.

Guide values for processing by injection moulding are given in Table 7.9.

Table 7.9 Processing data for EP

Drying prior to injection	When you have stored EP resins at low temperatures to counteract crosslinking, dry and heat material in a closed container prior to injection.
Melt temperature inside the plasticising cylinder	55 to 115 °C
Back pressure	10 to 15% of injection pressure
Injection pressure	Injection: 800 to 2500 bar Holding pressure: 40 to 60% of injection pressure Processing practically without material cushion
Mould temperatures	155 to 200 °C
Flow	Depending on the material medium to low flow
Shrinkage	Immediate mould shrinkage: 0.1 to 0.5%
Post shrinkage	Practically zero
Storage	Limited storage (generally weeks to a few months), material forms crosslinks

7.2 Elastomer Injection Moulding Compounds

This section will discuss issues relating to the injection moulding of rubbers, specifically to the moulding of liquid silicone rubber (LSR) which is of large commercial interest. Firstly however, a quick introduction to rubber materials is given.

There are two types of rubber, natural rubber and synthetic rubber.

Natural rubber was for a long time the only raw material of the rubber industry. It is obtained from the sap of the rubber plant of the rubber plantations in Asia (rubber milk or latex). The thickened material is rolled, dried and pressed.

Synthetic rubber is a product of organic chemistry. The basic material is oil, especially heavy petrol that is produced when cracking oil in the refinery. It is polymerised by means of various chemical combinations and additions. This results in synthetic latex milk. As with natural rubber a solid material can be extracted.

The raw material for liquid silicone rubber is quartzite or quartz powder. All rubber types can be processed on injection moulding machines.

7.2.1 Processing Procedures for Elastomeric Materials

7.2.1.1 Compression Moulding

With this procedure a blank, prepared in shape and weight, is put into the opened mould and pressed under high pressure. This is shown in Figure 7.1.

The disadvantages of this production method are:

- Long cycle times
- High material consumption
- Considerable additional treatment is necessary
- No accurate parts are obtained
- Only semi-automatic operation is possible.

7.2.1.2 Injection Moulding

In this method the injection moulding screw plasticises prepared cords whereby the elastomeric material is homogeneously heated up.

Advantages of heating the rubber:

- Shorter vulcanisation times, 12-16 s/mm wall thickness
- A higher accuracy of the parts
- A better quality of the parts
- Any additional treatment is not necessary
- 10-20% material is saved
- Flash-free parts or parts with minimum flash are obtained
- Less wear on the mould
- Automatic production.

7.2.1.3 Injection Moulding Machines for Rubber Compounds

Standard machines can be prepared for the processing of elastomeric materials. The machines must, however, have the following equipment:

- Cylinder for elastomeric material
- Mould blow unit
- Dosage delay
- Special control for cleaning and brushing devices in the mould (to remove bloom).

NB: Special 10 D cylinders with compressionless screws and a big feed bag for specially sensitive rubber materials are available for many injection units.

7.2.1.4 Peripheral Device

A temperature control unit is needed for heating the plasticising cylinder. Depending on the rubber material a temperature between 50 and max. 170 °C is needed inside the plasticising cylinder. In most cases such a temperature control unit works with oil.

7.2.1.5 Mould

Mould heating is achieved by means of electrical heating. It is recommended that both mould halves are controlled separately. The heating power for each mould half should be between 1,200 and 1,600 Watt. The mould temperatures vary according to the elastomeric material. Standard values range from 110 °C to 210 °C. In view of the high mould temperatures insulating platens are required with a thickness of 10 mm for the mould.

7.2.2 Silicone Elastomers – Liquid Silicone Rubber (LSR) Systems

Increasing expenses for machines, staff and especially raw materials forced manufactures of plastic materials and custom moulders to look for new solutions. Thus liquid silicone rubber was introduced by the chemical industry in 1978. The base material is quartzite or quartz powder (SiO_2), making it independent of crude oil (in contrast with synthetic rubber).

LSR is vulcanised into silicone rubber in the mould. Preforms made from LSR, in spite of the high material prices, are often considerably more cost-effective than those made from conventional types of rubber, because the processing costs are considerably lower.

Among the variety of plastics and elastomers, the silicones occupy a special position. Their significant difference from other plastics lies in their chemistry. Most plastics have a backbone of carbon-carbon bonds. In contrast to this, the silicones are distinguished by alternating silicon and oxygen atoms.

Since there are very many different possibilities for varying organic groups within the molecule chains, there are many different products in the silicone marketplace with widely differing properties. The primary products which are produced on a silicone base are silicone elastomers, silicone oils, which are used as anti-foaming agents or parting agents, chemical specialities such as cosmetics or polishes, silicone resins, textile products and paper coatings.

7.2.2.1 Classification

Silicone rubbers may be classified by crosslinking method, viscosity and vulcanisation temperature. Differentiation can be made between hot and cold (room temperature vulcanisation (RTV) rubber types. Within each of these two groups are found single-component and two-component systems, grouped by viscosity range. The viscosity of the rubber (fluid-mouldable, pasty, plastic-firm) determines the processing method and influences the characteristics of the vulcanised material. Figure 7.3 depicts the classification of silicones.

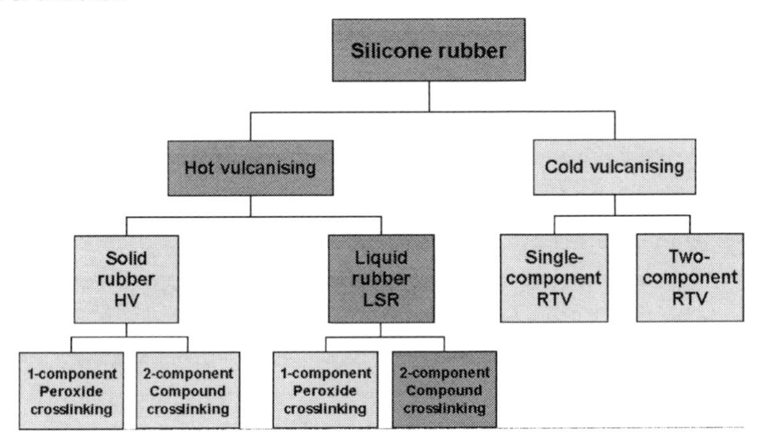

Figure 7.3 Classification of silicones

It is silicone rubber, hot vulcanisation, liquid rubber LSR-2K crosslinking compound, which is discussed in the following sections on LSR processing. All other silicone rubber types require different machine equipment and process conditions for LSR which are beyond the scope of this book.

7.2.2.2 Crosslinking

Compound crosslinking is based on the compounding of SiH to double bonds. The reaction takes place relatively quickly. A catalyst must be present for the initiation of the reaction. Platinum complexes, which are added in very low concentrations to the rubber, are used as catalysts. The crosslinking reaction begins at room temperature. With increasing temperature, the reaction speed increases rapidly. Figure 7.4 depicts a schematic view of compound crosslinking.

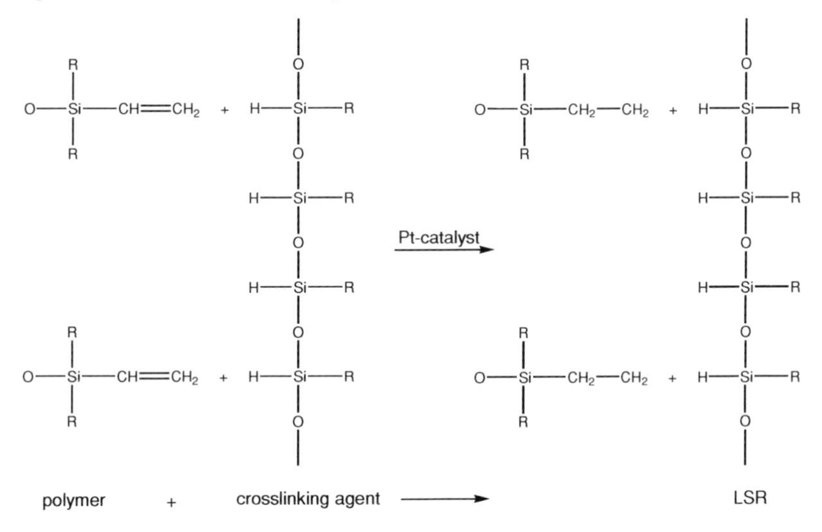

Figure 7.4 Compound crosslinking

Rubber systems with compound crosslinking are based on two-components, since a separation into two components is necessary for storage. One component must contain the catalyst, the other the crosslinking agent.

7.2.2.3 Processing of Two-Component LSR

Because of the particular material properties of LSR in comparison to other materials which are capable of being injected, a special processing technology is required. This technology is known by its abbreviation LIM, which stands for liquid injection moulding.

Injection moulding of LSR is possible with an LSR cylinder assembly. However, a multi-component dosage unit is required. A multi-component dosage unit mixes the components A and B and feeds the resulting compound to the injection moulding machine. Equal amounts of A and B are mixed and then fed into the injection cylinder. A typical processing system will include:

- Multi-component dosage unit with containers
- Component A: crosslinking catalyst
- Component B: components to be crosslinked
- Component C: colouring paste
- Mixer (not necessary for some machines if using a mixing screw)
- Feeding device for the injection unit
- Clamping unit with mould.

7.2.2.4 Advantages of Processing LSR Materials

There are a number of advantages in moulding LSR compared to processing other thermoset materials.

- Low injection pressure processing on small, low-cost injection moulding machines with low clamping forces possible
- The number of mould cavities can be increased
- Easy ejection of moulded parts
- High-precision parts can be produced
- Few burrs on parts
- Considerably shorter cycle times in comparison with conventional rubber materials
- Vulcanising times of 5-6 seconds per millimetre wall-thickness with a mould temperature of 200 °C
- Temperature range for end product: –60 to +300 °C.

7.2.2.5 Injection Process

You can process LSR with modern injection moulding machines. However, you need special LSR equipment for the machine. Today injection moulding of LSR is solely done with screw injection moulding machines. The screw plasticises the prepared compound. During the plasticising the material is kept at a temperature of 5 to 25 °C. This is followed by material injection by the screw with low pressure into the mould.

7.2.2.6 Configuration of Injection Moulding Machines and Processing Data

For injection moulding of LSR materials the following equipment is required:

- Special LSR plasticising cylinder with cooling jackets
- Hydraulically actuated needle-type shut-off nozzle with cooling jacket
- A special pressure relief valve for the material feed line from the dosage and mixer units
- A mould blow unit
- If necessary a special control circuit for cleaning or spraying the mould.

General processing data for LSR is given in Table 7.10.

Table 7.10 Processing data for LSR	
Cylinder temperature	5-25 °C Higher cylinder temperatures carry the risk of transient scorch on the cylinder wall. Moulded-on particles ripped into the mould can impair the preform's properties.
Screw speed	Average screw speed (100-200 rpm)
Back pressure	Back pressure = 0 Because of LSR's low viscosity, the back pressure considerably reduces the dosing rate or completely prevents dosing (screw return stroke)
Flowability	Very good, as LSR systems have low viscosity Therefore, the injection pressure and mould clamping forces required are also very low
Injection pressures	100-200 bar
Mould temperatures	180-250 °C Higher mould temperatures accelerate vulcanisation, e.g., a 10 g component vulcanises in 50 s at 150 °C, 5 s at 200 °C
Shrinkage	Processing shrinkage and post shrinkage are practically zero
Special features	Processing of LSR systems is possible on normal screw injection moulding machines with special LSR cylinder fittings (hydraulic needle shut-off nozzles, screws with non-return valves, but without compression, cylinder temperature controlled by fluids). The two LSR components are dosed in a special apparatus in a 1:1 ratio and mixed together and then fed into the drawing-in aperture of the injection unit. At temperatures under 20 °C, the 'shelf life' of the mixed components is very long (Dow Corning gives at least 30 hour), i.e., when work is interrupted overnight the dosing and mixing equipment and the cylinder do not need to be cleaned.

NB: To keep cylinder temperature at 5 to 25 °C, the cooling jackets will be connected directly to a cooling water distributor. To reach extremely low temperatures an external cooling device can be used. For injection a single-stage non-compression mixing screw with back flow prevention valve is used.

Liquid silicone rubber (LSR) for the production of elastic parts in injection moulding technology is gaining an ever increasing significance in the processing of elastomers. The characteristic for liquid silicone rubber is the low viscosity in comparison to solid silicone rubber and other elastomers.

7.2.2.7 Specific Properties

- High heat resistance up to constant temperatures of 180 °C
- Good low-temperature resistance and cold flexibility to –50 °C
- Good tensile strength and tear strength
- High resistance to weathering and ageing
- High electrical insulating properties as well as resistance to shock
- Constant mechanical and electrical properties within a wide temperature range
- High resistance to alcohol, polar solvents and weak acids
- Anti-adhesive and hydrophobic characteristics
- Excellent physiological properties.

The LSR raw-material producers provide a number of differing types of material in order to cover as broad a spectrum of applications as possible. The following typical commercial products are available at this time:

- Standard types with Shore hardness of 20 to 70 Shore A
- Electrical conducting types
- Oil-sweating, self-lubricating types
- Flame-retarding products
- Types for medical technology, physiologically harmless

- Oil-resistant types for the automotive industry
- Fluorosilicone rubbers.

Based on the wide number of excellent properties and the fact that LSR can be dyed or tinted without problems, there are application areas in nearly all branches of industry. A few examples here are the electrical and automotive industries, machine construction, medical technology and the food industry. The applications range from keyboard covers, seals, plug connectors and medical apparatus, to nipples for baby bottles.

7.2.2.8 Crosslinking Reaction

Liquid silicone rubbers vulcanise in heat (170 °C to 210 °C) following compound crosslinking. In this technology, the crosslinking agent and the polymer material react under the influence of a catalyst. In contrast to peroxide crosslinking, no fission products occur here.

In order to begin the reaction at a desired point in time, the monomers are held in separate containers A and B. The A component contains the catalyst, and the B component contains the crosslinking agent. The compound crosslinking begins immediately after mixing the two components, even at room temperature. However, the reaction at that point proceeds very slowly and unevenly, so that the vat time (the processing time after the mixing of the two components) amounts to three days at room temperature. Figure 7.5 depicts the vat time of a typical LSR as a function of temperature. Observe here the rapid decrease in processing time with increasing temperature.

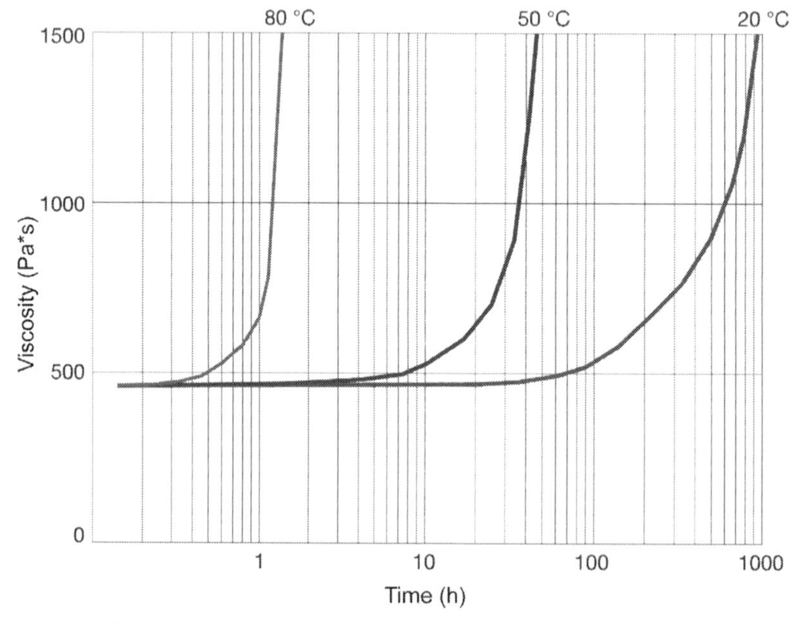

Figure 7.5 Vat time of LSR as a function of temperature

From a temperature level of approximately –20 °C and below, there is no discernible reaction.

This characteristic temperature dependence means increased monitoring during longer periods of production stoppage. Hot machine parts which come into contact with the mixed material for longer periods can become stuck with material which has vulcanised.

7.2.2.9 Structural Viscosity

The consistency of the non-vulcanised 2-component LSR is flow-capable and pasty. Although the pure polymer material has the characteristics of a Newtonian fluid in its flow behaviour (the viscosity is independent of the shear rate of the process), a mixture of polymer with extenders displays a structural viscous property. Structural viscosity means that the viscosity of the mixture decreases significantly with increasing shear rate. This may be observed in every thermoplastic.

This property has consequences which relate to the processing method because the material is subjected to widely-different shear stresses:

- low shearing (= 1 to 10 s^{-1}) of the individual components during conveying and dosing
- medium shearing (= 50 to 500 s^{-1}) during mixing and homogenising
- high shearing (= 500 to 50,000 s^{-1}) in the injection process phase and filling the mould.

Figure 7.6 depicts the viscosity characteristics curve as a function of shear rate and temperature for two different commercial silicone rubber components.

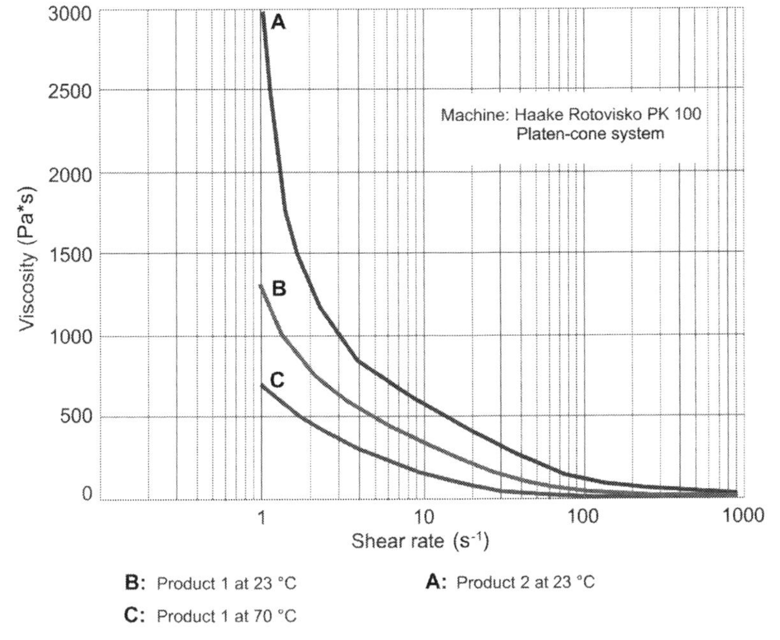

Figure 7.6 Viscosity characteristics curve for two commercial LSR products

The sharp decrease in viscosity with increasing shear stress is very clear. As the material temperature is increased, the viscosity increases relative to the same shear rate. Since the material experiences a very strong shear stress just at the injection process phase, the viscosity sinks dramatically at that point. The result of this is that the entire machine technology, and especially the mould, must be matched to a very 'thin' material.

7.2.2.10 Vulcanisation

In general, the vulcanisation of standard LSR materials increases rapidly from approximately 110 °C. This may therefore be established as the starting point of the compound crosslinking during processing. For this reason, mould temperatures during LSR processing lie between roughly 170 °C to 210 °C. Vulcanising periods with optimal heat economy and suitable mould geometry in a temperature range of 180 °C, are approximately 3 to 5 s/mm of wall thickness. This means a crosslinking speed, which is 3 to 4 times faster than with other rubber systems, and hence a correspondingly higher number of moulding cycles.

7.2.2.11 Internal Mould Pressure

As with any hot vulcanising material, liquid silicone rubbers also undergo an increase in volume due to the heating in the mould and the related vulcanisation. Internal mould pressures of 100 to 150 bar are the result. However, specific injection pressures of 1000 bar and more are required for filling the mould. The necessary injection pressure is highly dependent upon the sprue system. In order to produce parts that are as free of variations as possible, this data must be applied during the calculation of the clamping force. In addition, because of the low viscosity of the material which is introduced into the mould, all of the tolerances of the processing machine and the mould must be kept especially low in order to avoid over-injections.

7.2.2.12 Shrinkage

LSR parts display shrinkage during processing. This is a result of the significantly different expansion coefficients of the mould steel and the polymer material. In addition to the material-specific data, the degree of the shrinkage is highly dependent upon the mould temperature, the generated pressure level and the direction of injection.

7.2.2.13 Tempering

Compound crosslinking silicone rubbers do not form fission products during crosslinking. However, despite the high degree of complete chemical reaction, a 100% reaction cannot be achieved. This means that volatile polymer components remain behind in the finished part. For applications in the food industry or in medical areas, liquid as well as solid silicone rubber products must be tempered in accordance with regulations, for example the German BgVV and in the US the FDA.

This tempering takes place in ventilated kilns with fresh air supply at approximately 200 °C over a period of 2 to 4 hours.

7.2.2.14 Preparation of LSR

LSR is produced by the raw goods manufacturer, prepared, mixed and filled ready for use. The crosslinking agent required for vulcanisation is added to one portion of the material, the initiating catalyst is added to the other portion.

As a result, two components are produced, A and B. The two components are matched in such a way that the ready-for-use LSR systems may be mixed in a 1:1 relationship. Both components are shipped to the processor in 20 litre or 200 litre barrels.

7.2.3 Processing Components for LIM

7.2.3.1 Dosing Equipment for LSR

Since the raw material is shipped from the manufacturer in 20 litre or 200 litre barrels, it must be further processed at the production plant with dosing equipment which is specially designed for the purpose. This type of dosing equipment must fulfil the following requirements:

- Removal and transport of the material from the shipping drums
- Mixing the two material components in a 1:1 relationship
- Added dosage of colour and additives if applicable
- Conveying the processed mixture to the injection moulding machine.

Figure 7.7 depicts a schematic view of the system with a functional overview of a typical commercially available dosage system for liquid silicone. The removal of the A and B material components is performed by barrel presses. In these presses a device called the follower plate is pressed onto the material surface so that the material flows in the desired direction of conveyance under the applied pressure. Piston-pumps are installed as conveyor and dosing pumps. The two components pass through conduits to a mixing block in which they are brought together for the first time. Both pumps must be set to run synchronously in order to achieve the desired 1:1 mixing ratio. A static mixer, through which the material components must run before delivery to the injection moulding machine, is installed downstream for further homogeneous mixing. For pigmentation with colour paste or for dosing of additives, the dosing systems have separate small dosing units available. Portions of 0.3 to 5% can be added for mixing.

The dosing system for liquid silicone rubber, which was described previously, is designed for only one machine. In order to save costs on space and investments, central material supply units are also available. Here, multiple injection moulding machines are supplied from one dosing system. Modern material conveyance systems operate on the concept of central material supply with a decentralised supply by volume at every injection moulding machine. Figure 7.8 depicts this type of system in a schematic overview.

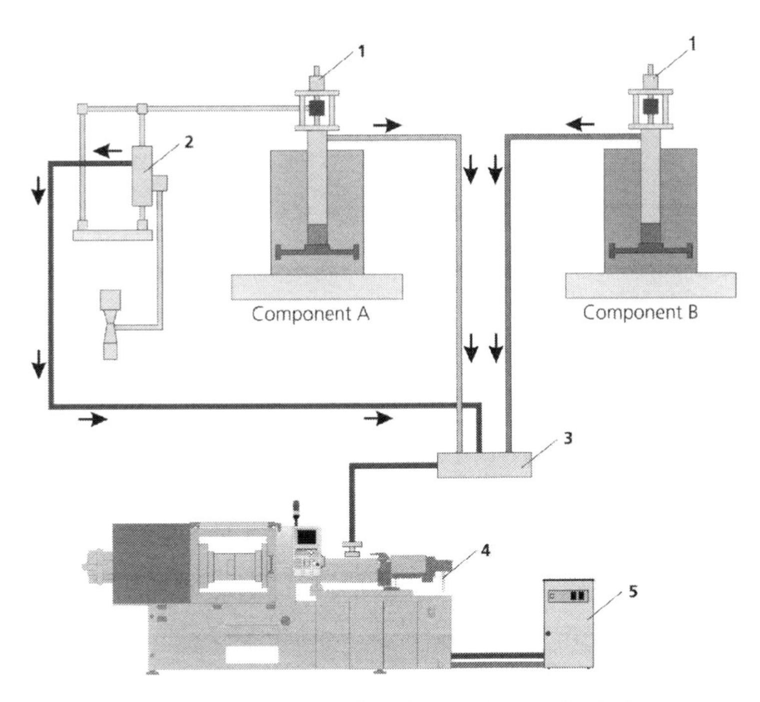

Figure 7.7 Mixing and dosing equipment for LSR
(1) Dosing apparatus for A and B components of the LSR, (2) Dosing apparatus for colour dosing, (3) Mixing station, (4) Injection moulding machine with LSR cylinder module

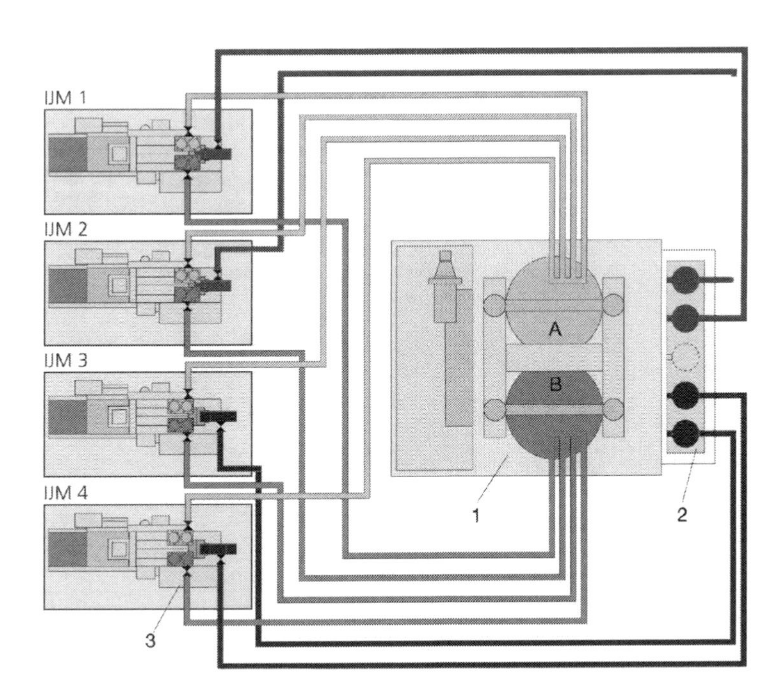

Figure 7.8 Central material delivery with decentralised volume dosing
(1) Central material supply with components A and B, (2) Supplemental frame with up to 5-colour or additive delivery pumps, (3) Volume gauge at every injection moulding machine

A central material delivery system consists of 200 litre barrel presses with delivery pumps. This delivery system supplies several stand-alone dosing systems that are located directly at the injection moulding machine. Decentralised dosing systems may consist, for example, of time-controlled needle dosing valves with electronic volume gauges. They ensure the exact mixture ratio during dosing. By means of additional electrically controlled dosing units, colours and additives may be freely dosed. Mixers and the inlet to the machine are downstream from the dosing system.

Since every injection moulding machine has autonomous volume dosing with controls, there is no limitation in regard to the number of machines to be supplied.

A means for exchanging barrels without interruption may be integrated into the system as an added option in order to avoid unnecessary down times for the machines.

7.2.3.2 Machine Technology

Injection moulding machines for the processing of LSR are differentiated by a few process-relevant peculiarities from machines that process other materials, and especially through a special injection unit. A standard injection unit can be converted relatively quickly and easily for LSR processing by the installation of an LSR cylinder module. Since LSR technology places great demands on the machine in order to ensure problem-free operation, the following points should be kept in mind during the selection of a suitable injection moulding machine.

- The machine control unit must be structured simply and flexibly in order to make the hook-up of different peripheral machines possible. In addition, secondary equipment and mould heating must be operated and monitored directly through the machine control unit.
- The most important motion process phases of the machine should be controlled to ensure the highest possible reproducibility of parts.
- A screw with position regulation allows, among other things, the absolutely precise maintenance of the defined stroke and pressure profile at injection.
- Ramp-controlled mould sealing provides for positioning on the clamping platen which is accurate to the millimetre, providing problem-free operation of the handling devices.
- Mould sealing, which is characterised by high rigidity and parallelism of the mould platen, as well as fast opening and closing speeds enable a rapid injection cycle. Due to the fast vulcanisation time of LSR, 4 to 5 second cycles are not rare.

7.2.3.3 Injection Unit

Silicone cylinder modules are available in different sizes since cylinder size and injection volume must be carefully adjusted to each other during processing. Because of this, a high degree of control accuracy of the screw motion is ensured for small or large injection volumes and hence the greatest possible reproducibility of parts from injection cycle to injection cycle. The injection unit of an LIM machine must perform the following tasks:

- Mixing of components as well as additional dosages of colour pastes or additives
- Precise volume dosage with sufficient speed within the heating period
- Reproducible and precise injection of a defined volume of LSR
- Pressure release from cold runners with open nozzles
- Avoidance of early crosslinking by tempering or cooling of the LSR in the cylinder and the machine nozzle
- Thermal separation by retracting the cooled machine nozzle from the heated mould.

7.2.3.4 Cylinder Module

Support elements and device drives in customary standard machine versions may be employed for the installation of a silicone cylinder module. A drive for a hydraulic needle sealing nozzle must also be provided. With the use of cold runner nozzles with needle sealing or with a cold runner head, the needle sealing nozzle on the machine can be omitted. Figure 7.9 depicts a LSR cylinder equipment kit.

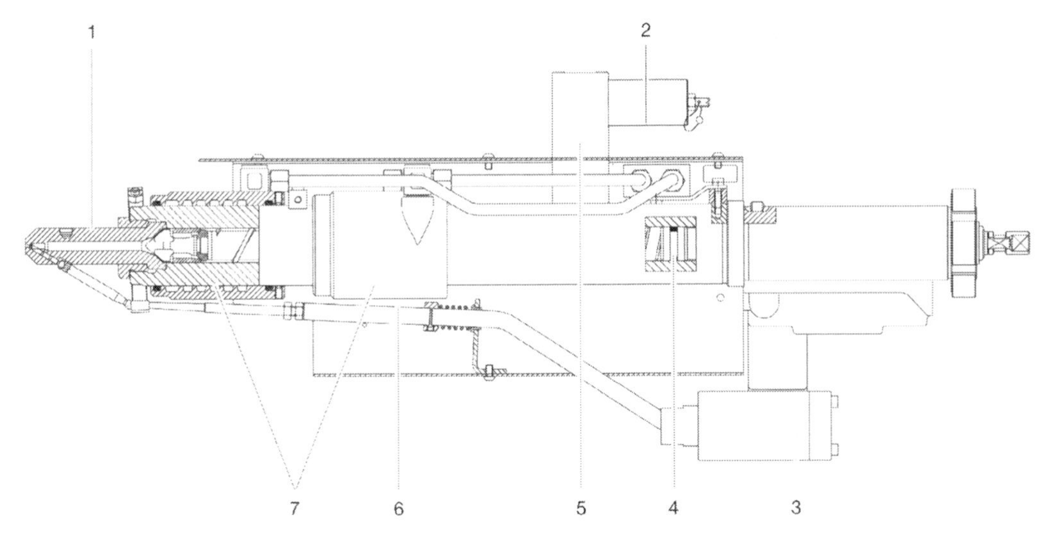

Figure 7.9 LSR cylinder equipment kit
1. Hydraulic needle sealing nozzle, 2. Connection with safety valve,
3. Hydraulic cylinder for sealing nozzle, 4. Seal from the drive side,
5. Insert, 6. Rod, 7. Heating and cooling mantle

The cylinder module is fitted with liquid thermal collars that are connected in series. Cooling is through a water battery or through a control unit. The temperature of the cylinder should be maintained at a constant level of approximately 20 °C, since crosslinking reactions may otherwise occur. The material inlet for the cylinder is placed at the front near the nozzle, since a high L/D ratio is not necessary for dosing. Mixing and homogenising of the individual components has already taken place before entry into the cylinder through the dosing system. Only a secondary homogenising of the components with any subsequent additives or colour pigments takes place in the screw.

The screw is designed as purely a mixing and conveyor screw without compression. Because of the low viscosity of silicone, the shaft of the screw is ground cylindrically at the connection of the inlet area and sealed with a radial seal from the drive unit. The coupling of the screw to the drive is with an appropriate coupling unit. An important element of this cylinder equipment kit is a high-pressure valve at the inlet to the cylinder. These valves serve on the one hand for control of material from the mixer of the dosing system and, on the other, they fulfil their function as a high-pressure valve.

7.2.3.5 Non-Return Valve

Because of the low viscosity of LSR, the non-return valve at the point of the screw has a special significance. At start-up and at the beginning of the injection process stage, this valve must close cleanly and without delay in order to maintain good reproducibility of the parts. Typical non-return valves for thermoplastic processing cannot be used since the typical seal rings cannot be positively moved because of the very low resistance of the silicone. The disc non-return valve for 2-K silicone has proven to be beneficial and is shown in Figure 7.10. It has the following features:

- Precise dosage characteristics
- Positive closing
- Low back-flow
- Precise injection volumes.

With the disc non-return valve, a seal ring is pressed onto the rotating spring in the closed position. During the plasticising process phase, the seal ring opens because of the pressure of the melt from the screw chambers and allows material into the screw pre-chamber. After the end of the plasticising phase, the spring forces the seal ring back into position. At the start-up of the screw, the non-return valve is thus already locked, and any leakage flow in the screw chambers is very low as a result.

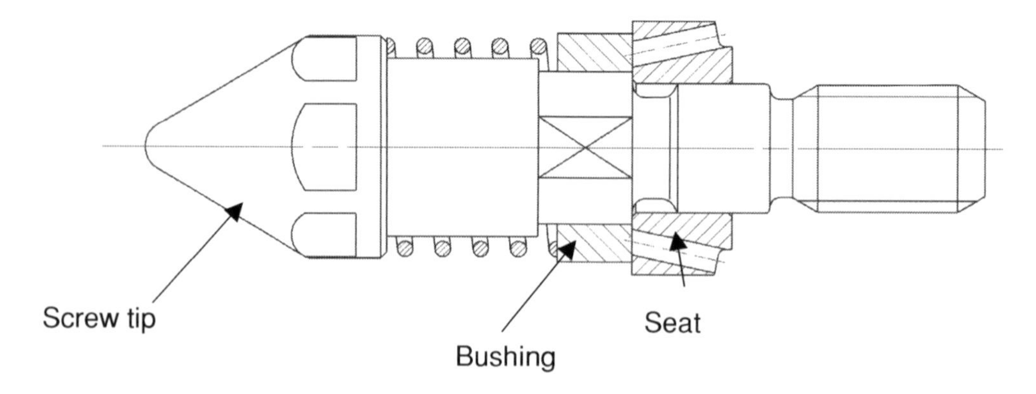

The bushing slides forward during screw rotation and plastic is pumped in front of the screw tip.

During injection the bushing slides back to make a seal on the seat.

Figure 7.10 Non-return valve for 2K silicone

7.2.3.6 Auxiliary Heating Control Circuits

Since LSR moulds are frequently heated electrically, internal heating circuits in the machine for the control and supply of mould heating are beneficial. The essential heating parameters may thus be managed directly with the other machine adjustment data and monitored by the control unit, thereby improving production accuracy. For example, ARBURG machines offer up to 16 such auxiliary heat-control circuits.

7.2.3.7 Interface for Brush and Cleaner Units

With LSR moulds, it is common practice to clean the form after every injection cycle or, to facilitate the removal process from the form, to brush the parts from the form. Interfaces for brush and cleaning units are customary for this.

7.2.3.8 Handling Interface

Because of the increase in automation in injection moulding manufacturing, more and more evacuation devices are being used in production. In order to ensure that these handling devices work reliably, they must be co-ordinated by the machine controls. For this, handling interfaces are required.

7.2.3.9 User-Programmable Inputs/Outputs

As moulded parts become more and more complex with many undercuts, the moulds themselves become more complex. As a result, hydraulic or pneumatic core-pullers must frequently be employed for removal from the mould. In order to control them, user-programmable inputs/outputs for the control unit are necessary. The core-pullers in the mould are thus controlled and monitored by the machine.

7.2.3.10 Air Blast Equipment with Pressure Reducer

The removal of LSR parts is frequently performed with pneumatic support. For this, air blast equipment is required, including a pressure reducer, which allows the compressed air to be introduced only at the specified pressure.

7.2.3.11 Control Unit for Vacuum Pump

As a result of the precise working of LSR moulds, there is very poor airflow. For extremely high injection speeds, and the extremely fast mould filling which is associated with them, or due to extremely tight tolerances, the air does not have adequate time to escape from the cavity. The result is either scorching or moulded parts which are not filled completely. In order to avoid this, the cavities are evacuated before the injection step. This signal for control of the vacuum pump must come from the machine control unit.

7.2.3.12 Hydraulic/Pneumatic Cold Runner Control

Cold runner systems that are ready to use are offered as standard in the marketplace. These function especially well with needle sealing nozzles that are controlled hydraulically or pneumatically. In order to allow integration with the process sequence of the injection moulding machine, interfaces were developed for control of this type of cold runner. The machine control unit also monitors the position of the needles.

7.2.3.13 Protective Screens Made of PC

In most instances, LSR moulds are heated electrically to temperatures of 170 °C to 210 °C. Because of the level of heat that is radiated from the mould as a result of this, a heat-resistant material must be installed for the protective screens of the machine. PC meets this requirement.

7.2.3.14 Simple Cold Runner Nozzle

The simple cold runner nozzle with hydraulic needle sealing system, as shown in Figure 7.11 makes possible either direct or indirect injection of silicone parts. Advantages in the application of this system are good thermal separation between the cold nozzle and the heated mould and the flat gate location on the moulded part.

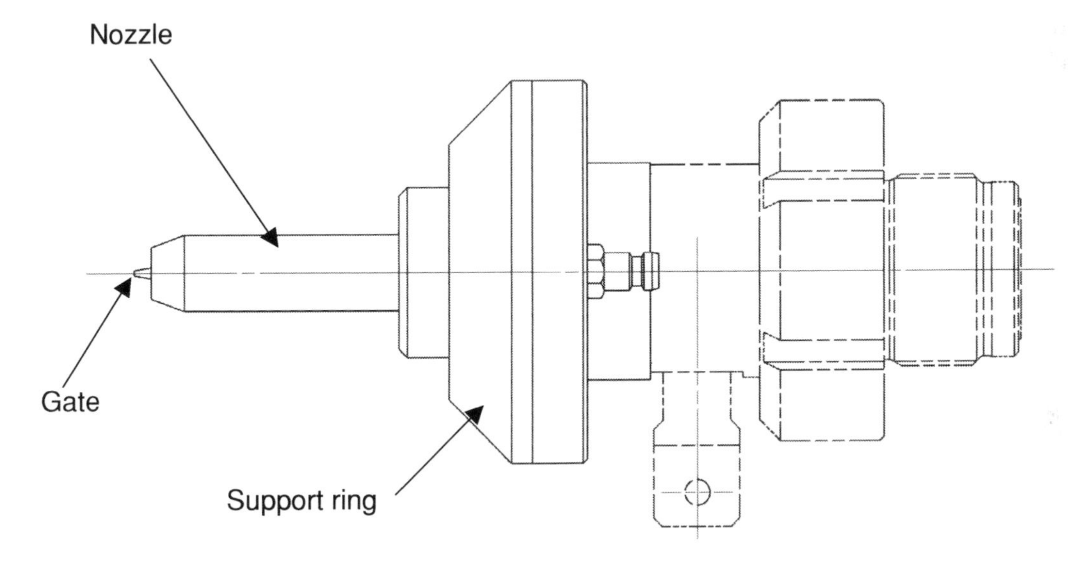

Figure 7.11 Single cold runner nozzle

7.2.3.15 Cold Runner Head

The cold runner head shown in Figure 7.12 is a machine-integrated system and hence applicable for different moulds. It is attached to the cylinder with a split coupling and enters into the mould through the fixed platen. The individual nozzles are fitted with a hydraulic needle sealing system, thus ensuring precise sealing of the sprue. As a result, the injection locations are absolutely clean and the surface quality is flawless. With the open design concept and the modular construction, manifold systems with up to six sprues may be easily realised.

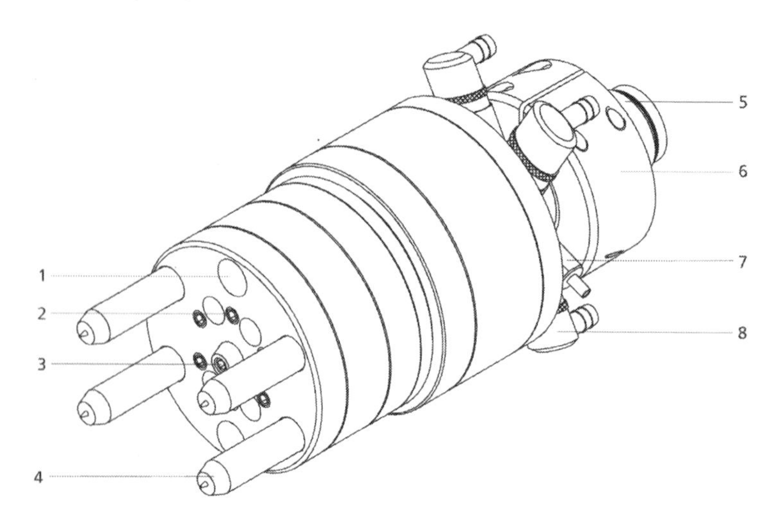

1. Insert cartridge
2. Volume flow limiter
3. Cone centring
4. Nozzle with shut-off needle
5. Cylinder connector
6. Split coupler
7. Hydraulic connection
8. Temperature control connection

Figure 7.12 Cold runner head

The mould can be completely separated from the cold runner head in connection with a quick-coupling system, thus ensuring an optimal separation between hot and cold during stationary periods. Along with the technical benefits of this system, its application has a cost-reducing effect in the construction of moulds, because different moulds can be matched to the same cold runner head. The following significant features characterise the cold runner head:

- A component of the machine and not of the mould
- Set-up and assembly as with a special nozzle
- Retraction from the mould possible at any time
- Opening and closing of the needle sealing nozzles through the machine controls
- Arrangements with up to 6 nozzles
- Each individual nozzle with a needle sealing system
- Volume flow may be regulated
- Good cleaning and maintenance features
- Simple and sturdy construction of the exchangeable needles and nozzles
- Optimal thermal separation between hot and cold modules
- Uniform force application to all nozzles through the central hydraulic cylinder
- Reduced mould costs, since installation with different moulds is possible
- Easy matching of new moulds to the same cold runner head
- Cycle-time savings
- Cost reductions through material savings
- Absolutely clean injection locations.

7.2.3.16 Mould Technology

The mould is heated electrically. It is recommended that control of both mould halves is carried out separately. The heating power for small moulds should be between 1,600 and 2,200 Watt per mould half, varying according to the mould size (e.g., 215 x 300 mm). The mould temperatures vary according to the LSR material. Standard values range from 180 to 250 °C.

Making LSR moulds requires expert know-how. It is recommended that only mould designers and makers who already have experience in this technique are used. In view of the high mould temperatures you need insulating platens with a thickness of 10 mm for the mould.

Because of the described material-specific characteristics of LSR, correspondingly high demands are placed on the moulds in regard to rigidity, processing precision and temperature control. Even flaws of less than 0.01 mm can form ridges and thus make cost-intensive reworking procedures necessary for the moulded parts. The primary distinguishing feature of the different mould technologies is the manner of injection. Figure 7.13 provides an overview of the injection systems in use today.

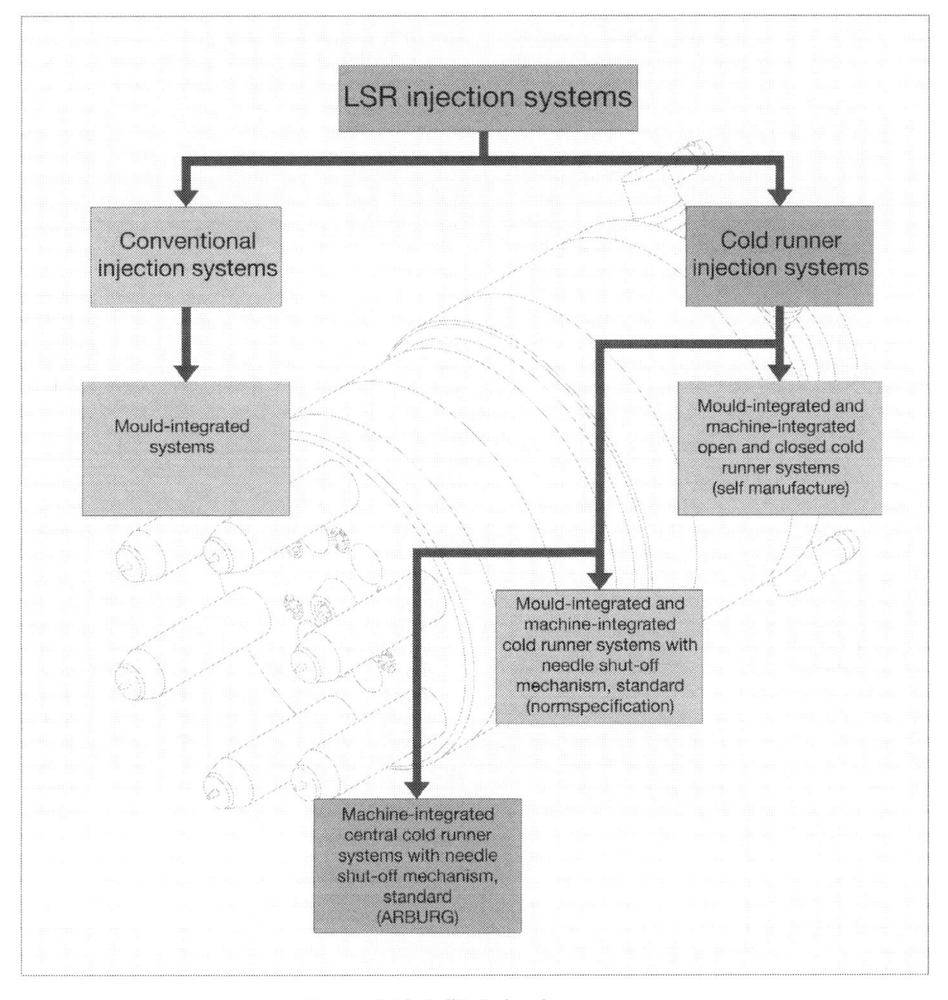

Figure 7.13 LSR Injection systems

As is the case with the use of other materials, the injection in LSR moulds has the task of guiding the material flowing from the machine nozzle into the cavity with as little pressure loss as possible and without interference.

Conventional Injection System

The conventional injection system, that is, the non-temperature controlled, mould-integrated sprue runner, provides the simplest way of achieving this. Injection systems such as those with thermoplastic processing are possible here.

The cone or pin-gated sprue, with or without a submanifold, offers the least resistance to the in-flowing material. It is applied primarily in simple moulds for thick-walled parts with high quality requirements. The significant disadvantage of this injection method is the high area for isolating, separating and recycling of the very loosely dimensioned sprue cone with the submanifold.

Point gates for simple or multiple sprues are frequently found in the mould parting line. The runners are guided up to just before the mould cavities with a somewhat larger cross-section, in order to flow into the narrow, short gate runner. The sprue can thus be separated easily from the form part, although not automatically.

Tape or film gates bind flat parts from one side. This allows a material flow that promotes uniform orientation in the part, and thus fewer tendencies for distortion. In order to set this sprue optimally, the part must be injected asymmetrically. Using the injection unit positioned horizontally where it may be displaced parallel to the fixed platen of the sealing unit, is especially suited for this application. An additional option for injecting this type of part can be realised by locating the injection unit at the parting line.

Tube-shaped parts that must have a high degree of cylindrical accuracy and be free of welding seams are joined with tunnel sprues. Tunnel sprues provide the significant advantage of automatic sprue separation from the moulded part at the ejection of the part. The sprue runners end shortly before the mould cavities and open into bores which form tunnel shapes leading toward a lateral wall surface of the mould cavity. Figure 7.14 depicts a mould construction with a conventional, mould-integrated runner system (pin-gated sprue with submanifold).

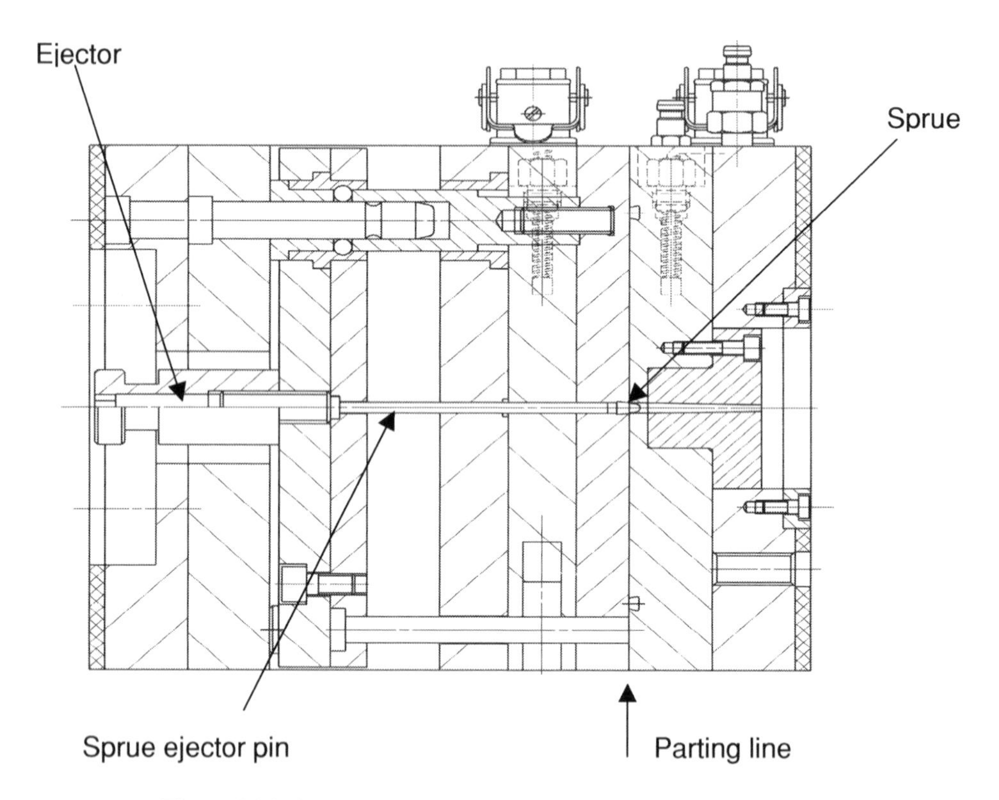

Figure 7.14 Conventional sprue system, mould-integrated

The significant disadvantage in the application of a conventional system is the fact that material that has once been vulcanised cannot be returned into the system for reprocessing. Based on the high price of the raw materials for liquid silicone, this is a very considerable economic factor.

As an alternative to the conventional injection system, cold runner systems are coming more and more into use in the processing of liquid silicone. These systems may be compared with the hot runner technology found in thermoplastic processing.

Cold Runner Sprue System

The cold runner sprue system may be divided into categories. First of all, there are mould-integrated and machine-integrated, open and closed cold runner systems, which are primarily self-manufactured. These systems are not generally available since they are utilised in special applications by individual process technicians.

The decisive problem with the application of an open cold runner system exists in thermal separation around the gate zone. It has proven to be very difficult to inject a moulded part directly with a cold runner nozzle. The vulcanisation process is not limited to just the part, but also continues in the gate zone of the cold runner nozzle. This vulcanised plug is fixed in the part at the next injection as an unsightly mark. As an alternative method for this cold runner system, processing may also be handled with a nozzle sealing system.

Mould-Integrated Cold Runner System

Manufacturers now offer standardised mould-integrated and machine-integrated cold runner systems with pneumatic needle sealing nozzles. The advantage of this type of standardised cold runner system is its simple installation in the injection moulds. The cold runner system may be considered as a standard mould in design. It is equally possible to use a cold runner system for multiple moulds. Figure 7.15 depicts a mould construction with the application of a mould-integrated cold runner system as a standard specification.

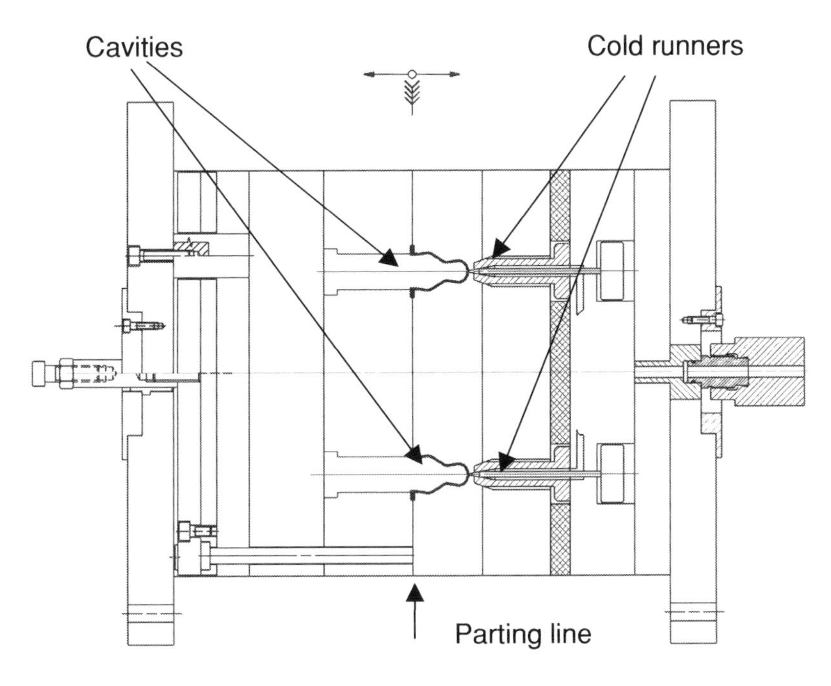

Figure 7.15 Cold runner system as mould standard specification

The application of this type of standard specification makes possible manufacturing of moulds with a minimum injection by the precise separation of the cold sprue from the hot cavity. The sprue markings are minimal thanks to the pneumatic needle sealing system. Cycle times can be reduced because of the high temperatures of the cavity. As an ARBURG standard for example, cold runner systems are offered with up to four gate points.

Machine-Integrated Cold Runner Systems

The third category of cold runner systems is represented by machine-integrated systems. This can utilise a simple cold runner nozzle, as well as the cold runner head with up to six gate points. The major advantage of these systems is that they are a fixed component of the machine. The needle sealing nozzles may thus be controlled from the machine control system. Machine-integrated cold runner systems may therefore be installed independently of the mould. LSR moulds can hence be produced much more cost-effectively, since it is only necessary to procure the cold runner once per machine. The simple application of a cold runner nozzle in an existing design is depicted in Figure 7.16.

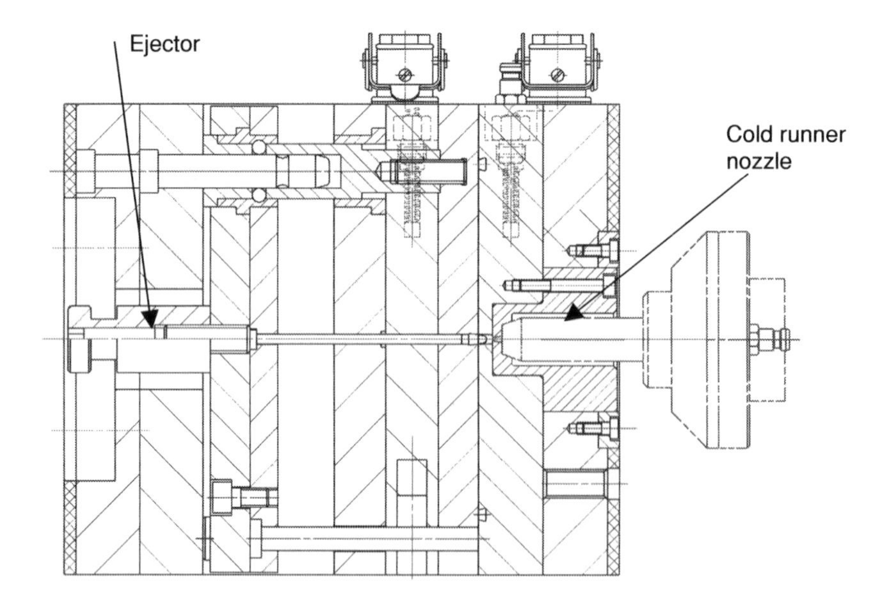

Figure 7.16 Mould design with a simple cold runner nozzle

The advantages of the application of this type of cold runner nozzle in a conventional injection system are clear:

- Material savings through direct injection into the cavity, or through injection onto a sprue manifold
- Optimal thermal separation between the cold sprue and the hot cavity
- Absolutely clean injection areas
- Retraction from the mould possible at all times
- Control of the needle sealing nozzle through the machine controls.

The cold runner head is installed with LSR moulds with several cavities. This cold runner head is mounted as a forward unit on the cylinder. Through machine contact, the head enters the mould through the fixed mould platen so that the nozzle tips reach the cavities. Because of the open concept and modular construction design, application of the head is flexible, and it may be fitted with up to six nozzles. Reducing or adding to the number of nozzles is simple and easy to perform with the use of blank adapters. Set-up and assembly on the machine are similar to a special nozzle so that installation work on the mould is eliminated. Figure 7.17 illustrates an installed cold runner head.

7.2.4 Application of LSR Parts in Food Production

LSR parts can be applied in a great variety of fields. However, for use in food production for example, the BgVV (Germany) has only allowed the application of LSR materials of 50 and 60 Shore hardness. The particular raw material supplier should be aware of the suitability of their materials for specific applications if there are concerns.

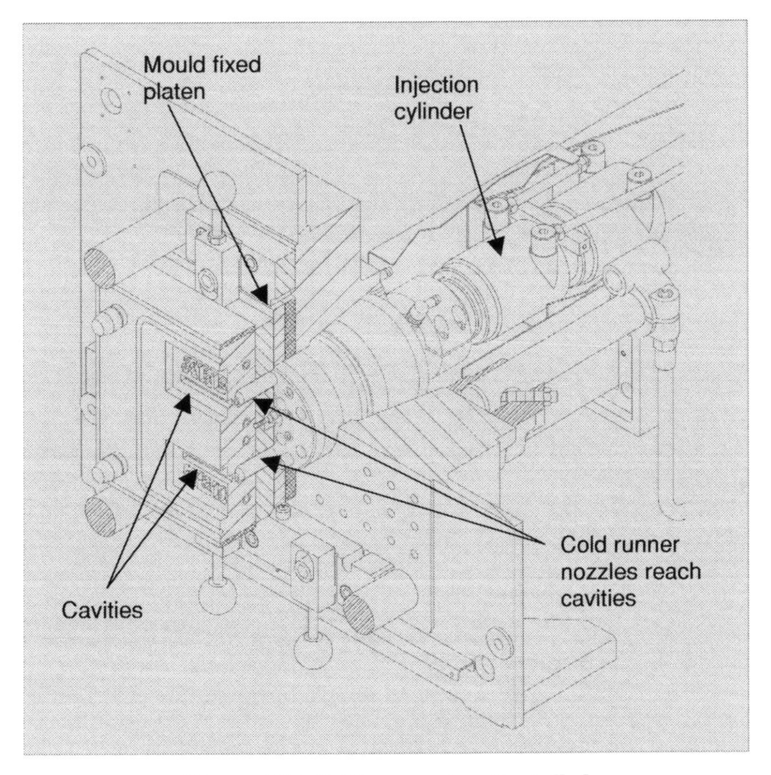

Figure 7.17 Cold runner head installed

7.2.5 Summary

Outstanding material properties and good processing factors make LSR more and more the material for applications with high demands today. Among the varied benefits are good electrical insulating properties, excellent heat, cold and light resistance as well as the physiological compatibility of many types.

The elimination of preparation techniques is certainly another reason for the rapid acceptance and expanded use of LSR. Especially for purely thermoplastic process operations, LSR provides an excellent potential for entering into the manufacturing of elastomers. This is without the necessity of making high investments in facilities and expertise for material preparation.

For LIM processing, only a correspondingly modified injection moulding machine and a dosing machine for conveying and mixing in a 1:1 ratio are necessary. In addition, heated moulds, which are especially constructed for the processing of LSR, are used.

Cold runner systems are the state-of-the-art in mould technology. Here, mould-integrated or machine-integrated cold runner systems with needle sealing nozzles are generally used.

Nonetheless, LSR is not a material which can be processed entirely without problems. Liquid silicone rubber is a modern high performance material that must be thoroughly managed in practical applications. An entire series of process factors must be observed in order to produce LSR moulded parts with quality and high value. Each process factor in the cyclical flow shown in Figure 7.18 must be mastered precisely in order to work successfully with liquid silicone rubber.

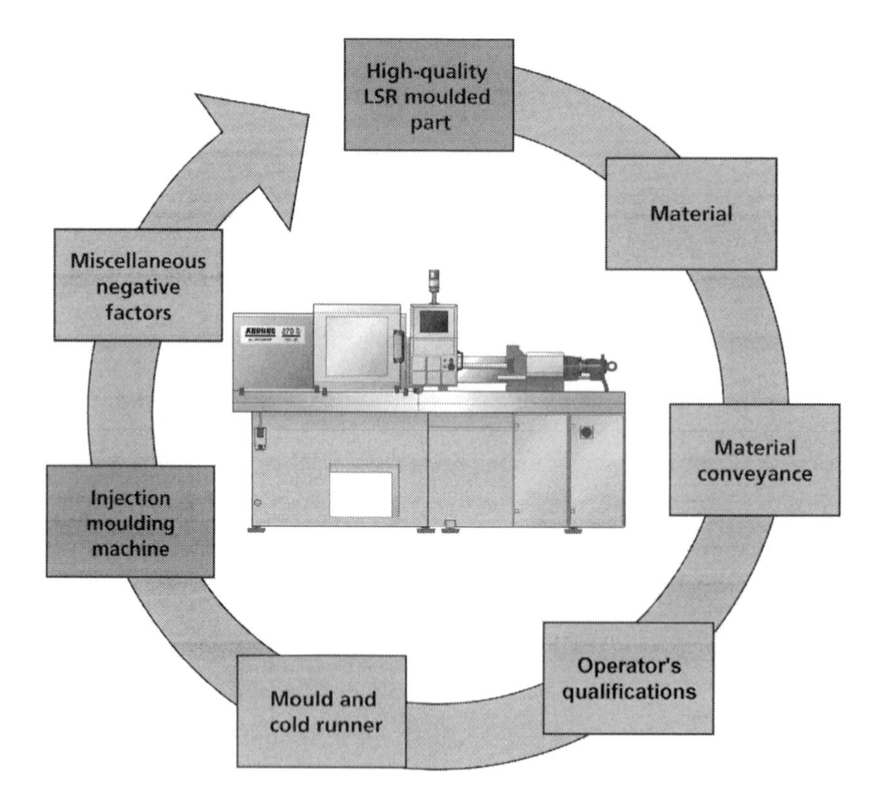

Figure 7.18 Process factors in the production cycle of a successful LSR part

Machine technology, mould construction and process management contain many special characteristics which must be ascertained, before a successful result is achieved.

7.3 Processing of Ceramic and Metal Powder Materials

7.3.1 Introduction

In recent years industry users have considerably raised their standards, asking for higher output, longer duration and lower servicing costs. The possibilities of conventional materials for machine construction have thus been exceeded.

In this situation ceramic and metal powder materials have opened new perspectives to solving engineering problems. In the 1960s industrial products were first made of aluminum oxide. Due to its excellent characteristics this material has won more and more ground in the technical industry, chemistry, electronics and machine manufacturing.

Some of the outstanding characteristics of ceramic materials are:

- High mechanical stability, even at high temperatures
- Good resistance to changes in temperatures (but avoid thermal shocks >100 °C, these will cause cracking)
- Good chemical resistance
- Very hard and wear resistant
- Low specific gravity.

In principle all materials available in a sinterable powder can be mixed with an appropriate binder and processed on injection moulding machines. Therefore in addition to the traditional oxide ceramics it is also possible for example to use metals, carbides and nitrides. Some typical materials are shown in Table 7.11.

Table 7.11 Some ceramic and metal materials and their colours	
Material	**Colour**
Aluminium oxide (Al_2O_3)	White
Zirconium oxide (ZrO_2)	Yellowish
Silicon carbide (SiC)	Dark grey
Silicon nitride (Si_3N_4)	Dark grey
Magnesium oxide (MgO)	Yellow-white
Stainless steel	Grey
Copper (Cu)	Red brown
Hard metal (WC-Co/Ni)	Dark grey

Since mixing and injection units can be subjected to increased wear when using powdered materials, it is recommended to keep the grain size of the powder as small as possible. However, the range in which the grain size of the material guarantees optimum production, and therefore the desired properties of parts, is relatively narrow (Table 7.12).

Table 7.12 Powder properties and performance		
Properties	**Coarse powder**	**Fine powder**
Surface of the moulded part	Rough	Smooth
Green strength (strength after binding but before sintering)	Low	High
Injection moulding performance	Poor	Good
Wear on machine and mould	High	Low

Fine powders produce less surface roughness, can be processed with little wear and result in higher green strength. The characteristics of various powder materials can be see in Tables 7.13 and 7.14.

Table 7.13 Material grain sizes	
Material	**Grain Size**
Metals (atomised)	<30 μm, X50~6-7 μm
Carbides	<1.5 μm, X50~0.7 μm
Silicates (e.g., porcelain)	<45 μm, X50~5 μm
Oxides (e.g., Al_2O_3)	<15 μm, X50~0.75 μm
X50 = magnification	

Table 7.14 Shrinkage	
Grain shape	**Shrinkage**
Round	Slight, isotropic
Irregular	High, isotopic
Flat	Anisotropic

7.3.2 Binder

The binder must have the following properties:

- Be dimensionally stable during debinding
- Have a good storage capacity
- Does not react with the powder materials
- Has a high green strength. (strength after binding)
- Good mould removal properties
- Thermal stability
- Decomposes easily and completely upon debinding.

The bonding between the powder and the powder particles should be as great as possible, so that the centrifugal forces arising during the injection process do not give rise to any separation of the two components, thereby resulting in inhomogeneously filled parts.

In order to achieve good injection moulding characteristics and isotopic sintering with a low rate of shrinkage, the use of spherical shaped particles should be favoured.

During mixing, the binder and powder are combined to form the most homogeneous compound feedstock possible. In order to guarantee that the powder particles are completely surrounded by the binder components, all powder agglomerations must be broken up by shear forces.

The binder component should be kept as low as possible, as this reduces shrinkage during sintering, thereby allowing closer tolerances on the finished part to be maintained. The binder component of injection moulding masses is between 35-55% by volume. If the binder component is too low, the quality of the injection moulding properties is reduced and the wear on the machine and injection equipment increases greatly.

Various possibilities are available for homogenising the powder and binder. For example, the material can be mixed with a twin screw extruder, a shear roller extruder, or a rapid mixer.

The use of powder injection allows the economical manufacture of components that could not be made by employing metal cutting or pressing methods. Injection moulding allows almost limitless design freedom for shaped parts. The production process includes moulding, debinding and sintering of the shaped parts. This means that the final tolerances of the components are determined by the following important factors:

- Binder content
- Powder characteristics
- Mixing process
- Injection parameters
- Distortion due to gravity
- Sliding properties of the sintering surface

7.3.3 Processing

To determine the most economical process one needs to consider the size of the part, its geometry and the quantity to be manufactured.

7.3.3.1 Compression Moulding

The ceramic powder is compacted inside a mould under high pressure. Bores, steps, bevels and threads can be pressed into the part in the pressing direction only, since the part has to be ejected after the moulding. The compacted parts can be further machined to have additional bores, undercuts etc. (high tech, high cost machining necessary).

7.3.3.2 Injection Moulding

Injection moulding is an economical method of processing ceramics. The material is plasticised by a screw. Dosage and plasticising take place simultaneously, while the previously moulded part still cools inside the mould. With automatic removal devices the entire process can run automatically.

Advantages of processing ceramics on injection moulding machines are:

- Wall thickness of part may vary
- Holes and undercuts possible
- A higher density possible
- Better surface and end quality
- Shot weights of less than 1 g possible.

7.3.4 Configuration of Injection Moulding Machines

Injection moulding machines can be used to process ceramics providing that the following equipment is available:

- Plasticising cylinder, ARBID version with electrical heating
- Dosage delay
- Thermostatic water valve (feeding yoke)
- Nozzle tip 4 mm bore
- 'U'-Version of clamping unit (swivelable or vertical)
- Injection unit with adaptor for parting line injection.

Note: The 'U'-Version of the clamping unit and the adaptor for parting line injection facilitate manual or automatic parts removal.

Peripheral devices: A cooling device for the mould is necessary. In most cases parts removal is (still) done manually. However, automatic removal devices are increasingly being used.

Mould: Moulds for powder injection can have features normally used in manufacturing plastics, such as sliding bars, pull cores and cavity pressure transducers. The major difference is the wear characteristics, powders are far more abrasive and moulds may require hardening or the use of alloy materials. The mould needs sufficient cooling and will be connected to a cooling device. It is recommended that both halves are controlled separately. The temperatures vary according to the ceramic material. Standard values range from 6 to 10 °C. The mould should have exchangeable inserts. Mould cavities must be high gloss polished. Mould designs should incorporate large ejector pins and sufficient beveling to facilitate easy part removal. Apply insulating platens to both mould halves to prevent condensation.

Wear: There will be wear on the plasticising cylinder and the mould, especially when processing inorganic materials.

7.3.5 The Process Requirements

To facilitate processing on injection moulding machines the material needs to be free flowing and must allow for good plasticising. Once moulded the material requires debinding. The process sequence is shown in Figure 7.19.

7.3.5.1 Debinding and Sintering

Debinding is carried out in furnaces suitable for the binder system. The binder may be removed by way of catalysis, dissolving or thermal decomposition. Debinding can be effectively achieved by a suitable furnace temperature and atmosphere. Removing the binder transforms the moulded part into a porous and sensitive moulded body, referred to as a 'brown compact'. In this condition, the part is only kept stable by minimum binder residues and weak van der Waals forces.

In order to firmly bind the particles of the brown compact to each other, the part is sintered by additional heat up to 2000 °C. This process is the same for the pressed sintered parts. The final moulded part is formed by way of diffusion and/or the forming of liquid phases and grain growth.

Injection moulded sintered parts can approach the theoretical material density up to a value of 99.9%. With suitable powder they will shrink isotropically during sintering and have homogeneous characteristics. Any injection moulding flaws that arise cannot be remedied by sintering. However in most cases it may only be necessary to refinish fits or regrind cutting edges. Fine powder and polished surfaces guarantee excellent surface qualities in the injection moulding of metal or ceramics.

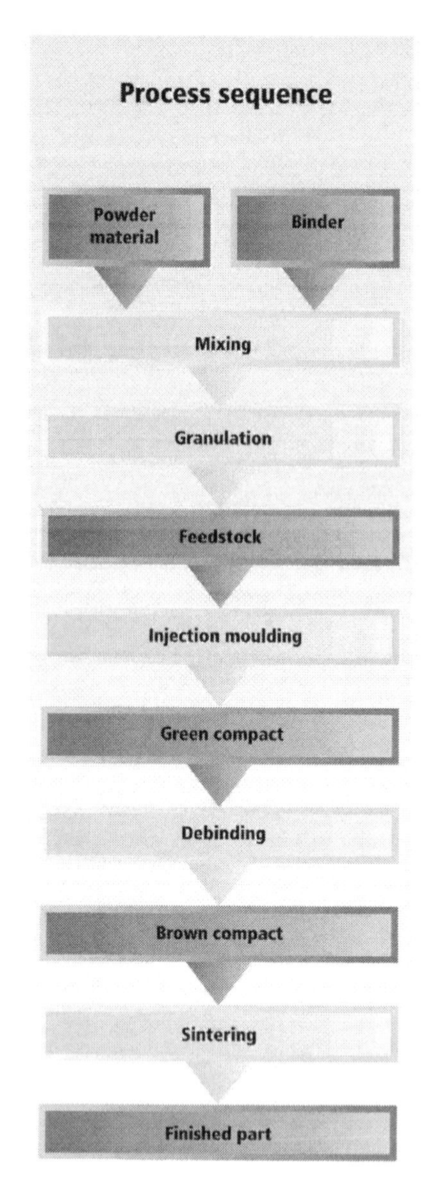

Figure 7.19 Process sequence for injection moulding of powder materials

7.4 A Growing Market for Moulders

Injection moulding technology can be employed successfully to process a wide variety of materials. In all cases the basic requirements are adequate equipment for the injection moulding machine and a suitable material. When these requirements are met, injection moulders have a good chance to win a larger share of the market.

Our knowledge of the specific properties of materials and adequate processing methods has greatly increased, but new applications are continually being developed.

8 Guide Values and Processing Instructions for the Most Important Thermoplastic Injection Moulding Compounds

This chapter will illustrate recommended drying and processing settings for the most common plastics. This will include recommended settings for temperature profiles, injection pressure and speeds, holding pressures, back pressures and cooling times. It will also discuss the injection moulding process parameters and their effect on the process. However, before some materials can be moulded, they must first be dried. This chapter therefore begins with a guide to drying materials prior to moulding. A summary is shown in Table 8.1. Incorrect drying can lead to moulding faults, which will be further discussed in Chapter 9.

8.1 Pre-Drying Material

Table 8.1 Drying guide for thermoplastics. Maximum drying temperature and maximum allowable moisture content

Product	Maximum drying temperature (°C)	Maximum allowable moisture content (granulate) (%)	Drying time (h)
PA	100	< 0.15	4
PBT	120	< 0.02	4
PBT + PC/ASA	120	< 0.02	4
POM	120	< 0.15	4
PES	200	< 0.05	4
PSU	160	< 0.05	4
PAEK	200	< 0.05	4
LCP	80	< 0.02	3
SB	80	< 0.2	3
ABS	90	< 0.2	3
PP + Talc	120	< 0.03	3
ASA/PC	110	< 0.1	3

Note: ASA = acrylonitrile-styrene-acrylate

Many plastics are hygroscopic (absorb water) and must be dried before moulding to remove this residual moisture. There are a number of different types of material dryer available on the market. Often they come equipped with the ability to delivery the material straight to the hopper to prevent further moisture build up. Attention should be paid to the maximum allowable water content as failure to remove moisture above these levels can cause moulding problems. It should be noted that Table 8.1 serves as a rough guide only in terms of drying temperature and time. This is because these can vary depending on the type of dryer used. Attention must also be paid to hopper size as material left exposed to the atmosphere for extended periods begins to reabsorb water. This may occur even when left for just an hour, depending on atmospheric conditions and material type. Hoppers should not therefore be overloaded with hygroscopic materials. Once the material has been correctly dried, attention can be turned to plasticising.

8.2 Examples of Moulding Parameters of Selected Materials

A guide to the basic parameters of nozzle temperature, mould temperature, injection pressure, holding pressure and back pressure for selected materials is shown in Table 8.2. It can be seen that injection moulding parameters vary widely from material to material. With this in mind, it is important to understand what these injection parameters are and how they affect the injection moulding cycle. The next section will therefore discuss the injection moulding process.

Table 8.2 Control values for processing						
Material	**Nozzle-side cylinder temperature[1, 2] (°C)**	**Mould temperature (°C)**	**Injection pressure (Bar)**	**Holding pressure (Bar)**	**Back pressure (Bar)**	**Remarks, see footnotes**
PS	160-230	20-80	650-1550	350-900	40-80	
SB	160-250	50-80	650-1550	350-900	40-80	
SAN	200-260	40-80	650-1550	350-900	40-80	
ABS	180-260	50-85	650-1550	350-900	40-80	
PPO mod.	245-290	75-95	1000-1600	600-1250	60-90	
PVC - hard	160-180	20-60	1000-1550	400-900	40-80	3, 5, 8
PVC - soft	150-170	20-60	400-1550	300-600	40-80	3, 5, 8
CA	165-225	60-80	650-1350	400-1000	40-80	3, 4, 8
CAB	160-190	60-80	650-1350	400-1000	40-80	3, 4, 8
CP	160-190	60-80	650-1350	400-1000	40-80	3, 4, 8
PMMA	220-250	20-90	1000-1400	500-1150	80-120	4
PC	290-320	85-120	1000-1600	600-1300	80-120	4
PES	320-390	100-160	900-1400	500-1100	80-120	4
PE - soft	210-250	20-40	600-1350	300-800	40-80	
PS - hard	250-300	20- 60	600-1350	300-800	60-90	
PP	220-290	20-60	800-1400	500-1000	60-90	
PA 6,6	270-295[9]	20-120	450-1550	350-1050	40-80	4, 8
PA 6	230-260[9]	40-120	450-1550	350-1050	40-80	4, 8
PA 6,10	220-230[9]	20-100	450-1550	350-1050	40-80	4, 8
PA 11	200-250[9]	20-100	450-1550	350-1050	40-80	8
PA 12	200-250[9]	20-100	450-1550	350-1050	60-90	
PA amorph.	260-300	70-100	900-1300	300-600	60-90	
POM	185-215	80-120	700-2000	500-1200	40-80	3, 8
PET	260-280	20-140	800-1500	500-1200	80-120	
PBT	230-270	20-60	800-1500	500-1200	80-120	
PPS	300-360	20-200	750-1500	350-750	40-80	
FEP	340-370	150				5
ETFE	315-365	80-120				5

1. If no other empirical values are available: nozzle temperature = set nozzle-side cylinder temperature. Cylinder temperatures falling in direction of material throat, drop of 5-10 °C for each heating zone; max. temperature difference between nozzle-side and throat 20 °C. For more than 2 heating zones, set nozzle-side heating zone and the following to same temperature.
2. For heat-sensitive compounds set higher temperatures only for short cycle times (shorter dwell time in cylinder).
3. Heat-sensitive.
4. Process only dry granules.
5. Do not use shut-off nozzles, only open nozzles.
6. Injection without non-return valve recommended.
7. Work only without non-return valve.
8. Work only with low back pressure.
9. To improve material feed behaviour: set temperature at same level or slightly rising towards throat.

8.3 Injection Moulding Process Parameters and Quality of Moulded Parts

The properties of an injection moulded part depend upon the working material and upon the processing conditions. In the production of a series of parts, a certain deviation in quality features such as weight, dimensional consistency and surface characteristics may always occur. The size of this deviation will vary from machine to machine and from material to material. Furthermore, external influences or negative factors have an effect on the quality of an injection moulded part. Examples of such negative factors may include changes in the viscosity of the melt, temperature changes in the mould, viscosity changes of the hydraulic fluid and changes in the characteristics of the plastic.

The causes through which these negative factors may arise are, for example, machine start-up after a long period of non-operation, changes in material properties in the processing of a new lot or a different colour, and environmental influences such as the ambient temperature at the time of processing.

The design purpose of injection process regulation is to make these negative influences ineffective, and thus to attain an even higher reproducibility of the parts. The decisive factor for all quality features, which are concerned with dimension and weight, is the internal pressure of the mould. Constant maintenance of this pressure curve in every cycle guarantees uniformity of the quality of injection moulded parts. If the mould internal pressure curve is maintained at a constant, all of the negative factors mentioned above are compensated.

During injection moulding without injection process regulation, a specified pressure curve is established for injection and holding pressure, which can also be maintained with assurance with a regulated machine. However, the mould internal pressure curve that arises can only be assumed. Pressure losses through the runner manifold as well as the mould-specific filling behaviour cannot be identified.

With the application of injection-process regulation, the mould internal pressure is first measured and compared with a nominal value. If there is a deviation, a hydraulic valve that applies pressure to the injection cylinder is actuated. It is thus possible to follow the nominal value precisely and independently of negative factors. The switch over from injection to holding pressure also occurs as a function of internal pressure. Thus, no pressure spikes can occur since the switch over takes place when a specified threshold value is reached.

The following benefits are achieved through the application of injection process regulation:

- Significant reduction in start-up cycles – the required consistency in quality characteristics is achieved after just a few cycles.
- Better reproducibility of the parts – the deviation spread of the various dimensions lies significantly below that of a non-regulated machine.
- Cycle time reduction – by the ability to visualise the internal pressure signal, the sealing point can be determined much more easily and accurately.
- Re-starts – if the same internal pressure curve is applied at a re-start, the resulting parts are exactly alike.
- Improved quality of the parts through effective speed and pressure profiles – internal pressure profiles without spikes make possible the production of parts with low residual stresses. Switch over as a function of internal pressure prevents over-injection of the part, regardless of the selected dosage stroke.

The enormous significance of a mould internal pressure curve is characterised by the large number of parameters that can influence the appearance of the curve. Only the most important influencing factors are mentioned below:

- In the injection phase: the injection speed, the flow resistance as a function of the type of plastic, the material temperature and the mould wall temperature.
- In the pressure holding phase: the material temperature, the mould temperature, the level of the holding pressure and the duration of the holding pressure.
- In relation to the maximum mould internal pressure: the injection speed, the material temperatures, the switch over point (from injection pressure to holding pressure) and the material flow.

The appearance of the internal pressure curve additionally contains important criteria that influence the following quality data:

- In the injection phase: the appearance, the surface characteristics, the orientation and the degree of crystallinity of the moulded part.
- In the pressure holding phase: the formation of ridges, the weight, dimensions, shrinkage, shrink holes and sink marks and the orientation are influenced.

The properties and the quality of a component are predominantly determined by the moulding process in the mould. The dominating limiting quantities here are the pressure and temperature cycles in the mould cavity. It would be ideal if pressure and temperature were uniform at any point in the cavity, and if the temporal pressure and temperature cycle also remained the same from batch to batch. Then shrinkage would be the same in all component batches, there would be no internal stresses and no tendency towards warping in the component, and one component would fall out in just the same condition as another.

This ideal pressure and temperature distribution within the mould, which was as uniform as possible, is practically impossible to achieve with injection moulding, as a pressure drop is bound to occur while the mould is being filled, due to the flow resistance. Temperature differences will also arise because filling takes a finite time, even if this is usually very short. To get close to an ideal state, i.e., to aim for the most uniform possible filling process, the flow resistance during the filling of the mould plays a decisive role. The lower the flow resistance, the faster the mould is filled, and the smaller are the local pressure differences in the mould.

These factors have corresponding consequences for the design of the component and mould, and the process parameters chosen.

As regards the influence of the mould geometry, the following is generally valid: the flow resistance should be kept as low as possible, e.g., by avoiding sharp edges in the component (pressure losses due to abrupt turning by the compound flow).

For the process parameter settings the following points can be generally applied:

- Screw injection speed as high as possible
- Compound temperature as high as possible.

High temperatures result in low viscosity for the compound flowing in, low pressure losses, and thus low pressure differences and short filling times.

In practical machine setting, there are naturally limits here, which are discussed in more detail when the individual process parameters are dealt with. Here are just a few examples:

It will not always be possible to take the injection speed right up to the machine's performance capability limit. As the injection speed rises, the tendency to free jet formation and thus to the occurrence of surface faults increases. If assistance cannot be provided here by suitable mould design, the machine must be operated at a low injection speed – it may be that two or three speed stages will be available for injection. The compound temperature must not become so high that heat damage occurs. The more sensitive the compound, the better to select a larger safety margin from the upper temperature limits. But it should also be pointed out from the start that setting should not fall into the other extreme.

Too much caution can bring about the exact opposite of the desired effect: low temperatures increase the viscosity, and thus cause higher flow losses due to friction – which heats up the compound again as it is injected into the mould. In this way, cylinder temperatures that are too low can actually lead to higher compound temperatures in the mould than in materials where the cylinder temperature was set higher.

The higher the mould temperature is set, the longer the cooling off lasts and the longer the cycle time is. Therefore a temperature should be chosen which is only as high as the desired quality demands in order to be able to produce components as economically as possible. A summary of the most important factors in producing quality components is shown in Figure 8.1.

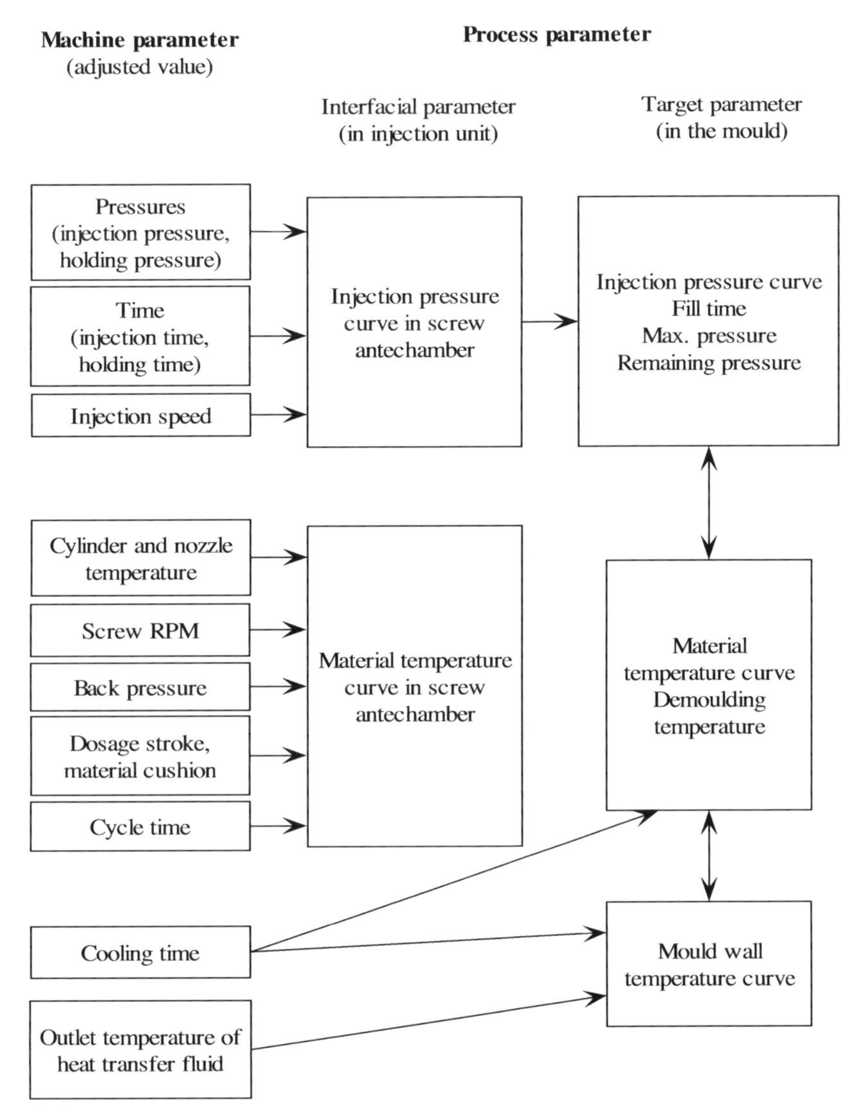

Figure 8.1 Important parameters for part formation
(Arrows indicate which parameters affect other parameters)

8.4 Injection and Mould Cavity Pressure

It was discussed in Section 8.3 that the pressure curve in the mould (as well as in the compound temperature and the mould wall temperature) is decisive for the quality of the component. Therefore, the connection between injection pressure and internal mould pressure will now be indicated. With this, comes the possibility of exerting a better influence on the pressure cycle in the mould during injection and thus on the component's quality.

Drive pressure (e.g., in the hydraulic stroke drive of the screw) and injection pressure (e.g., in the screw forward cavity) are practically proportional to one another (independent of the injection conditions) and display the same cycle (if we disregard the friction losses as the screw moves forward) – see Figure 8.2. The internal mould pressure, on the other hand, is lower than the injection pressure, conditioned by the flow losses during injection, which depend on the viscosity of the melt, the injection speed, and the geometry of the flow path.

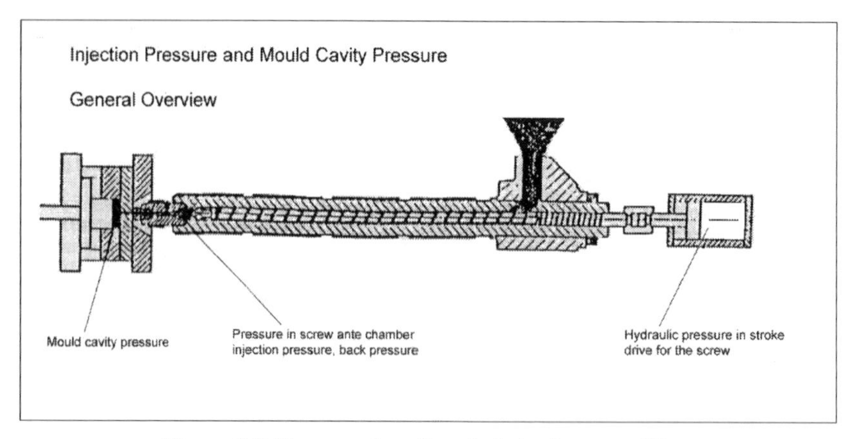

Figure 8.2 Pressure locations in injection moulding

Figures 8.3 and 8.4 show the injection pressure and the internal cavity pressure curves. The internal cavity pressure can be measured by sensors within the mould, and can be indicated, or visually displayed, using an oscilloscope or a pen recorder. The pressure cycle in the vicinity of the gate is the most informative factor here.

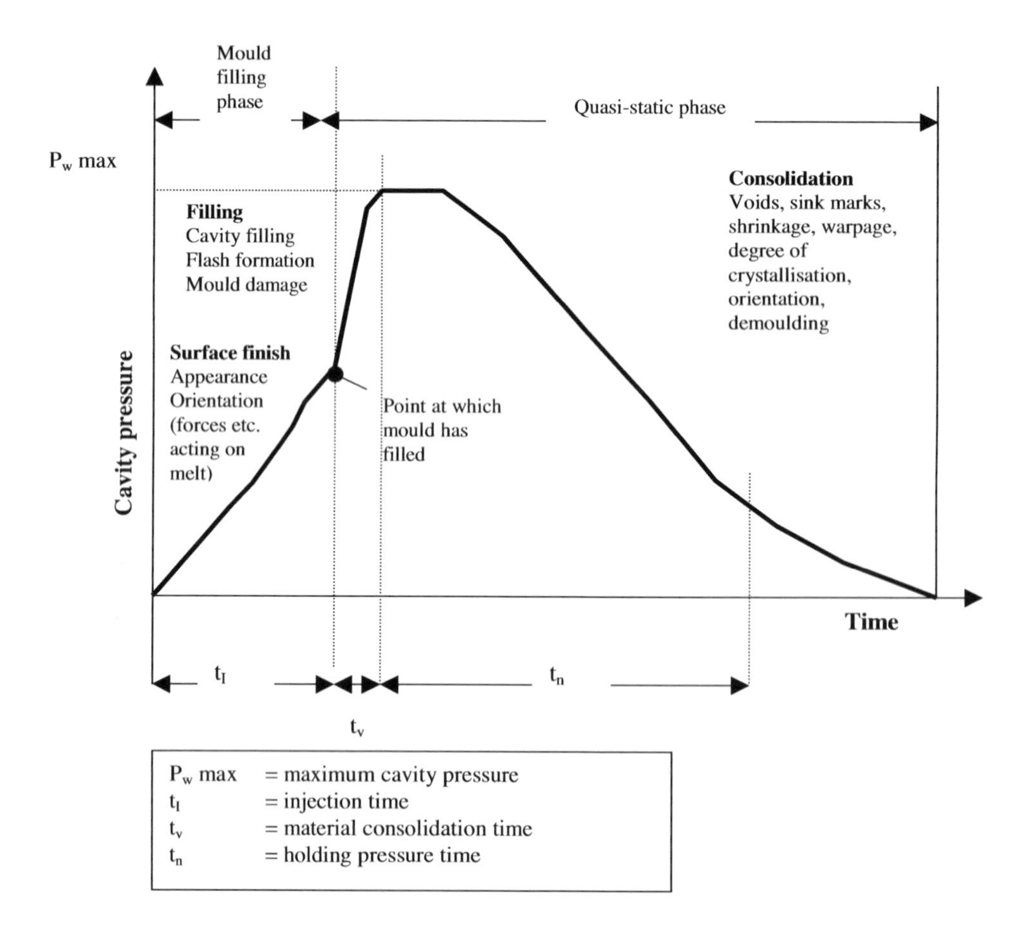

Figure 8.3 Correlation between quality characteristics and pressure profile in the moulding

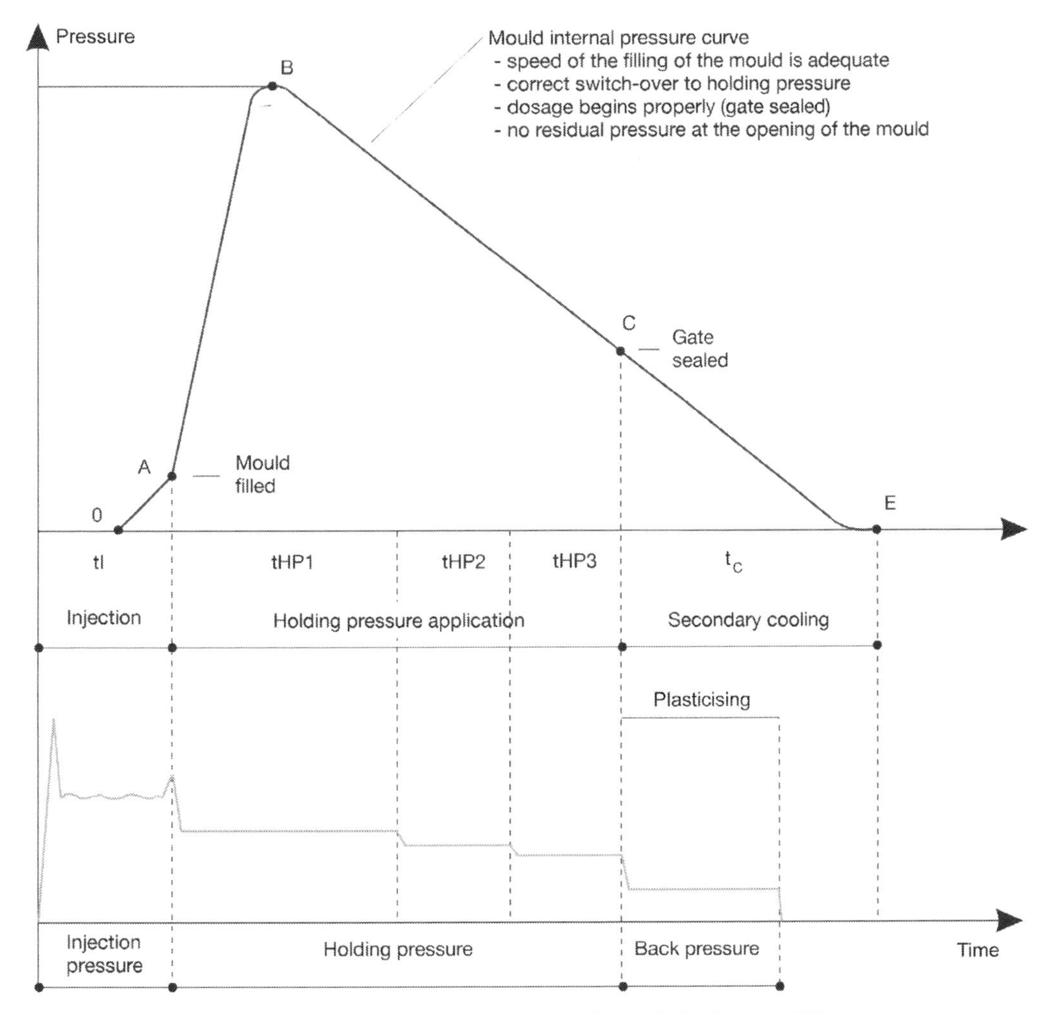

Figure 8.4 Mould cavity pressure curve in relation to injection moulding pressures

Key to Figure 8.4:

0-A Filling phase, or more correctly dynamic filling phase, because, even though solidification is beginning, compound is pressed down into the cavity up to the curing point, C.

A-E Quasi-static phase. The mould is volumetrically filled. Flow movements of the compound (reserve thrusts from the screw forward cavity) occur only transiently until compacting (up to B) and balancing out of shrinkage during solidification (up to C).

A-B Compression phase

B-C Pressure holding phase (up to sealing point)

C-E Shaping phase (cooling off to temperature at which components are ejected from mould)

HP Hydraulic Pressure

t_c Cooling time

As can be seen from Figure 8.4, the internal mould pressure follows the injection pressure, with a time delay. At A, the cavity is volumetrically filled, and between A and B the compound is packed in the mould. The maximum internal mould pressure, at B, is not reached until some time after the maximum injection pressure is obtained.

Even if the injection pressure stays the same, the internal mould pressure drops slightly, as a result of shrinkage of the compound, down to the curing point, C. From here the pressure drops rather faster, because now no more compound can be pushed back, right to the residual pressure, Pw remainder, when the mould is opened, at E.

8.5 Injection Pressure and Injection Time

The injection pressure and holding pressure selected must be as high as necessary to fill the cavity sufficiently fast, completely and efficiently, but, on the other hand, as low as necessary to produce low-stressed injection moulded components and avoid difficulties when the components are ejected from the mould.

The injection time (injection time and holding time), i.e., the duration of effect of the injection pressure must be selected to be just long enough to solidify (seal) the gate. If the injection time is too short, compound can flow back out of the cavity, sink marks occur, and in general there are larger tolerance variations. Overlong times are uneconomic and increase the internal stresses of the injection moulded component, especially close to the gate. The correct injection time can be determined by weight measurement.

With injection times greater or equal to solidification time, the injection moulded component weight remains practically the same (does not increase). With injection times less than solidification time, the injection moulded component weight decreases. If the flow stops before the mould has fully filled, the final moulding weight will be lower. The occurrence of sink marks is also a sure indication that the injection time (duration of effect of pressure) is shorter than the solidification time.

With amorphous thermoplastics holding pressure reduction is necessary. This can avoid difficulties in ejecting the parts and is necessary to get low-stress injection moulded parts.

With semi-crystalline thermoplastics a constant holding pressure is recommended in order to ensure an undisturbed crystallisation process.

8.6 General Information on Filling Speed

8.6.1 Initial Injection Speed

The smaller the flow path cross-section is in relation to the screw/piston surface, the higher the filling speed is. A larger injection cylinder in the same injection unit thus produces a higher filling speed for the same initial injection speed. The initial injection speed, and with it the filling speed, should be selected to be as high as possible, so that the mould is filled as quickly as possible with compound with as uniform a temperature as possible. Then the temperature and pressure variations in the mould are slight, and low-stress free components can be obtained. In this way, the component should be filled as uniformly as possible, with the flow head moving away from the gate. Free jet formation is to be avoided by suitable design.

For thin-walled parts, the optimum filling speed is higher than for thick-walled parts, so as to obtain uniform filling of the moulding through the flow head. Too low a filling speed causes a greater temperature variation between those parts of the preform near the gate and those far from it, due to increased cooling off of the compound while the cavity is being filled. The higher viscosity of the colder compounds also requires higher injection pressures, which in turn require stronger locking pressures.

Too long a filling speed can also lead to surface faults. If compound which has already solidified onto the mould wall is displaced by a subsequent filling, cross-grooves occur vertically to the direction of flow (gramophone record effect). The effects of too high a filling speed are described in Chapter 9.

8.7 Filling Speed and Orientation

During the filling of the mould, orientation effects can arise, especially through friction influences. The molecules initially lying randomly in the compound are now stretched and orientated in the direction of flow. Such orientations lead to nonuniform shrinkage and nonuniform preform properties (anisotropy). The higher the shear rate, the greater the orientation of the polymer chains. Also, the higher the filling speed and the greater the viscosity of the compound flowing in, the higher the shear rate. Thus, higher filling speeds are bound to lead to an increased tendency to orientation.

However, the higher the compound temperature is, and therefore the lower the viscosity value, the less negative effect a high filling speed will have. A high compound temperature, in connection with a high mould wall temperature, will cause the oriented molecules to lose their orientation after the filling process has ended (relaxation). This reduces orientations, along with their negative effects. Therefore, before any reduction in the filling speed, a check should be made on whether orientation phenomena can be reduced by increasing the compound temperature and the mould temperature. Here also, preference should be given to compound temperatures and mould temperatures that are as high as possible (as already stated when dealing with the most favourable injection pressure).

In terms of the flowability of the materials themselves, sometimes material suppliers illustrate the relationship between wall thickness and injection speed as a flow path/wall thickness ratio (L/s). If a ratio of 100:1 is given, this means for a wall thickness of 1 mm, then the length of flow from the gate will be 100 mm. Because flow is dependent on wall thickness a variety of mould wall thickness may be quoted. If the material is required to flow further, e.g., 125:1, more pressure will be required to fill the cavity and more orientation in the material will result. The relationship between pressure and wall thickness is shown in Figure 8.5. Therefore, ideally moulds should be designed with consideration of flow path lengths and wall thickness ratio in mind.

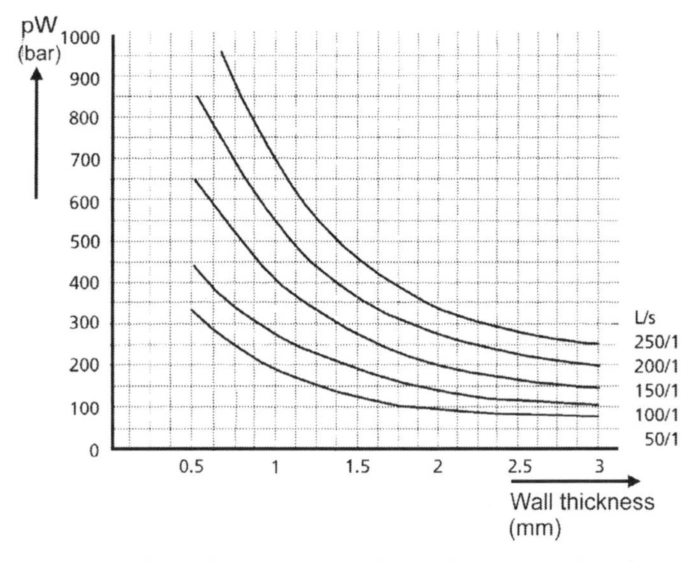

Figure 8.5 Injection pressure/wall thickness relationship

8.8 Effects of Too High Filling Speed

A high filling speed causes a high shear rate between the compound in the core and the compound on the mould wall. The shear stress arising under such conditions can lead to impairment of the plastic (shear fracture). Particularly high shear stresses arise if the compound has to turn sham corners, especially with abrupt changes in the cross-section. This should be taken into account in the design of the moulding.

Many plastics display particularly high shear fracture sensitivity, e.g., fluoroplastics such as Teflon. But, PMMA or PC also display more pronounced tendencies to shear fracture than, for example, the polystyrenes.

If the filling speed is high, the air must be removed from the cavity sufficiently quickly. If this is not the case, the compression, and thus the heating of the air increases (Diesel effect, see Chapter 9). This can lead to heat damage of the material, or can even cause burns. So care should be taken to ensure that the air removal system in the parts of the mould filled last works well. Under certain circumstances, it is sufficient to reduce the mould clamping force to the necessary level, if the air can be extracted through the parting plane.

A high filling speed can cause nonuniform mould filling due to:

- Free jet formation at gate ('sausage injection moulding'), which results in surface faults,
- Splitting of the compound flow, which leads to unnecessary joint line formation,
- Tearing loose of batches already solidified, which causes a deterioration in the surface finish and usually in the mechanical properties as well.

These faults can usually be avoided by suitable construction design, even at high speed.

Finally, high filling speeds can lead to uncontrolled and excessive heating of the compound, due to strong constrictions in runner cross-sections that are too small. This is particularly the case if the working compound temperature is too low, and thus the melt viscosity is higher. In certain circumstances, heating due to constriction losses can become so great that, when the set cylinder temperatures are reduced, the compound temperature in the mould cavity does not fall, but rises. This matter will be referred to again in connection with the cylinder temperatures.

8.9 Setting the Initial Injection Speed

The initial injection speed is set, on the overwhelming majority of the injection moulding machines in operation today, through a throttle in front of the drive cylinder of the piston or screw. A higher consistency of initial injection speed from batch to batch can be obtained by using a quantity controller instead of the throttle. If the initial injection speed is set using a throttle, the connection between the throttle position and the initial injection speed is not linear. For this reason – and also to guarantee a repeatable speed setting in case of a change of machine, the initial injection speed for each machine should be determined independently of the throttle setting (by measuring the initial injection time).

A prerequisite for the initial injection speed set actually being achieved is that a sufficiently high pressure head must be available during the initial injection. This is true for both kinds of initial speed setting (throttle or quantity control). Thus, the first pressure stage for initial injection must be selected sufficiently high. The higher the flow resistance during filling, the higher the pressure selected should be. The initial injection pressure can be raised or lowered to check whether the injection pressure head is sufficiently high. If the initial injection speed no longer changes when the pressure changes (measure initial injection time, at at least 0.01 s), the pressure head is satisfactory. If possible, the initial injection pressure should be set so high that it lies 10-15% above the value at which the pressure influence on the speed becomes noticeable. A guide to injection times relative to the viscosity of materials is given in Table 8.3. Table 8.4 is a guide to the viscosity type of a number of common materials.

Table 8.3 Recommended values for injection time for low, medium and high viscosity materials

Injection volume (cm^3)	Injection time in seconds		
	Low viscosity	Medium viscosity	High viscosity
1-8	0.2-0.4	0.25-0.5	0.3-0.6
8-15	0.4-0.5	0.5-0.6	0.6-0.75
15-30	0.5-0.6	0.6-0.75	0.75-0.9
30-50	0.6-0.8	0.75-1.0	0.9-1.2
50-80	0.8-1.2	1.0-1.5	1.2-1.8
80-120	1.2-1.8	1.5-2.2	1.8-2.7
120-180	1.8-2.6	2.2-3.2	2.7-4.0
180-250	2.6-3.5	3.2-4.4	4.0-5.4
250-350	3.5-4.6	4.4-6.0	5.4-7.2
350-550	4.6-6.5	6.0-8.0	7.2-9.5

Table 8.4 Viscosity of various plastic types

Viscosity	Plastic Types
Low	PE soft, PA 4.6, PA 6, PA 66, PA 6.10, PA 11, POM, PET, PBT, PPS, TPE
Medium	PS, SB, SAN, ABS, PPO mod., PVC soft, CA, CAB, CP, PE rigid, PP, PA 12, PA amorphous
High	PVC rigid, PMMA, PC, PSU, PES, PEI, PAI, PVDF, FEP, ETFE

8.10 Plasticising

8.10.1 Compound Temperature

The slighter the spatial temperature variation of the compound in the mould and the temporal temperature variations from batch to batch are, the higher and the more uniform will the quality of the component be. The optimum compound temperature is a parameter specific to each material. As mentioned earlier, the compound temperature distribution in the mould becomes more uniform and the tendency toward orientations in the components is reduced. However, the temperature of the compound must in no case be so high that damage occurs through decomposition. For this reason, it should not be forgotten that, as well as the cylinder and nozzle temperature settings, the following setting values and factors also affect the compound temperature:

(1) Screw speed
(2) Back pressure
(3) Dwell time of compound in cylinder (essentially determined by the cycle time)
(4) Friction (friction losses) during initial injection
(5) Mould wall temperature.

(1), (2), and (3) affect the compound temperature in the screw forward cavity.

(4) and (5), like the first factors, influence the compound temperature in the mould.

Too low a compound temperature, as well as causing the drawbacks for the component referred to already, also brings the risk of nonuniform and/or incomplete plasticising (melting) of the compound. It should be taken into account here that a low cylinder temperature does not necessarily mean a low compound temperature. Strong friction as a result of a higher viscosity leads to higher pressure losses, even during initial injection, and thus a steeper rise in temperature. Apart from the danger of local thermal overloads in the compound, there is also the risk of mechanical damage.

8.10.2 Cylinder Temperatures

The temperature ranges provided by the plastic manufacturers can be regarded as the most suitable practical guides. If no manufacturer's data are available, the guidelines in Table 8.2 should be used. As a first step, these temperatures can be selected as setting values for the nozzle-side cylinder heating band. The influences referred to in the previous section are not taken into account here.

The following data is of assistance in determining the actual compound temperature in the screw forward cavity. The nominal temperature set is the temperature of the cylinder wall, or of the measuring point, which is close to the cylinder wall. If the additional heating caused by friction is disregarded, the plastic (as a poor heat conductor) will reach the wall temperature only after a long dwell time. The friction during plasticising causes a rise in temperature the higher the screw speed and the back pressure.

Methods of measuring the actual compound temperature will be discussed in Section 8.10.6.

The setting values for the cylinder heating bands should, as far as possible, be graded in descending order going towards the material feed area. The cylinder temperature at the material feed area should not be too low, so that the frictional heat does not become too great there. Thus, in practice, a 20 °C temperature difference between the first cylinder heating area and the last has proved itself for almost all thermoplastics.

For PA 66, an even smaller temperature spread gives better material feed behaviour and more uniform plasticising. For unplasticised PVC and PMMA in certain circumstances, it is important that the cylinder temperatures do not decrease going towards the material feed area, but actually increase in order to keep the frictional heat low. With this in mind, the cooling of the cylinder yoke for the plasticising cylinder must also be mentioned. Supporting body temperatures of approximately 30-40 °C (warm to the touch) are to be recommended. Too low a temperature causes condensation formation, which can lead to material feeding difficulties and component faults due to moisture absorption by the granules.

8.10.3 Nozzle Temperature

The effect of the nozzle heating should be that the plasticised compound maintains its temperature while flowing through the nozzle. The nozzle temperature should be set accordingly – usually to the same level as the cylinder heating on the nozzle-side. The nozzle temperature – at least on small and medium-sized injection moulding machines – is essentially determined by the cylinder heating on the nozzle-side, because the nozzle's heat capacity is very small in relation to that of the cylinder. Thus cases can occur in which it is not absolutely necessary to control the nozzle temperature, and it is enough to balance out the heat losses of the nozzle, using a power controlled heater band (cheaper than control).

The power setter for the nozzle heater band should then be set, as stated above, so that the compound does not alter its temperature as it flows through the nozzle. But controlling the nozzle heating can lead to satisfactory results only with heat balance, and thus with sufficiently long and continuous production. When the machine is started up, and during any interruptions to production, the nozzle temperature control should be transferred to a control system. It is not possible to do without a nozzle temperature control system today, especially for the thermoplastics that are more difficult to process. It has already been stated that the nozzle-side cylinder temperature should be taken as a guideline for the nozzle temperature. The optimum temperature can be slightly lower or higher, in accordance with the material concerned.

If an additional temperature measuring point is established in the nozzle, it is relatively easy to determine the optimum setting temperature. It is the temperature at which the nozzle temperature measured during injection undergoes practically no change. If the temperature is set too high, the nozzle cools off while the compound is flowing through, and if the temperature is set too low the nozzle temperature rises again. A nozzle temperature lower than the last cylinder heating will often turn out to be effective (even if it is not optimum), in order to avoid a leak at the nozzle. Deviating from the optimum value is in this case usually justified, because this causes the thickness of the material cushion to change during initial injection and, when two pressure stages are being used, this can have significantly more negative effect (strong influence on pressure cycle in the mould).

8.10.4 Temperature Profile Guideline for Plasticising Cylinder

There are four different temperature profiles that can be used for injection moulding: horizontal, ascending, descending and finally ascending with descending at the nozzle. An example of each of these set ups is shown in Figure 8.6.

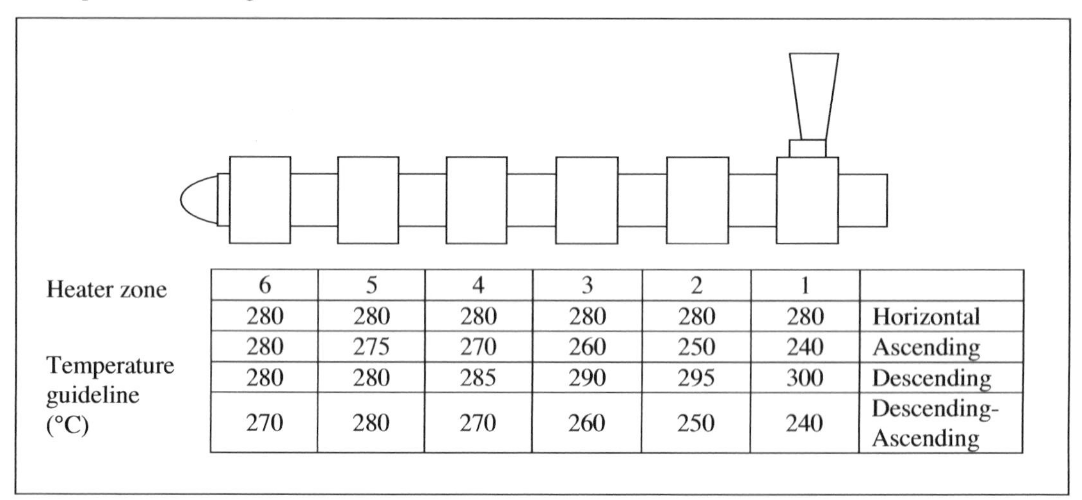

Heater zone	6	5	4	3	2	1	
Temperature guideline (°C)	280	280	280	280	280	280	Horizontal
	280	275	270	260	250	240	Ascending
	280	280	285	290	295	300	Descending
	270	280	270	260	250	240	Descending-Ascending

Figure 8.6 Examples of types of temperature profiles

Horizontal

This temperature profile is recommended when fast heating of material is required, e.g., fast cycle times, long (entire) screw stroke or screws with a deep flight. This profile setting ensures that the material is evenly heated and is completely molten before injection. When using glass-filled material this profile can reduce wear and tear of the cylinder assembly.

Ascending

This is the most common temperature profile. It is especially recommended for heat sensitive materials like POM and PES. This also includes flame-retardant material like flame-retardant PA and PVC.

Descending

This special temperature profile is necessary in exceptional cases, e.g., when working with PA 66.

Ascending and descending at the nozzle

This temperature profile is normally used with an open nozzle for prevention of thread formation or drooling. With this profile setting there can be drawbacks as the colder material (which may be hardened) at the nozzle may be injected into the mould to form the product.

The most suitable temperature profile setting can only be achieved through practical experience with the particular material in question.

8.10.5 Recommended Temperatures for Cylinder and Mould

There are recommended temperatures for the cylinder and mould for thermoplastic and thermoset materials as shown in Tables 8.5 to 8.7. Special instructions specified by the manufacturer must be observed.

Table 8.5 Temperatures for cylinder and mould: amorphous thermoplastics

Injection material	Nozzle-side cylinder temperature (°C) [1,2]	Feed yoke temperature (°C) [1,2]	Mould temperature (°C)	Transition temperature (°C ca.)	Remarks, see footnotes
PS	160-230	30-35	20-60	90	
SB	160-250	30-35	20-60	85	4
SAN	200-260	30-35	40-80	100	4
ABS	180-260	30-35	40-85	105	4
PVC rigid	160-180	30-35	20-60	80	3, 5, 6, 7, 8, 10
PVC soft	150-170	30-35	20-40	55-75	3, 5, 8, 10
CA	185-225	30-35	30-60	100	3, 4, 8
CAB	160-190	30-35	30-60	125	3, 4, 8
CP	160-190	30-35	30-60	125	3, 4, 8
PMMA	220-250	35-45	60-110	105	4
PPE (PPO) mod.	240-290	35-45	70-120	120-130	3, 4, 11
PC	290-320	35-45	60-120	150	4, 11
PAR	350-390	45-65	120-150	190	4, 11
PSU	320-390	45-65	100-160	200	4, 11
PES	340-390	45-65	120-200	260	4, 11
PEI	340-425	45-65	100-175	220-230	4, 11
PAI	340-360	45-65	160-210	275	4, 11
PA amorph.	260-300	35-45	70-100	150-160	4

If no other practical values are available: Set nozzle temperature – nozzle-side cylinder temperature. Cylinder temperature reduction approaching the feed zone by max 5-10 °C per heating zone; max. temperature difference between nozzle and feed side of 20 °C. Set the nozzle-side heater band and the following one to the same temperature with more than 2 heater zones.

1. Only set the upper temperature with a high shot count with a thermally sensitive material (shorter dwell time in the cylinder).
2. Thermally sensitive!
3. The granulate must be dried before processing!
4. Do not use shut-off nozzles, only open nozzles!
5. Injection without check valve recommended!
6. Only operate with screw tips without check valves!
7. Only operate with low back pressure!
8. To improve the feed performance, set the same or a slightly higher cylinder temperature approaching the feed side.
9. A corrosion protected cylinder unit (Arbid) is recommended.
10. An abrasion proofed cylinder unit (Arbid) is recommended for processing of reinforced materials (e.g., fibre glass).
11. With thermostat cooling water valves:
    30...35 °C 2...3 scale markings
    35...45 °C 3...4 scale markings
    45...65 °C 4...5 scale markings

Table 8.6 Temperatures for cylinder and mould: crystalline thermoplastics					
Injection material	**Nozzle-side cylinder temperature (°C) [1,2]**	**Feed yoke temperature (°C) [1,2]**	**Mould temperature (°C)**	**Crystalline melt temperature (°C) ca.**	**Remarks, see footnotes**
PE soft	210-250	30-35	20-40	105-115	11
PE rigid	250-300	30-35	20-60	125-140	
PP	220-290	30-35	20-60	156-168	
PA 4.6	210-330	45-65	60-150	295	4, 8, 9, 11
PA 6	230-260	45-65	40-100	215-225	4, 8, 9, 11
PA 66	270-295	45-65	50-120	250-265	4, 8, 9, 11
PA 6.10	220-260	45-65	40-100	210-225	4, 8, 9, 11
PA 11	200-250	34-45	40-100	180-190	8, 9, 11
PA 12	200-250	35-45	40-100	175-185	9, 11
POM	185-215	35-45	80-120	165-175	3, 8
PET (PETP)	260-260	45-65	50-140	255-258	3, 4, 11
PBT (PBTP)	230-270	45-65	40-60	220-225	3, 4, 11
PPS	300-360	45-65	20-200	280-288	4, 11
PFA	350-420	45-65		300-310	5, 10
FEP	340-370	45-65	150-200	285-295	5, 10
ETFE	315-365	45-65	60-120	270	5, 10
PVDF	220-300	35-45	70-90	171	5, 10
PEEK	350-380	45-65	150-180	340	
PEEKK	390420	45-65	150-180	363	
PEK	400430	45-65	150-180	365	
PAA	250-290	45-65	120-150	235-240	
PPA	320-350	45-65	135-165	310	
LCP	280-450	45-65	30-160	270-380	
Thermoplastic elastomers					
TPE-A	200-260	30-35	20-50		
TPE-E	200-250	30-35	20-50		
TPE-S	180-240	30-35	20-50		
TPE-U	190-240	30-35	20 40		
TPE-O	110-180	30-35	15-40		

If no other practical values are available: Set nozzle temperature – nozzle-side cylinder temperature. Cylinder temperature reduction approaching the feed zone by max 5-10 °C per heating zone; max. temperature difference between nozzle and feed side of 20 °C. Set the nozzle-side heater band and the following one to the same temperature with more than 2 heater zones.

1. Only set the upper temperature with a high shot count with a thermally sensitive material (shorter dwell time in the cylinder).
2. Thermally sensitive!
3. The granulate must be dried before processing!
4. Do not use shut-off nozzles, only open nozzles!
5. Injection without check valve recommended!
6. Only operate with screw tips without check valves!
7. Only operate with low back pressure!
8. To improve the feed performance, set the same or a slightly higher cylinder temperature approaching the feed side.
9. A corrosion protected cylinder unit (Arbid) is recommended.
10. An abrasion proofed cylinder unit (Arbid) is recommended for processing of reinforced materials (e.g., fibre glass).
11. With thermostat cooling water valves:
    30...35 °C 2...3 scale markings
    35...45 °C 3...4 scale markings
    45...65 °C 4...5 scale markings

Table 8.7 Temperatures for cylinder and mould: thermosets				
Resin type		Mould (°C)	Plasticising cylinder	
			Nozzle-side (°C)	Feed-side (°C)
Phenol-formaldehyde	PF	145-175	75-95	45-60
Urea-formaldehyde	UF	140-165	80-95	45-55
Melamine-formaldehyde	MF	150-180	75-100	50-60
Melamine-phenol-formaldehyde	MP	150-180	70-100	45-55
Unsaturated polyester	UP	140-165	80-105	40-65
Diallyl phthalate	DAP	150-190	85-110	50-60
Epoxy	EP	155-200	85-115	50-60
Silicone	SI	180-250	5-20	5-20

8.10.6 Measuring the Compound Temperature

There is still no completely satisfactory solution as regards the measurement of the compound temperature (as distinct from measuring the compound pressure). The simplest method is injection into the air, or into a small container made of a heat insulating material, and measurement of the compound temperature with an insertion thermometer. Naturally, this measurement cannot be continuous – the cycle must be interrupted. That is the decisive disadvantage of this process, because in this way the heat balance of the machine is disturbed. The accuracy of measurement here depends on the quantity of the compound injected out. The smaller this is, the more inaccurate is the measurement, because a small quantity cools off more rapidly.

With small quantities, the heat transfer to the thermometer, which is usually cooler, causes a noticeable cooling off of the compound, and thus gives a false reading. In this case, the temperature of the thermometer should be brought up to somewhere near the compound temperature before measurement. This can be done during the cylinder heating.

It is scarcely possible to obtain repeatability of the measurement result to within less than around 5 °C with this method. Significantly higher repeatability, and thus significantly higher accuracy also, can be obtained by incorporating a thermometer into the cavity in front of the screw.

Here the thermometer must be mounted in such a way that no (colder) plastic can be deposited on it, or that such plastic will be continually rinsed off. The disadvantage of this process lies in the necessity of building in structures that lie in the compound flow and increase the flow resistance, and also increase the risk that dead corners will form. This causes more difficulties on smaller machines than on larger ones. This is because the quantity of compound which is still in front of the screw tip after initial injection (which should certainly be kept as small as possible) is generally increased by the addition of these structures. The effects involved are more negative with smaller injection equipment or a smaller sprue fraction than with larger injection equipment.

Temperature measurements on an ultrasonic basis are certainly also possible, but the costs involved are still very high. With this method, the compound temperature is measured continually as the compound flows through the nozzle.

8.10.7 Screw Speed

An adjustable screw speed is the prerequisite for suitable processing of materials. The optimum screw speed is a parameter that depends on the material. For example, optimum rpm speeds for a range of plastics will be given by the manufacturers, from which the optimum speed for the screw used can be determined.

'Optimum' should here be understood as follows. The melt is satisfactorily homogenised, the friction being not high enough to arouse fears of any damage to the plastic due to heat stress or mechanical stress. If the screw speed is too high, not only does the friction heat rise, but the homogeneity of the melt is usually impaired. With fibreglass-reinforced compounds, the wear on screw and cylinder also increases significantly.

It is not only the higher plasticising yield, i.e., a shortening of the plasticising time, which favours raising the screw speed. In borderline cases, the time from the start of plasticising to the opening of the mould is no longer determined by the cooling off time required for the preform, but by the plasticising time. Here, it would be better to use a cylinder fitting with larger screw diameter, so that processing could go on at a lower speed.

If there is no data on the optimum range of screw speed or rpm, or if these ranges are very wide, the screw speed is matched to the cooling off time. The cooling off time required before the opening of the mould (determined by the permissible temperature at which components can be ejected from the mould, or the permissible residual pressure in the mould) is obtained, and the screw speed is then selected in such a way that plasticising is completed shortly before the cooling off time is completed (before the opening of the mould). In this way, the time when the compound is standing still in front of the screw tip is kept short, and any possible interference factors arising should be of little significance.

8.10.8 Back Pressure

The back pressure is intended to ensure that a sufficiently high level of homogeneity is attained by the melt during plasticising. This involves uniform temperature distribution, and the absence of air blisters or gas blisters.

The level of back pressure required depends on the melt viscosity of the plastic and/or its heat sensitivity (see Table 8.8). Care must be taken to set the back pressure (pressure in the compound in front of the screw tip) through the hydraulic pressure.

When starting up the machine, use the lower values from the manufacturer's data. For dry pigmented granules, the back pressure must be increased long enough for the colour distribution in the injection equipment to be satisfactorily uniform.

If the back pressure is too low, this will become noticeable through big variations of the compound pack during holding pressure from batch to batch (as a result of air blisters). If the component's weight is measured, too low a back pressure shows up through the underweight components which appear from time to time.

As the back pressure increases, the plasticising time increases (and the plasticising yield sinks), and in the same way the back heating increases because of the increasing and longer-lasting friction effect. Too high a back pressure leads to a greater extension of the plasticising time, and usually drawing-in difficulties (material feeding down from the hopper to the screw) also occur. The risk of compound damage through overheating also increases.

8.11 The Injection Stage

8.11.1 Filling to Packing

The importance of the pressure cycle to the quality of the moulding has already been discussed.

The injection pressure available ensures that the injection velocity (speed of filling) can be carried out as fast as possible. However, at some stage filling switches to packing. The switch from injection pressure (filling) to holding pressure (packing) is called the switch over position. This position on the injection ram can be set in a variety of ways for example by:

- Screw position
- Hydraulic (line) pressure
- Melt pressure
- Cavity pressure.

To begin with, the switch over position can be set at about two-thirds of the total shot volume and altered accordingly from there.

Table 8.8 Recommended values for back pressure					
Injection material	**Specific weight (g/cm^3)**	**Viscosity[2]**	**Back pressure (bar)**	**Screw circumference speed (m/min)**	**Remarks (footnotes)**
Amorphous thermoplastics					
PS	1.05	m	40-80		
SB	1.04	m	40-80		
SAN	1.08	m	40-80		
ABS	1.03-1.07	m	40-80		
PVC rigid	1.38-1.40	h	40-80	4-6	1
PVC soft	1.20-1.35	m	40-80		1
CA	1.26-1.32	m	40-80		
CAB	1.16-1.22	m	40-80		
CP	1.19-1.23	m	40-80	4-6	
PMMA	1.18	h	80-120	4-6	
PPE mod.	1.06-1.10	m	60-90		
PC	1.20-1.24	h	80-120	6-10	
PAR	1.2	h	80-120		
PSU	1.27	h	80-120		
PES	1.37	h	80-120		
PEI	1.87	m	40-80		
PAI	1.38	h	40-80		
Semi-crystalline thermoplastics					
PE soft	0.91-0.93	n	40-80		
PE rigid	0.94-0.96	m	60-90		
PP	0.9	m	60-90		
PA 46	1.18	n	40-80		
PA 6	1.13	n	40-80		
PA 66	1.14	n	40-80		
PA 610	1.06	n	40-80		
PA 11	1.04	n	40-80		
PA 12	1.02	m	60-90		
PA amorph.	1.12	m	60-90		
POM	1.41-1.42	n	40-80	10-15	
PET	1.3-1.37	n	60-90		
PBT	1.26	n	40-80		
PPS	1.34	n	40-80		
FEP	2.14-2.17	h	80-120	5-10	1
ETFE	1.70	h	80-120	5-10	1
PAA	1.43-1.64	n	40-80		
PPA	1.26-1.56	n	40-80	10-15	
PAEK	1.27-1.49	m			
LCP		n	40-80		
TPE					
TPE-E					
TPE-U	1.14-1.26	n			
Thermosets and elastomers					
Thermosets	1.2-2.0	h	40-80	4-6	
Elastomers		m	20-60	2-5	
LSR	1.86-1.88	n	20-60		
Powder					
Metal, ceramic		m	10...50		
Water	1.00				
Carbon fibre	1.75-1.9				
Fibre glass	2.49-2.52				

Notes:
1. Only operate with low back pressure
2. n = low, m = medium, h = high

Table 8.9 Typical values for injection pressures						
Injection material	Specific weight (g/cm^3)	Viscosity [3]	Injection pressure (bar)	Holding pressure (bar)	Mould cavity pressure	
					Relationship with (x) highest holding pressure stage	Expected cavity pressure (bar)
Amorphous thermoplastics						
PS	1.05	m	650-1550	300-700	0.75-0.5	150-350
SB	1.04	m	650-1550	350-800	0.75-0.5	200-400
SAN	1.08	m	650-1550	350-900	0.75-0.5	250-450
ABS	1.03-1.07	m	650-1550	400-900	0.75-0.5	300-550
PVC rigid [1, 2]	1.38-1.40	h	1000-1550	500-900	0.6-0.4	250-500
PVC soft [2]	1.20-1.35	m	400-1550	300-600	0.75-0.5	150-300
CA	1.26-1.32	m	650-1350	300-650	0.85-0.7	250-450
CAB	1.16-1.22	m	650-1350	300-900	0.75-0.5	250-450
CP	1.19-1.23	m	650-1350	400-700	0.75-0.5	200-350
PMMA	1.18	h	1000-1400	500-1150	0.6-0.4	350-550
PPE mod.	1.06-1.10	m	1000-1600	600-1200	0.75-0.5	350-600
PC	1.20-1.24	h	1000-1600	600-1300	0.6-0.4	350-650
PAR	1.2	h	1000-1600	600-1300	0.6-0.4	350-650
PSU	1.27	h	900-1400	500-1100	0.6-0.4	400-600
PES	1.37	h	900-1400	500-1100	0.6-0.4	400-600
PEI	1.87	m	750-1550	400-750	0.85-0.7	350-650
PAI	1.38	h	750-1550	500-1050	0.85-0.7	450-750
Semi-crystalline thermoplastics						
PE soft	0.91-0.93	n	600-1350	300-800	0.85-0.7	200-600
PE rigid	0.94-0.96	m	600-1350	300-800	0.75-0.5	200-600
PP	0.9	m	800-1400	500-1100	0.75-0.5	300-650
PA 46	1.18	n	650-1550	550-1050	0.85-0.7	450-750
PA 6	1.13	n	450-1550	400-750	0.85-0.7	350-550
PA 66	1.14	n	650-1550	550-1050	0.85-0.7	450-750
PA 610	1.06	n	450-1550	350-750	0.85-0.7	300-500
PA 11	1.04	n	450-1550	400-800	0.85-0.7	350-550
PA 12	1.02	m	550-1550	400-1000	0.75-0.5	350-550
PA amorph	1.12	m	900-1300	450-800	0.75-0.5	350-450
POM	1.41-1.42	n	800-2000	700-1500	0.85-0.7	550-1050
PET	1.34-1.37	n	800-1500	550-1050	0.85-0.7	450-750
PBT	1.29	n	800-1550	500-1000	0.8-.0.7	400-700
PPS	1.34	n	750-1500	400-750	?	350-600
FEP [1]	2.14-2.17	h	1000-1500	500-1000	0.6-0.4	300-600
ETFE [1]	1.70	h	1000-1500	500-1000	0.6-0.4	300-600
PAA	1.4-1.64	n	1000-1500	350-800	0.85-0.7	300-700
PPA	1.26-1.56	n	700-1500	350-800	0.85-0.7	300-700
PAEK	1.27-1.49	m	800-1500	450-800	0.85-0.7	400-700
LCP		n	400-1500	350-1000	0.85-0.7	300-800
TPE						
TPE-E						
TPE-U	1.14-1.26	n	400-1000	300-600	0.85-0.7	200-450
Thermosets/elastomers/other						
Thermoset	1.2-2.0	h	800-2500	300-1000	0.6-0.4	200-600
Elastomer		m				
LSR	1.86-1.88	n	300-800	120-359	0.85-0.7	80-250
Metal, ceramic		m				
Water	1.00					
Carbon fibre	1.75-1.9					
Fibre glass	2.49-2.52					

Notes:
1. Do not use shut-off nozzles, use open nozzles only
2. Operate without check valve only!
3. n = low, m = medium, h = high

A residual amount of material is generally left in the screw after each shot. This is called the melt cushion. A melt cushion is used to ensure the transmission of injection pressure from the hydraulics to the mould. Otherwise the screw may bottom out before an adequate holding pressure can be retained on the moulding in progress. If an adequate cushion is not obtained the shot volume should be increased accordingly. The size of the cushion depends on the size of the machine and may vary from 3 mm-10 mm. The length of time that holding pressure is held should be kept as low as possible to minimise induced stresses. This minimum time can be found by simply reducing the holding time until sink marks appear. The lowest time in which sink marks do not appear is the optimum setting here. Typical values for injection pressure settings are shown in Table 8.9.

8.11.2 The Mould

8.11.2.1 Mould Temperature: General Information

The mould temperature or mould wall temperature is one of the most important process parameters (in addition to the pressure cycle and the compound temperature), which is of decisive importance for the quality and dimensioning of the components. The mould temperature influences the shrinkage and thus the dimensioning of the compound in the mould, the surface finish and the orientations in the injection equipment and also, not least, the cycle time – through the cooling off time – and thus the component costs.

This is as much a matter of the level of mould temperatures as of their uniformity and repeatability. Economic quality improvement in injection moulding is not possible without good and uniform temperatures in the mould. Even with a more expensive injection process control or adjustment system, the negative influence of unsatisfactory mould temperatures can not usually be balanced out. If it is a question of narrowing the tolerances of the components, the first step is to check the mould temperature data, and, if necessary, these must be improved before, for example, doing any research into injection pressure control on the basis of the internal mould pressure.

Naturally, a prerequisite for evaluating the mould temperature data is the measurement of the mould temperature. Unfortunately, even today this is not often done automatically.

8.11.2.2 Level of Mould Temperature

The optimal mould temperature level is a parameter specific to the material. If no data from the plastics manufacturer are available, the guidelines in Tables 8.5-8.7 can be used.

High mould temperatures cause the component to cool slowly, which is necessary, for example, with the majority of semi-crystalline thermoplastics, in order to obtain components that are to size and have constant dimensions. The crystallisation of these compounds must be completed in the mould, i.e., it must be over before the components are ejected from the mould. Otherwise, after-crystallisation occurs over the course of time, which in every case causes alterations in dimensions, and frequently also leads to warping of the component.

High mould temperatures improve the flow behaviour of the compound in the mould, and the injection pressure requirements are lower. The surface finish of the components also improves. High mould temperatures break down orientations that arise during the filling of the mould, and there are thus fewer orientations in the moulded component.

The upper limit for the mould temperatures is naturally determined by the maximum temperature at which the components can be ejected from the mould, which is specific to each material, e.g., for amorphous thermoplastics it lies at least 10 °C below the freezing temperature. High mould temperatures lead to slower cooling, which means longer cooling times, and therefore longer cycle times.

So in the selection of the mould temperature level, a choice often has to be made between higher quality and a more favourable price for the components. If you allow yourself to be guided here by the principle 'Only as good as necessary, not as good as possible', a compromise between the two, giving a 'semi-optimum' temperature level, will very often lead to an economically acceptable solution.

8.11.2.3 Uniformity of Mould Temperatures

Uniformity has two aspects here – the spatial temperature distribution in the mould and the temporal temperature behaviour in the production cycle.

The level of the mould temperature influences the shrinkage, and thus the later dimensions of the component. Local temperature differences can cause parts to warp. Mould temperatures that do not remain the same from batch to batch lead to dimensional variations through varying shrinkage.

Uniform temperature distribution in the mould is essentially dependent upon the mould temperature system. Adequate and uniform temperatures are of importance, not only in relation to the warping tendency, but also for economic reasons from the point of view of the unit time. It is necessary to wait until even the hottest part of the component has cooled enough before ejecting the component from the mould. Thus, a uniform intensity of temperature becomes a prerequisite for economic manufacture. With cores, when adequate temperature patterns often cost more to achieve, there are often signs of omission in the mould design in this connection. The mould certainly becomes cheaper then. But it means that a substantial increase in the cycle time – often up to 100% and more – must usually be taken into account.

The temperature systems used are essentially responsible for maintaining the same mould temperatures throughout the manufacturing cycle. The temperature systems must be matched to the mould as regards their production capacity, i.e., they must be in a position to supply or extract the necessary amounts of heat sufficiently quickly.

In order to guarantee a repeatable, uniform temperature distribution in the mould, even after a change of mould or a refit, the inlet and outlet paths of the temperature control fluid at the mould must be unambiguously marked. If the connections are mixed up, this will certainly alter the temperature conditions, which, admittedly, need not always have a negative effect, but which can often lead to substantial deterioration, especially in cooling cores.

8.11.2.4 Mould Temperature Patterns

During the cooling of the component, the heat is drawn from it by the mould. The amount of heat involved depends on the component weight, the type of plastic (specific heat content), the ejection temperature, and the cycle frequency. The mould passes this heat to the environment (by heat conduction to the machine and by radiation or convection to the ambient air), if no special temperature control system is connected up. The higher the mould temperature, the greater the amount of heat released to the environment. So, when the machine starts up, the mould initially heats up to the temperature at which the amount of heat supplied by the solidifying and cooling plastic can be released to the environment.

This temperature is, in practice, scarcely likely to match the optimum mould temperature for the injection moulding process. The temperature thus reached will remain the same only for as long as the machine carries on running continuously, as long as no cycle interruptions occur, and as long as the heat removal conditions, e.g., the ambient temperature, do not alter. Satisfactory constancy and repeatability for the temperature level thus reached cannot therefore be obtained without special temperature control equipment.

If the desired mould temperature is lower than the automatically set temperature referred to above, additional heat must be drawn from the mould by cooling: if the required mould temperature is higher, more heat must be supplied to it through heating.

These are also referred to as hard temperature control, if heat is drawn from the mould, and soft temperature control, if heat is to be supplied. Separate temperature control for the two halves of the mould is also important from this point of view. The parts of the mould on which the component is shrunk, such as cores, for example, will accept more heat, and thus will be more strongly heated, if they are not temperature controlled in a suitable fashion. Parts of the mould from which the component shrinks away accept less heat and will therefore remain colder without temperature control. In extreme cases, it can even happen that one half of the mould, with the core parts, has to be cooled, while the other half has to be heated, so as to obtain the same mould temperature in both halves of the mould.

Again this clearly shows how important a good temperature control system for the mould is (it must be able to exchange the requisite amounts of heat fast enough). The temperature control device must also be suitable for the specific system (so as to be able to supply and extract these quantities of heat).

If hard temperature control is possible and necessary, its quality has an essential part to play in whether the process is economically viable, because inadequate cooling means the cycle time is extended. The ejection losses here can be up to 100%.

At this point, a warning should be given that trial injections should not be carried out without monitored mould temperature control. Not only would this have the consequence that later, in production conditions, the dimensions obtained could in certain circumstances be completely different, but also that the cycle times could be altered for the worse.

8.11.2.5 Temperature Control Devices

Fluids are usually used for mould temperature control, because both heat extraction and heat supply can be controlled equally well. Electrical heating for soft temperature control is certainly very rapid, but still not fast enough on its own: it would cause excessively high temperature variations. In all cases, electrical heating is used in conjunction with fluid temperature control, the latter being there to provide – if required – for sufficiently rapid heat extraction (so as to keep the temperature variations low). This combined temperature control system does indeed obtain the best results, but of course costs are significantly higher, and only rarely economically acceptable. So in what follows only pure fluid temperature control will be considered.

The cheapest fluid temperature control can be carried out using circuit water. Of course, it is only suitable for hard temperature control. For this purpose, a water distribution system is provided in all parts of the injection moulding machines, to which the temperature control circuits of the mould can be connected. The cooling intensity is set by the water throughput using valves. So as to obtain values with a certain amount of repeatability, throughput gauges and thermometers (in the counterflow) are required in each circuit.

For higher hard control requirements and for all cases involving soft temperature control, fluid temperature control units (forming units, heating units and cooling units) are used, which pump up a temperature control medium (water and oil) through the temperature control system of the mould. The cycle temperature of the temperature control medium is adjustable, and is adjusted for a pre-set value.

Units using water as a temperature control medium generally operate in a temperature range between about 20 °C and 95 °C. At temperatures of over 100 °C, the units must be under constant pressure in order to keep the water liquid, for with steam the heat transfer ratios are considerably poorer. The lower temperature limit is defined by the temperature of the cooling water available. If special cooling units are installed, cycle temperatures down to –10 °C can be reached.

Oil is used as a temperature control medium for cycle temperatures of over 95 °C. As a rule, a temperature range of 30 to 250 °C can be covered using these units, which can also be used for cycle temperatures below 95 °C. They are less suitable for hard temperature control, of course, as the heat absorption capacity of oil is smaller than that of water. Thus more heat can be transferred using water than oil, for the same amount of circulated fluid.

The mould temperature can be controlled using the devices described so far. The temperature control units can be used to regulate the flow temperature, but not the mould temperature. To set the desired mould temperature, it is necessary to measure the actual mould temperature and to alter the flow temperature (or the fluid quantity) until the values desired are obtained. It should be kept in mind here that temperature changes, as a whole, work through relatively sluggishly, and therefore it is necessary to wait sufficiently long after an alteration for a state of equilibrium to be re-established. It is especially important to bear in mind that the flow temperature is not identical to the mould temperature. In hard temperature control (cooling), it will always lie below it, whole in soft temperature control (heat supply), the flow temperature is higher than the mould temperature.

As in any kind of control, there are disturbing influences, such as, for example, cycle interruptions, which exercise their full effect. There is no heat supply from the injection equipment, and the open mould has a larger heat-radiating surface. So, when production resumes, the old temperature balance, and with it the original dimensions of the component, will not be reinstated until some few cycles have been completed.

Mould temperature regulation can be used to reduce the effects of these interference factors considerably. A temperature sensor in the mould acts as an actual value pick-up, and signals the actual mould temperature to the temperature control unit. The flow temperature is now no longer kept at a constant value, but is adjusted by means of a regulator until the actual temperature in the mould corresponds to the theoretical temperature for the mould set on the temperature control unit.

8.11.2.6 Matching Temperature Control Units and Mould Maintenance

Satisfactory results can only be obtained from mould temperature control if the performance of the temperature control unit is suitable for the quantities of heat to be exchanged in the mould. For example, if the temperature control unit not only operates well, but also sufficiently rapidly, so that only slight temperature variations occur in the mould.

If the dimensions of the temperature control unit were too large, that would not do any damage, but it would not be economic. If the dimensions are too small, this will increase the temperature variations in the mould, in every case. The temperature control units must therefore have available a sufficiently high pump delivery (l/min), with a sufficiently high transmission pressure level. It must be possible to overcome the flow resistances of the temperature control circuit with a suitable delivery.

Unfortunately, the transmission pressure on a number of temperature control units on the market today is insufficient. A safety valve is frequently built into such units, which opens a parallel circuit within the unit, once a limiting pressure has been exceeded. Unfortunately this is not always noticeable to the user. Only a fraction of the flow delivery then flows through the mould, and the temperature control suffers accordingly. Naturally, the flow resistances of the temperature control circuits should be kept as low as possible. To this end, there should be sufficiently thick hoses, as short as possible, between the temperature control unit and the mould. The temperature control channels in the mould must be of suitable dimensions.

It is also important that the heat transfer ratios on the walls of the temperature control flues do not deteriorate over the course of time. The temperature control channels must be suitably maintained and must be checked for cleanness after each mould change at least. If water is used for temperature control, then special attention must be paid to rust deposits and also, at higher temperatures, to scale formation.

8.11.2.7 Measuring the Mould Temperature – Checking the Uniformity of the Mould Temperature Control

Most information concerning the injection process cycle can be supplied by a temperature measurement point mounted directly in the moulding nest wall (mould wall temperature). During the cycle, this temperature value can be observed during the compression phase and a lower value during ejection. The lower value can give a good guide for the ejection temperature.

A measurement point position like this is not suitable for regulating the mould temperature, because of the temperature variations referred to. The temperature gauge should therefore be sufficiently far away from the mould wall, so that at the measuring point the temperature variations have already been sufficiently dampened. But it should also be an adequate distance away from the temperature control channels, so as to exclude reverse effects from this side. A middle position between the mould wall and the temperature control channels is recommended. Because the temperature falls towards the clamping surface, the measuring point should naturally not lie too close to this. It is in any case advisable to insulate the mould halves from the clamping surface, so as to reduce the interference factors. For high mould temperatures, economic considerations also require this.

If no fixed temperature measuring point is incorporated in the mould, it is possible to use probe thermometers as an aid. To carry out any measurements in the cavity, involves interrupting the cycle. The negative effects, which then arise on the accuracy of measurement and the production cycle must not be left out of consideration. But in no case should the measurement of the mould temperature – at least at one specific comparison point or reference point – be neglected.

A sufficiently uniform spatial temperature distribution can be obtained in the mould if the inlet and outlet temperatures of the temperature control medium do not differ by more than 5 °C from one another. In order to achieve this, a sufficiently high throughput volume of the temperature control medium is required.

For temperature control (regulated flow temperatures), the variations in the cycle temperatures should be less than 5 °C, in order to reach a suitable temporal constancy in the mould temperatures.

8.12 Recommended Values for Holding Pressure Time and Remaining Cooling Time

The cooling time consists of both the holding time and the residual cooling time. The holding pressure time and cooling time can be calculated as follows:

- For holding pressure with either a single holding pressure time (t_n) or holding pressure times ($t_{n1} + t_{n2}$)
- For injected parts with medium wall thicknesses of d = 1 mm to d = 4 mm and with mould temperatures less than 60 °C (over 60 °C: 30% extra)

the following empirical formula for the necessary cooling time (t_{kn}, in seconds) with start up applies:

$$t_{kn} = d\,(1 + 2d)$$

We recommend dividing the determined cooling time according to Figure 8.7 into the remaining holding pressure times and cooling time (t_k). Respective values should be taken from Table 8.10.

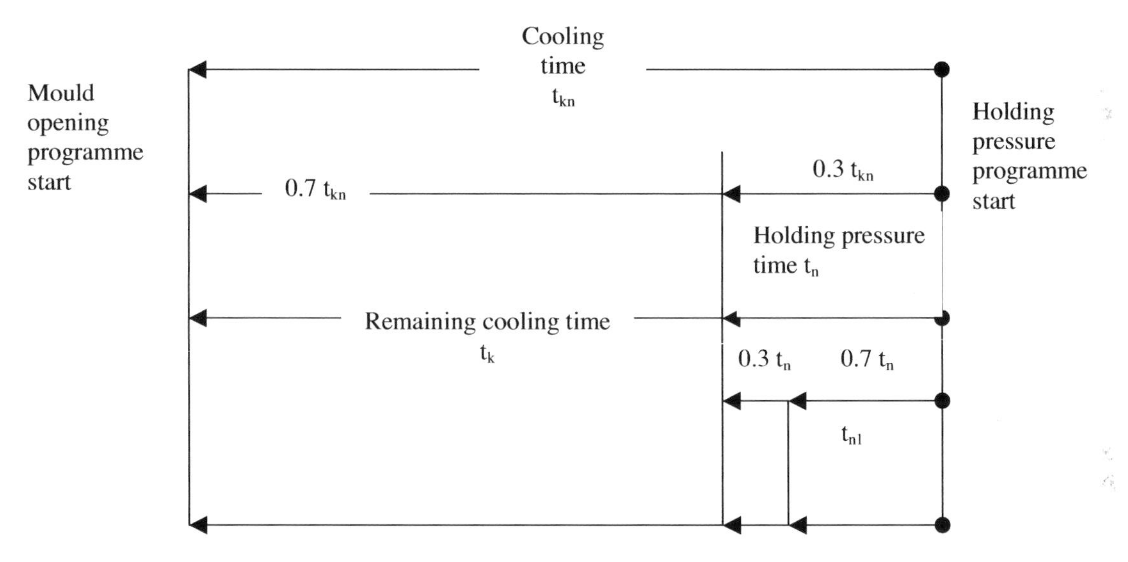

Figure 8.7 Holding time and cooling time

The remaining cooling time should be minimised so that the surface temperature of the component is just under the heat distortion temperature of the moulding material. This can be provided by the material supplier.

8.13 Cycle Time

The total cycle time consists of:

- Machine and mould movement time
- Time for mould opening and ejection, mould closing, injection unit movement.
- Residual cooling time
- Holding pressure time
- Injection time
- Plastication time

The total cycle time obviously has a considerable economic impact on the effective production.

Table 8.10 Holding and cooling values											
Mould temperature under 60 °C						Mould temperature above 60 °C					
d	t_{kn}	t_n	t_{n1}	t_{n2}	t_k	d	t_{kn}	t_n	t_{n1}	t_{n2}	t_k
1.0	3.0	0.9	0.6	0.3	2.1	1.0	3.9	1.2	0.8	0.4	2.7
1.1	3.6	1.1	0.7	0.4	2.5	1.1	4.6	1.4	1.0	0.4	3.2
1.2	4.1	1.3	0.9	0.4	2.8	1.2	5.3	1.6	1.1	0.5	3.7
1.3	4.7	1.4	1.0	0.4	3.3	1.3	6.1	1.9	1.3	0.6	4.2
1.4	5.4	1.7	1.1	0.6	3.7	1.4	7.0	2.1	1.4	0.7	4.9
1.5	6.0	1.8	1.2	0.6	4.2	1.5	7.8	2.4	1.6	0.8	5.4
1.6	6.8	2.1	1.4	0.7	4.7	1.6.	8.8	2.7	1.8	0.9	6.1
1.7	7.5	2.3	1.6	0.7	5.2	1.7	9.8	3.0	2.1	0.9	6.8
1.8	8.3	2.5	1.7	0.8	5.8	1.8	10.8	3.3	2.3	1.0	7.5
1.9	9.2	2.8	2.0	0.8	6.4	1.9	11.9	3.6	2.5	1.1	8.3
2.0	10.0	3.0	2.1	0.9	7.0	2.0	13.0	3.9	2.7	1.2	9.1
2.1	11.0	3.3	2.3	1.0	7.7	2.1	14.2	4.2	3.0	1.3	10.0
2.2	11.9	3.6	2.5	1.1	8.3	2.2	15.5	4.7	3.3	1.4	10.8
2.3	12.9	3.9	2.7	1.2	9.0	2.3	16.8	5.1	3.5	1.6	11.7
2.4	14.0	4.2	3.0	1.2	9.8	2.4	18.1	5.5	3.8	1.7	12.6
2.5	15.0	4.5	3.1	1.4	10.5	2.5	19.5	5.9	4.1	1.8	13.6
2.6	16.2	4.9	3.4	1.5	11.3	2.6	21.0	6.3	4.4	1.9	14.7
2.7	17.3	5.2	3.6	1.6	12.1	2.7	22.5	6.8	4.7	2.1	15.7
2.8	18.5	5.5	4.0	1.6	13.0	2.8	24.1	7.3	5.1	2.2	16.8
2.9	19.8	6.0	4.2	1.8	13.8	2.9	25.7	7.7	5.4	2.4	18.0
3.0	21.0	6.3	4.4	1.9	14.7	3.0	27.3	8.2	5.7	2.5	19.1
3.1	22.4	6.8	4.7	2.1	15.6	3.1	29.1	8.8	6.1	2.7	20.3
3.2	23.7	7.2	5.0	2.2	16.5	3.2	30.8	9.3	6.5	2.8	21.5
3.3	25.1	7.6	5.3	2.3	17.5	3.3	32.6	9.8	6.8	3.0	22.8
3.4	26.6	8.0	5.6	2.4	18.6	3.4	34.5	10.4	7.2	3.2	24.1
3.5	28.0	8.4	5.8	2.6	19.6	3.5	36.4	11.0	7.7	3.3	25.4
3.6	29.6	8.9	6.2	2.7	20.7	3.6	38.4	11.6	8.1	3.5	26.8
3.7	31.1	9.4	6.5	2.9	21.7	3.7	40.4	12.2	8.5	3.7	28.2
3.8	32.7	9.9	6.9	3.0	22.8	3.8	42.5	12.8	9.0	3.8	29.7
3.9	34.4	10.4	7.2	3.2	24.0	3.9	44.7	13.5	9.4	4.1	31.2
4.0	36.0	10.8	7.5	3.3	25.2	4.0	46.8	14.1	9.8	4.3	32.7

8.14 Setting the Injection Moulding Machine

The following twelve-step guide should help you on your way to set an injection moulding machine:

1. Set the melt and mould temperature as per specific material instructions.
2. Set up switch over position (roughly two-thirds of potential stroke) to prevent damage to press or mould and set an initial holding pressure at 0 bar. (The idea at this stage is not to completely fill the tool but to produce short shots.)
3. Set the screw rotation and back pressure as per specific material instructions.
4. Set the injection pressure and injection speed to maximum.
5. Set the holding time and cooling time.
6. Set mould open time (usually 2-5 secs).
7. Gradually increase shot volume to **almost** a full shot. You may use small increments of perhaps 5% at a time.
8. Switch the machine to automatic.
9. Set the ejectors and minimise the mould opening time.
10. Increase the injection volume to around 99% filled.
11. Increase the holding pressure in steps to optimise.
12. Minimise the holding pressure time.

9 Troubleshooting

9.1 Introduction

Like most manufacturers, the first ARBURG injection moulding machine was born of the necessity to perfect their own products. As a consequence, practical requirements were systematically incorporated in the design of the ARBURG ALLROUNDER from the very outset. Injection moulding machines continue to be modelled on the special needs of the customers. The objective is to offer machines that function ideally under the though conditions of day-to-day operation.

In other words, all machines must work accurately and smoothly, providing simple and reliable access to complex technical procedures using technology that is easy to operate. Many modern injection moulding machines meet these requirements: day in, day out, all over the world.

The user-friendliness of a machine, and thus its efficiency, depends not only on a perfected concept but also on the availability of the comprehensive services needed to round off a thorough customer support system. These services may include rapid telephone diagnosis, effective support by a customer service organisation and computer-controlled spare parts service. These facilities may also be backed by leasing and credit arrangements. Training courses may be available as well advice in the areas of mould design and machine planning.

With the benefit of many years of experience, manufacturers such as ARBURG have process engineers able to assist in the design and production of moulds. Their sound knowledge can save time and money when considering equipping and installing new machines. This kind of expert system is increasingly necessary to compete in the world of injection moulding.

9.2 Troubleshooting Guide

Sometimes, problems occur in producing parts of the desired quality. In most cases, the surface quality of thermoplastic injection moulded parts is the main criterion for their quality. Due to the complex interrelationship between the moulded part and the mould, the moulding compound and the processing, it is often very hard to recognise the origin of problems and thus to take immediate action. This chapter is based on the experience and knowledge of many experts. It was written during a three-year team project, which involved intensive work by 30 companies.

This troubleshooting guide is designed to help in analysing surface defects in the injection moulding and to provide hints on avoiding and or reducing defects. It consists of descriptions, pictures and notes about the different defects, which helps in classifying the problem. It provides a short explanation of possible physical causes for the defect. Furthermore, important notes on general faults are supplied, as well as notes on boundary conditions which should be considered.

There are also flow charts containing hints on avoiding or reducing defects. Remedy and hints are given, concerning the process, the moulded part, the mould design and the moulding compound. A blank Data Acquisition Record can also be found here to help in evaluating the optimisation process. Finally some case studies of real-life problems are included.

9.2.1 Detection and Classification of Defects

Injection moulding defects are classified into seventeen sections:

1. Sink marks (Section 9.2.3)
2. Streaks (Section 9.2.4)
   - 2.1 Burnt streaks (Section 9.2.4.1)
   - 2.2 Moisture streaks (Section 9.2.4.2)
   - 2.3 Colour streaks (Section 9.2.4.3)
   - 2.4 Air streaks and air hooks (Section 9.2.4.4)
   - 2.5 Glass fibre streaks (Section 9.2.4.5)
3. Gloss/gloss differences (Section 9.2.5)
4. Weld line (Section 9.2.6)
5. Jetting (Section 9.2.7)
6. Diesel effect (burns) (Section 9.2.8)
7. Record grooves effect (Section 9.2.9)

8. Stress whitening/stress cracks (Section 9.2.10)
9. Incompletely filled parts (Section 9.2.11)
10. Oversprayed parts (flashes) (Section 9.2.12)
11. Visible ejector marks (Section 9.2.13)
12. Deformation during demoulding (Section 9.2.14)
13. Flaking of the surface layer (Section 9.2.15)
14. Cold slugs/cold flow lines (Section 9.2.16)
15. Entrapped air (blister formation) (Section 9.2.17)
16. Dark spots (Section 9.2.18)
17. Dull spots near the sprue (Section 9.2.19)

In order to eliminate surface defects as soon as possible and as economically as possible, knowledge about the causes of the defects are essential. With this knowledge the processor can decide which remedy is most useful or what should be changed. Additionally, this chapter gives some important notes on economical and quick optimisation of the process and on avoiding defects. It should be noted that this guide only contains short explanations of the causes without any detailed description of the physical background.

9.2.2 Flow Charts for Troubleshooting

Getting rid of surface defects can be a hard task, due to the different physical causes. In order to help the processor, this chapter contains flow charts which systematically show how to eliminate the defect (see Flow Charts 9.1-9.22). The aim is to reach the desired quality by varying the process parameters.

Please note that for one defect, the diagram slides into different branches, according to the questions. Only one parameter should be changed as a remedy, in order to avoid mutual influence. Afterwards several cycles should be completed to ensure stable working conditions. In some cases various solutions are possible, but tendencies (+) or (-) are given. Should one parameter variation fail, go through the questions again and apply, if possible one remedy after another.

Of course, these diagrams, although taking practical conditions into account, only offer suggestions and cannot consider all eventualities that may occur during the process. In addition, they can easily be adapted to special conditions, or completed with your own knowledge. The diagrams will also help to decide whether the defect can be eliminated by changing the machine settings, or whether the mould or the part has to be changed.

Inquiry*

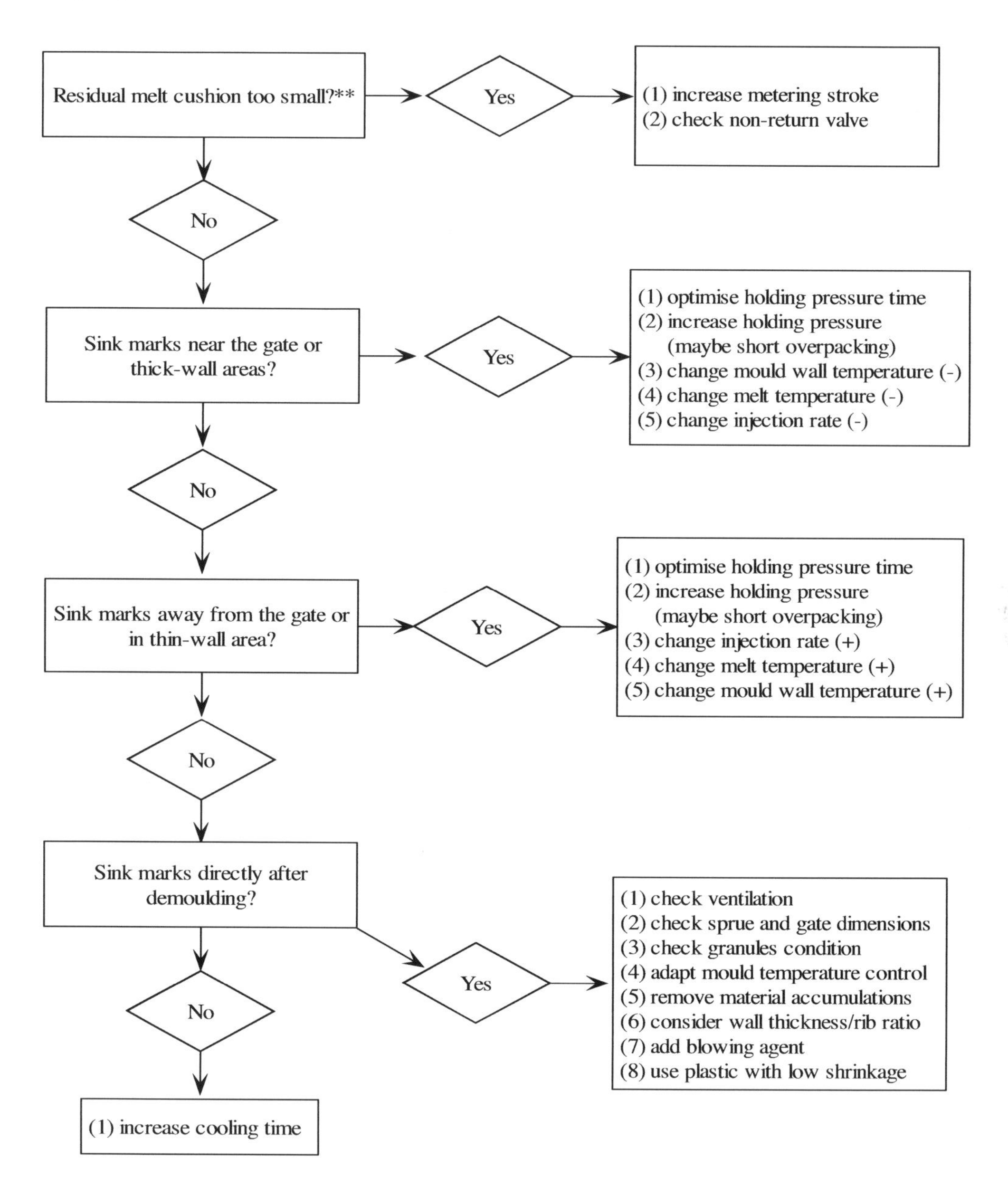

* Important! Check if there are voids in the moulded part after removing sink marks

** Residual melt cushion should be at least 2-5 mm

Flow Chart 9.1 Sink marks

Inquiry

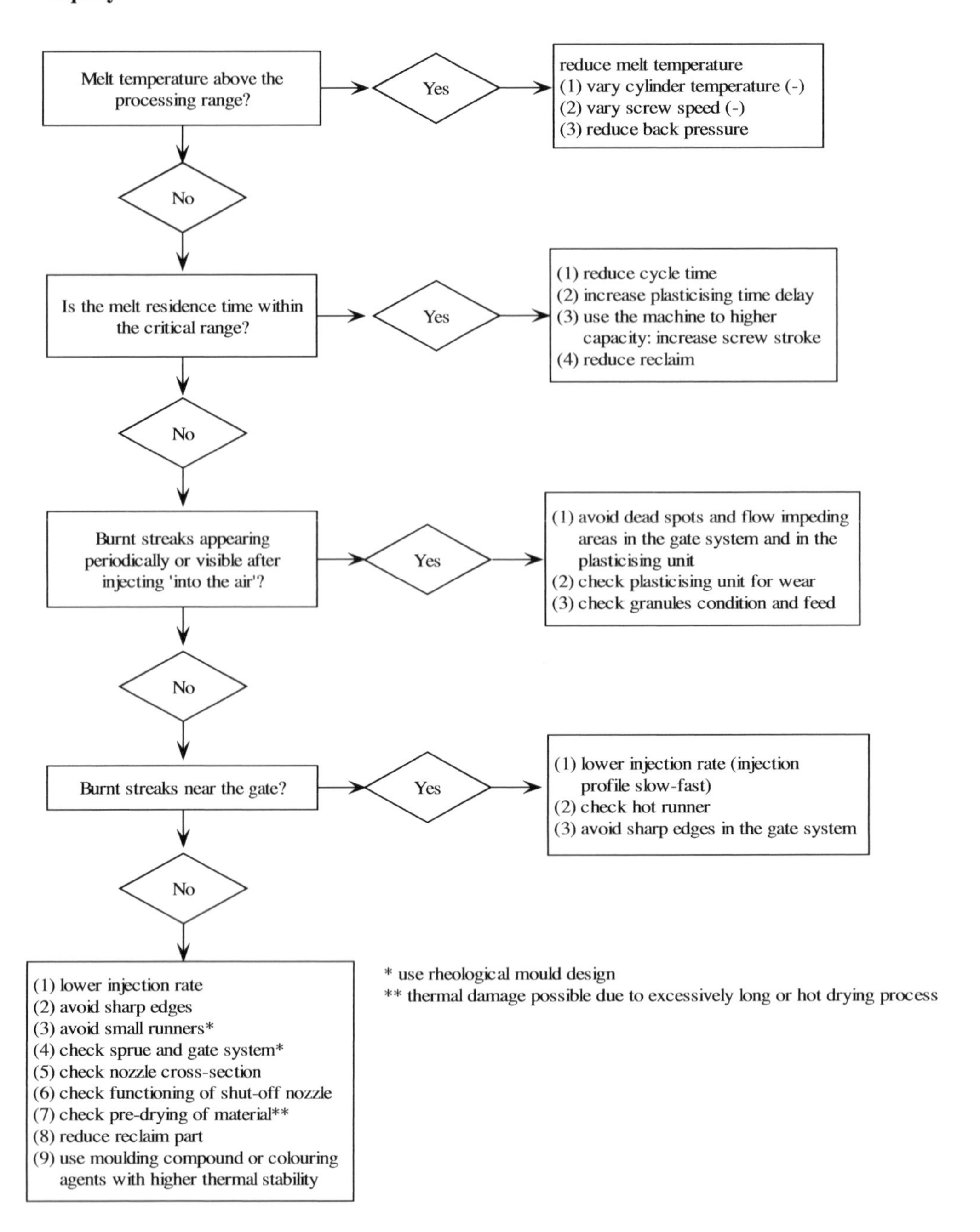

Flow Chart 9.2 Burnt streaks

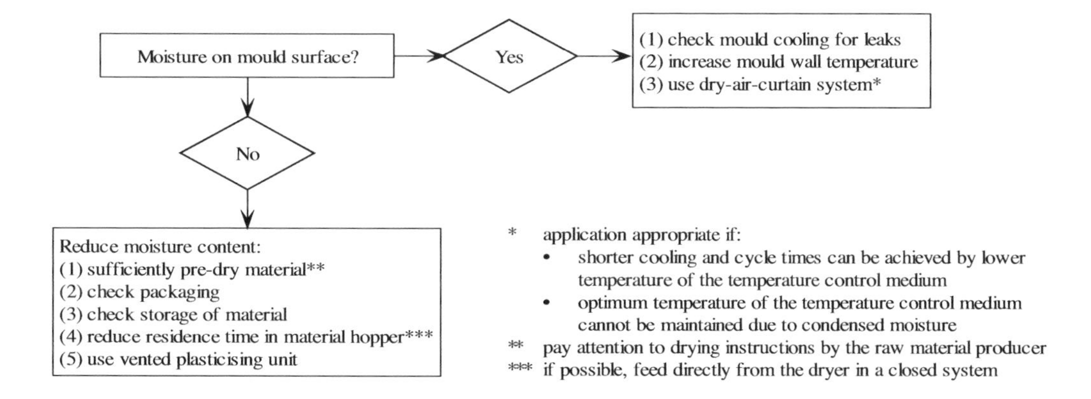

Flow Chart 9.3 Moisture streaks

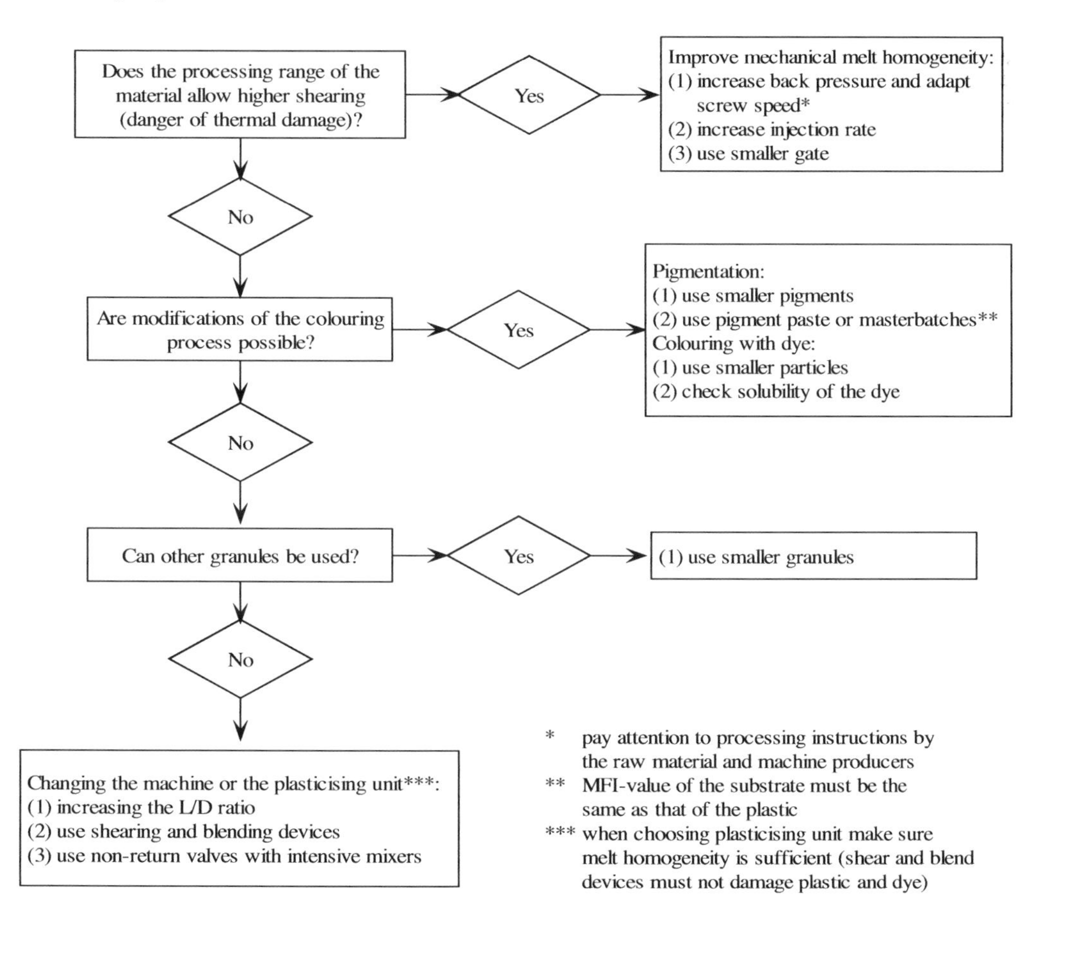

Flow Chart 9.4 Colour streaks

Inquiry

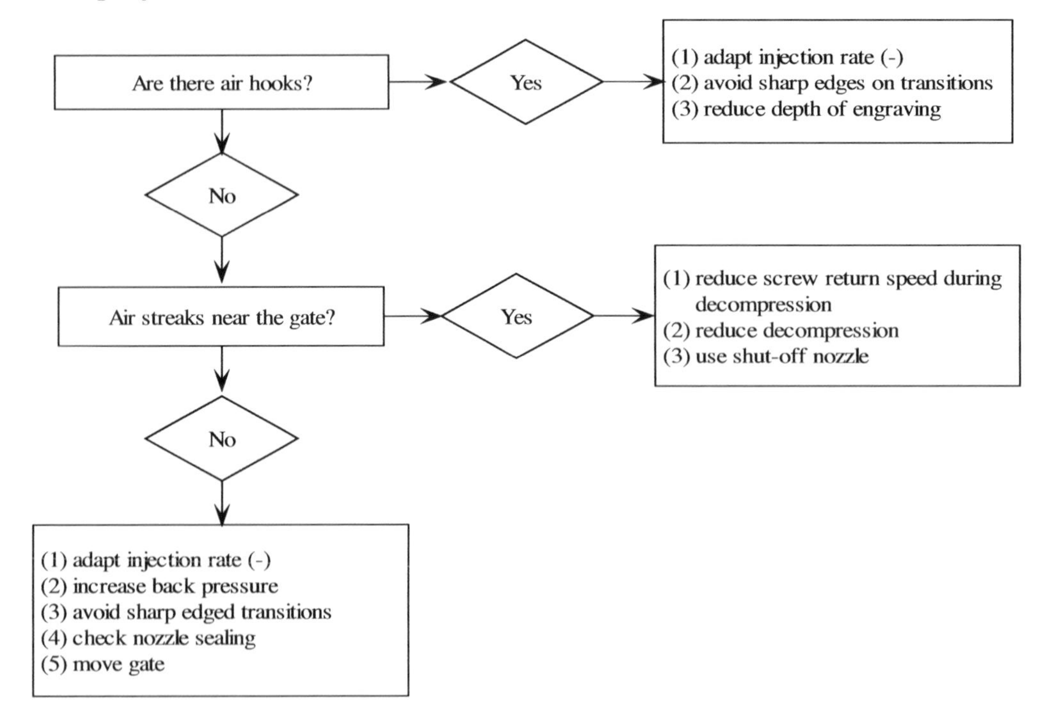

Flow Chart 9.5 Air streaks/air hooks

Inquiry

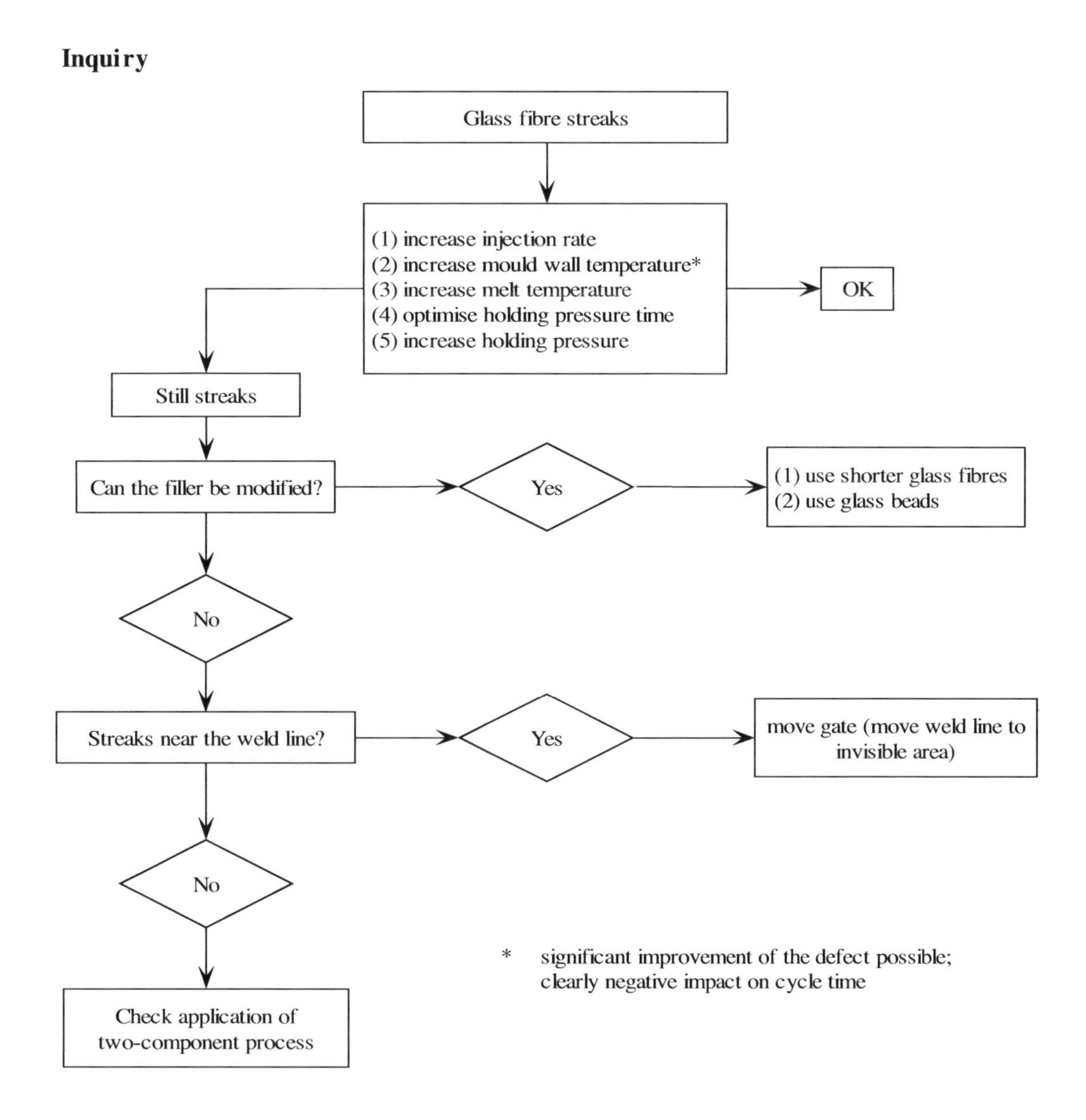

Flow Chart 9.6 Glass fibre streaks

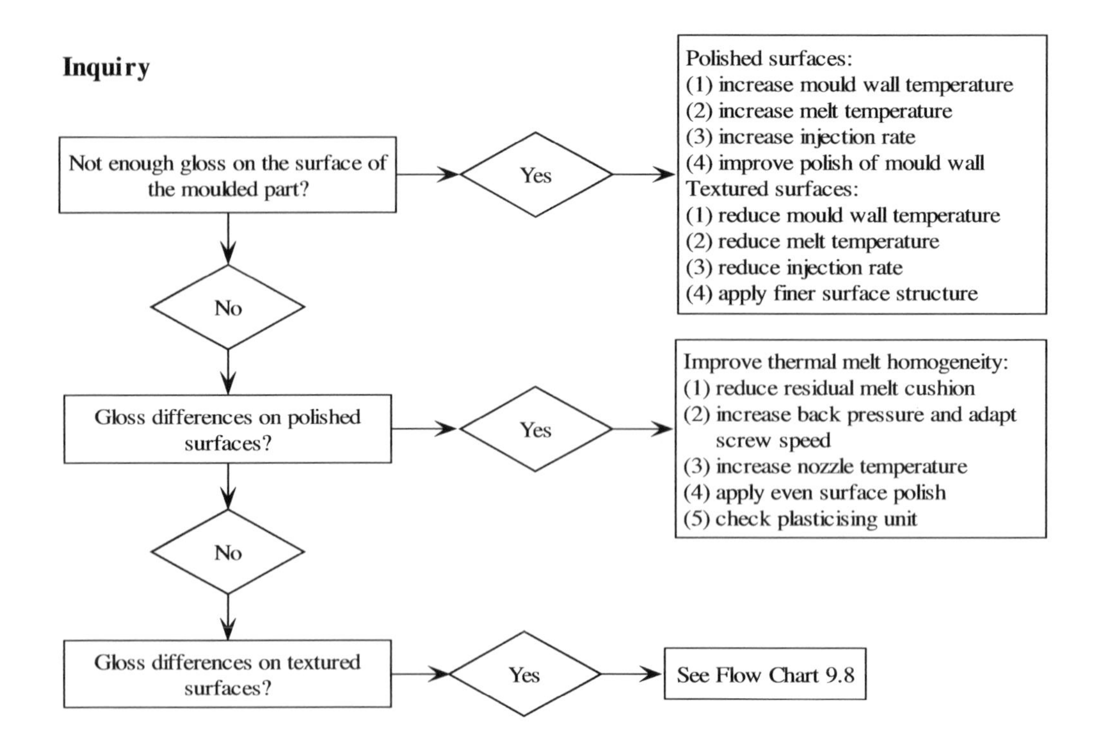

Flow Chart 9.7 Gloss/gloss differences (1)

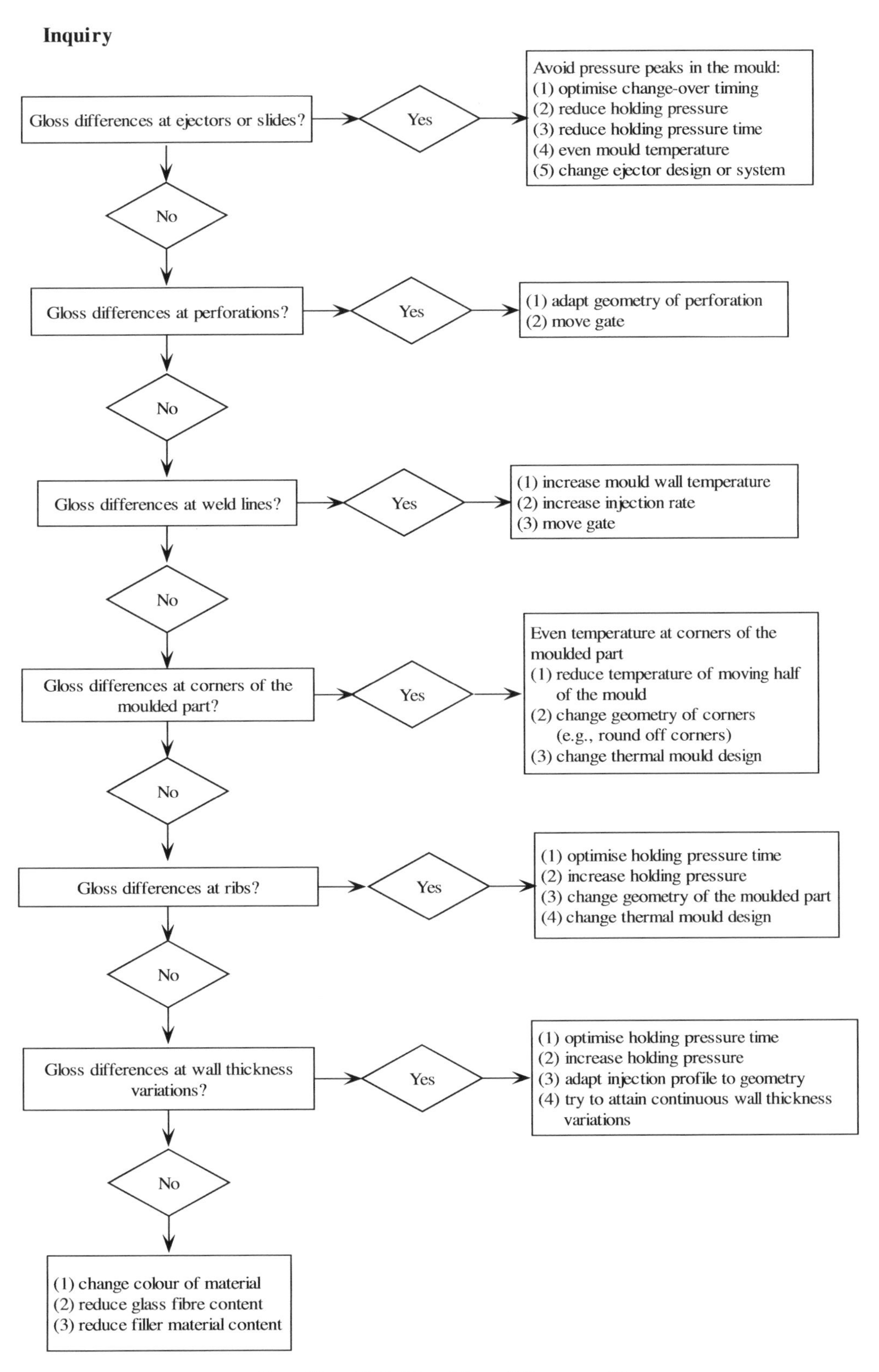

Flow Chart 9.8 Gloss/gloss differences (2)

Inquiry

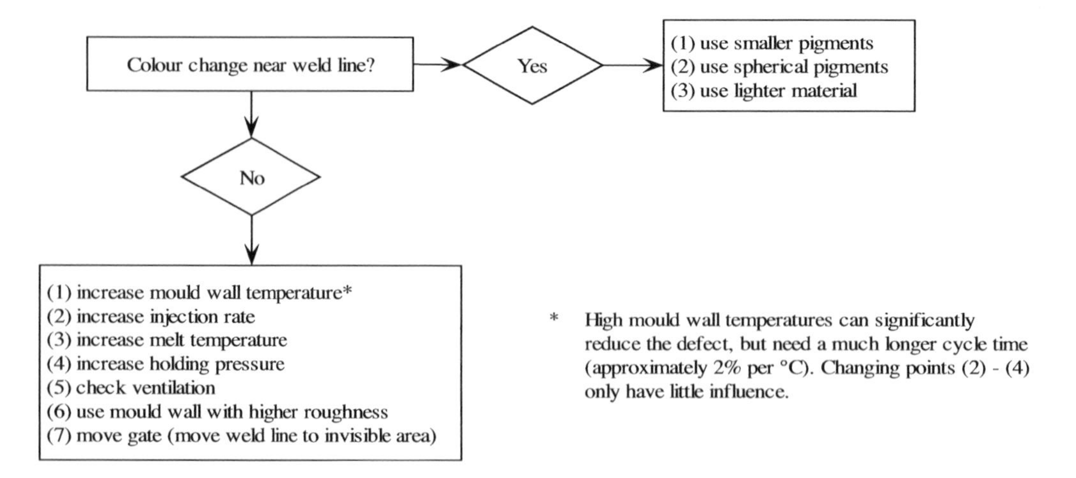

* High mould wall temperatures can significantly reduce the defect, but need a much longer cycle time (approximately 2% per °C). Changing points (2) - (4) only have little influence.

Flow Chart 9.9 Weld line

Inquiry

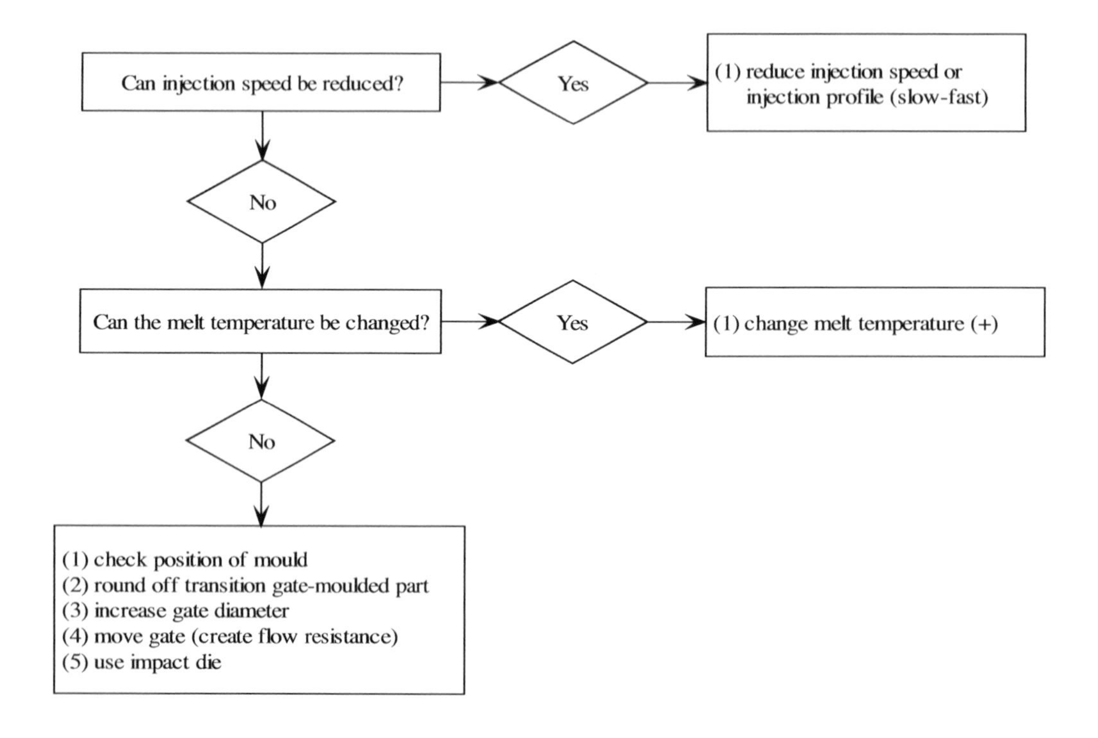

Flow Chart 9.10 Jetting

Inquiry

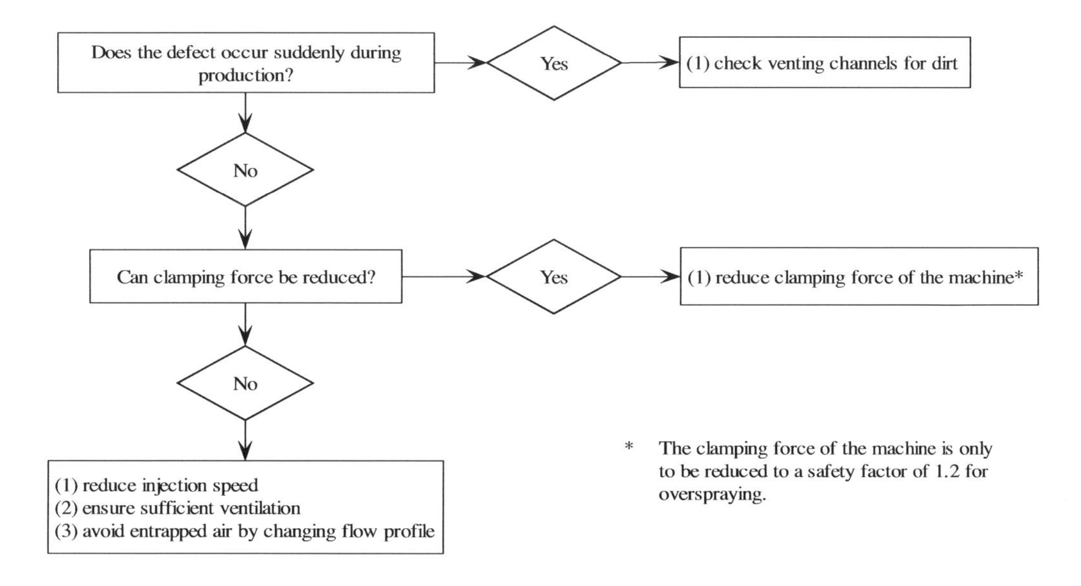

Flow Chart 9.11 Diesel effect

Inquiry

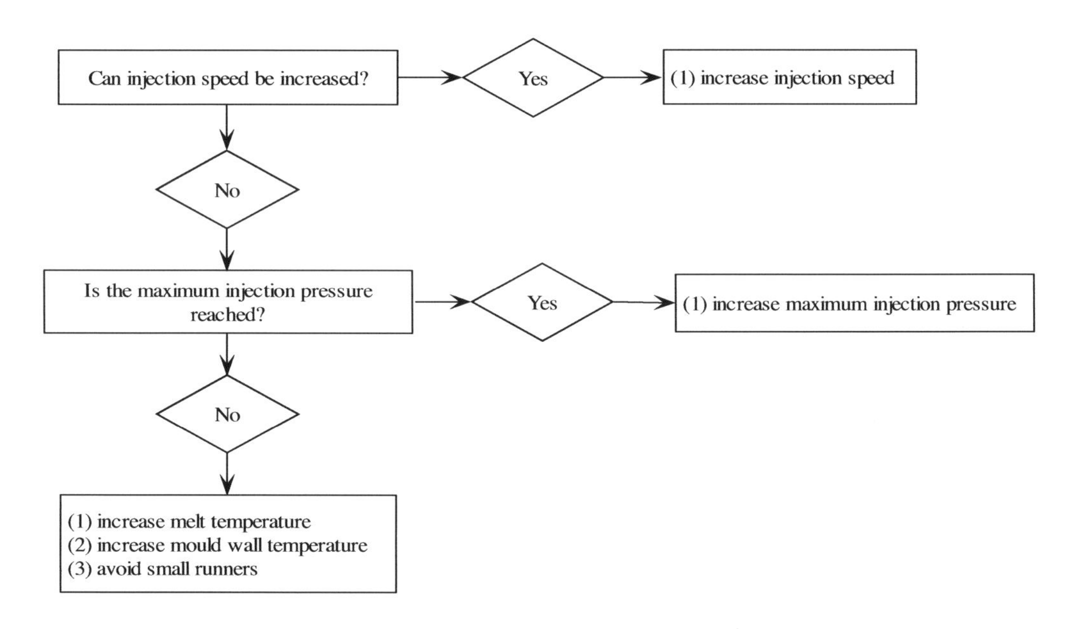

Flow Chart 9.12 Record grooves effect

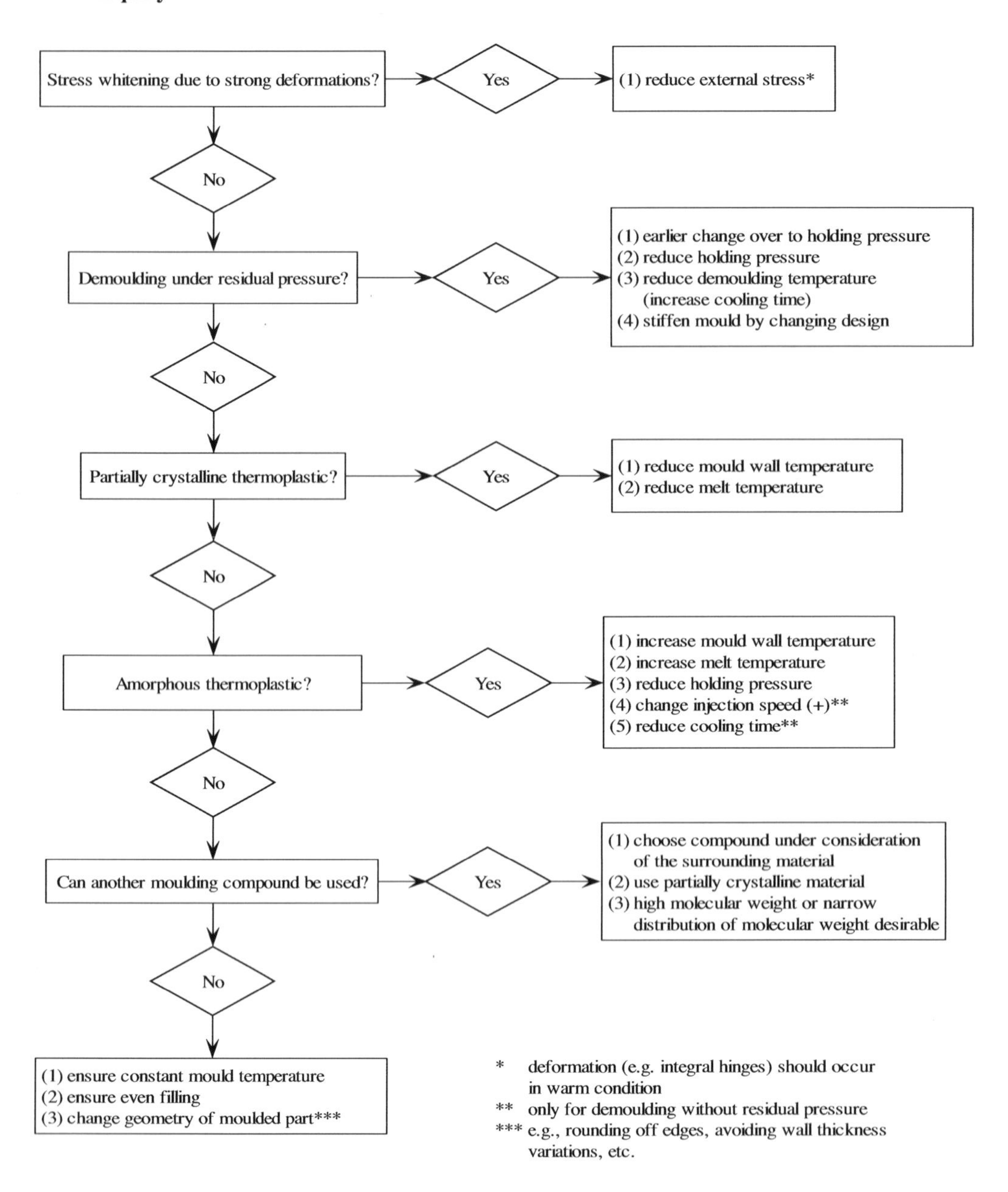

* deformation (e.g. integral hinges) should occur in warm condition

** only for demoulding without residual pressure

*** e.g., rounding off edges, avoiding wall thickness variations, etc.

Flow Chart 9.13 Stress whitening/stress cracks

Inquiry

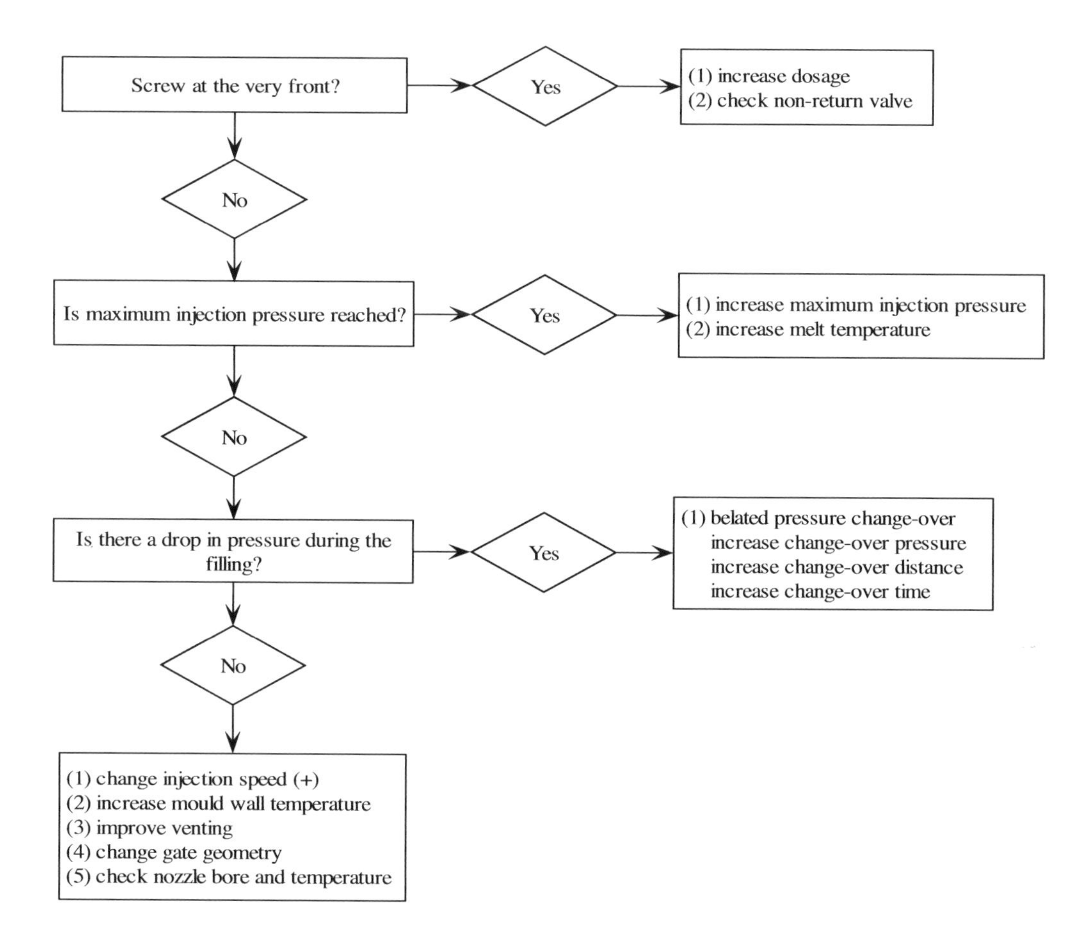

Flow Chart 9.14 Incompletely filled parts

Inquiry

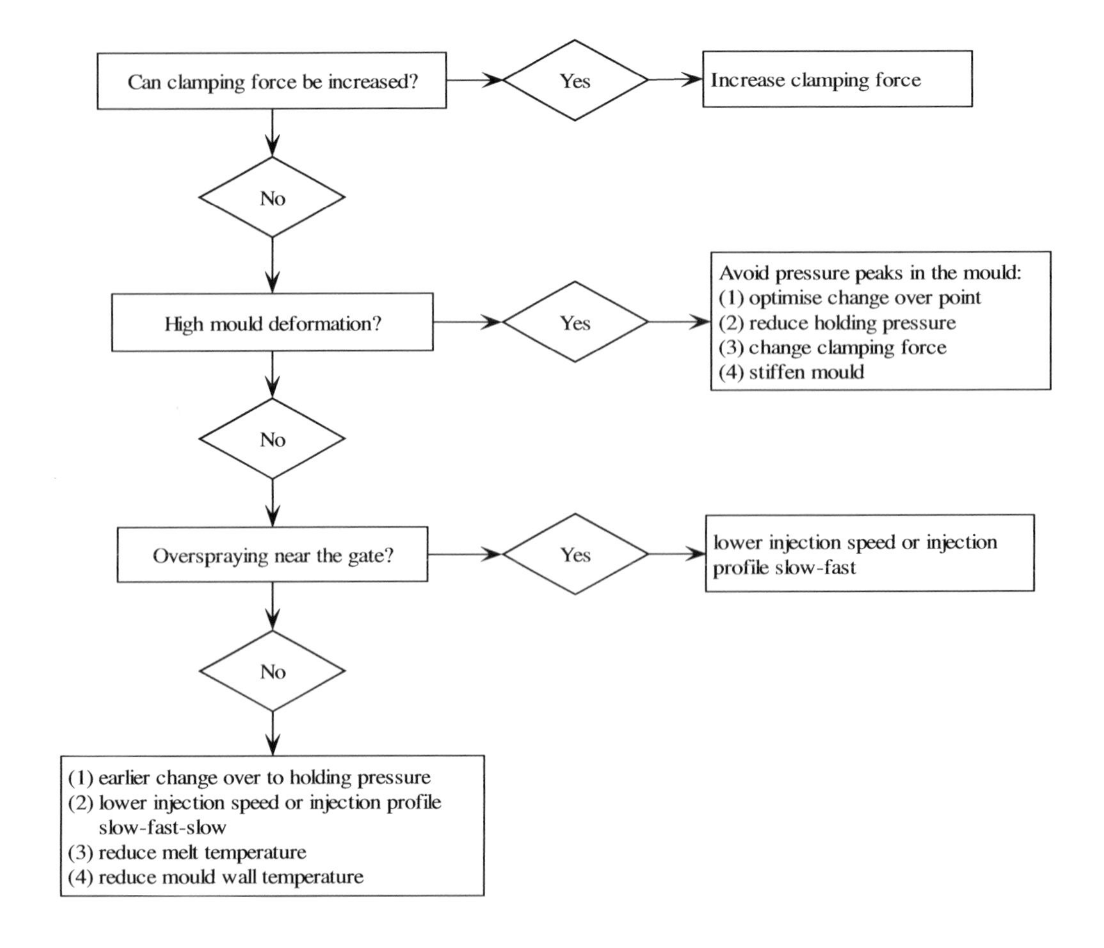

Flow Chart 9.15 Overspraying (flashes)

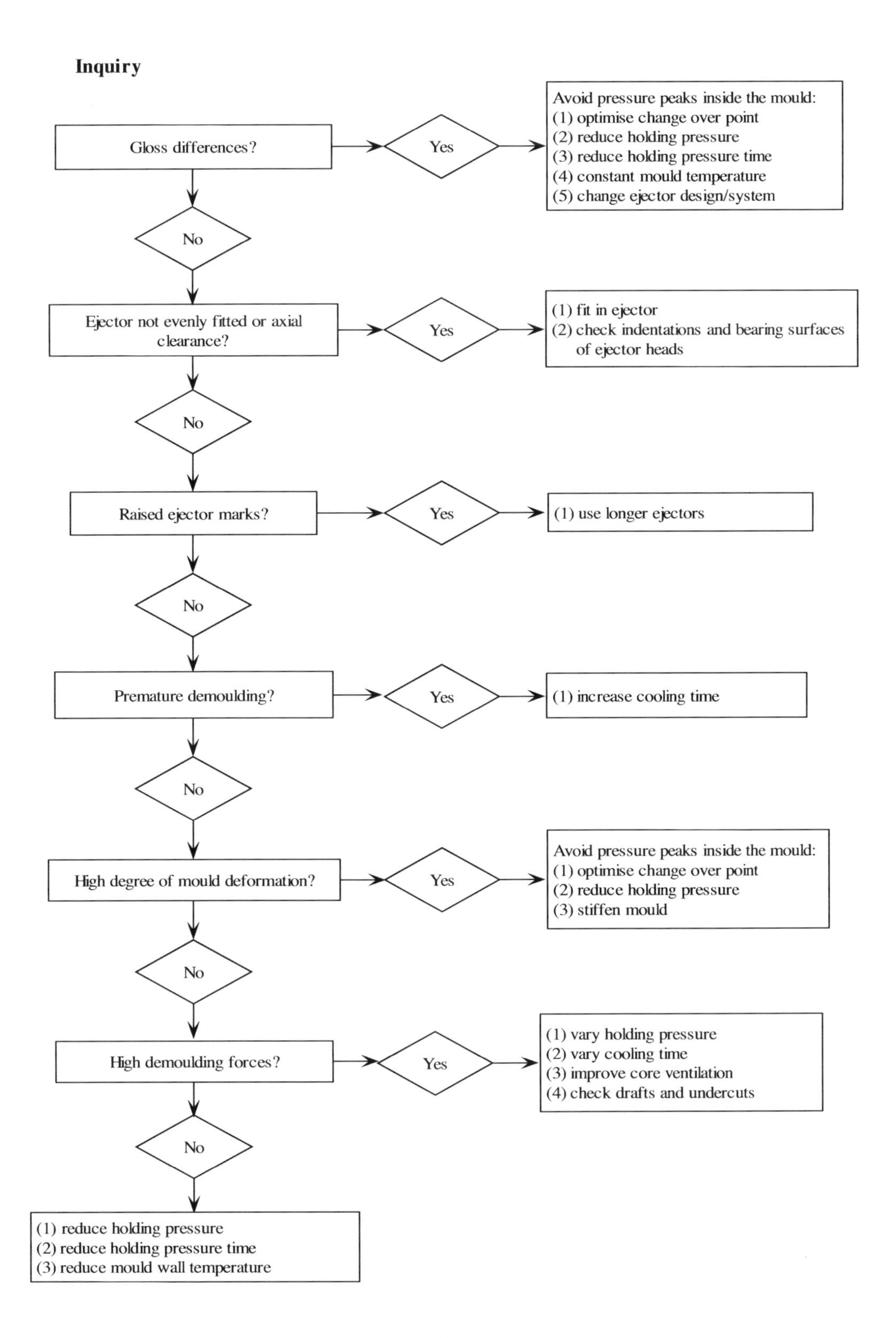

Flow Chart 9.16 Visible ejector marks

Inquiry

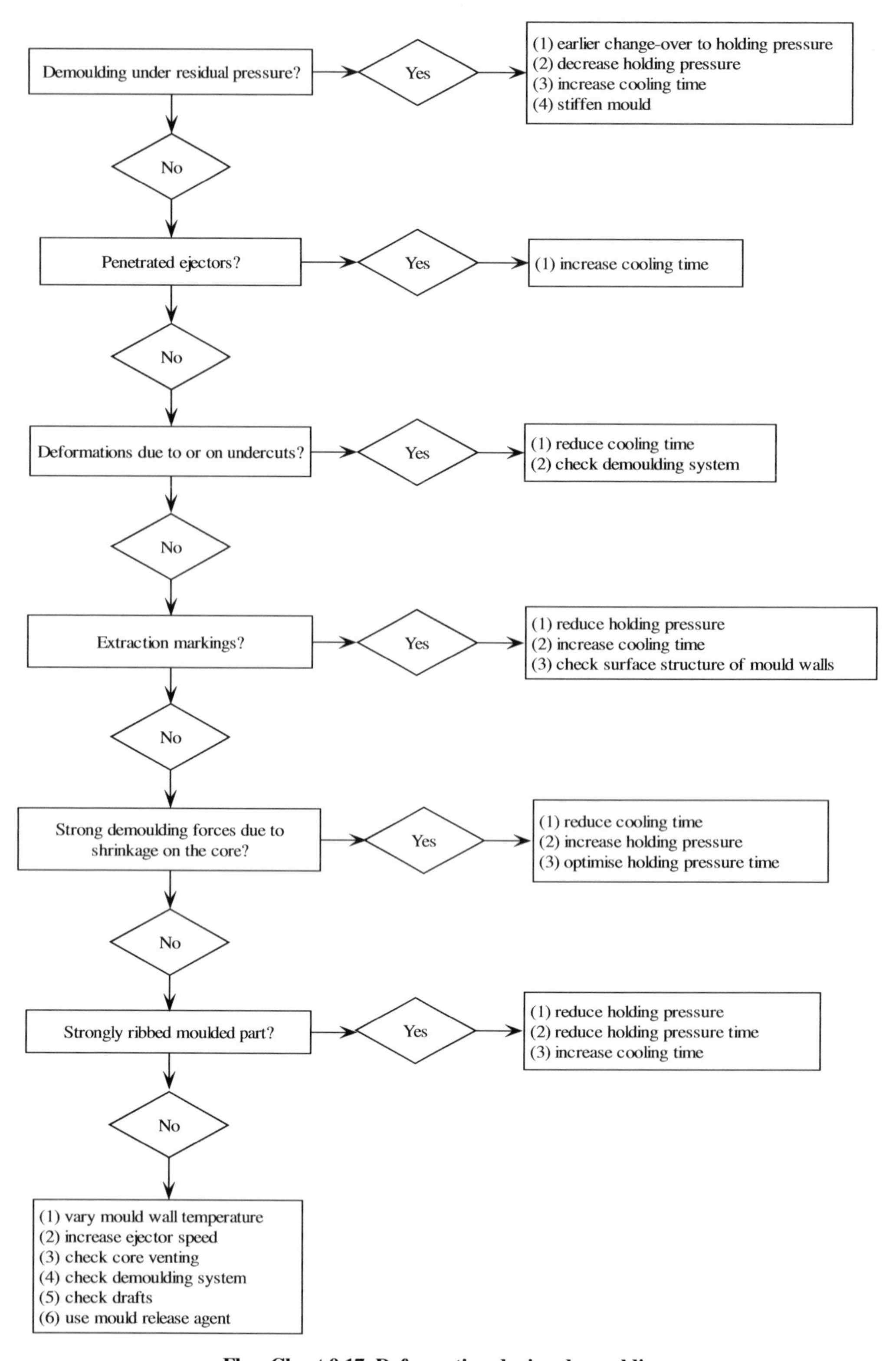

Flow Chart 9.17 Deformation during demoulding

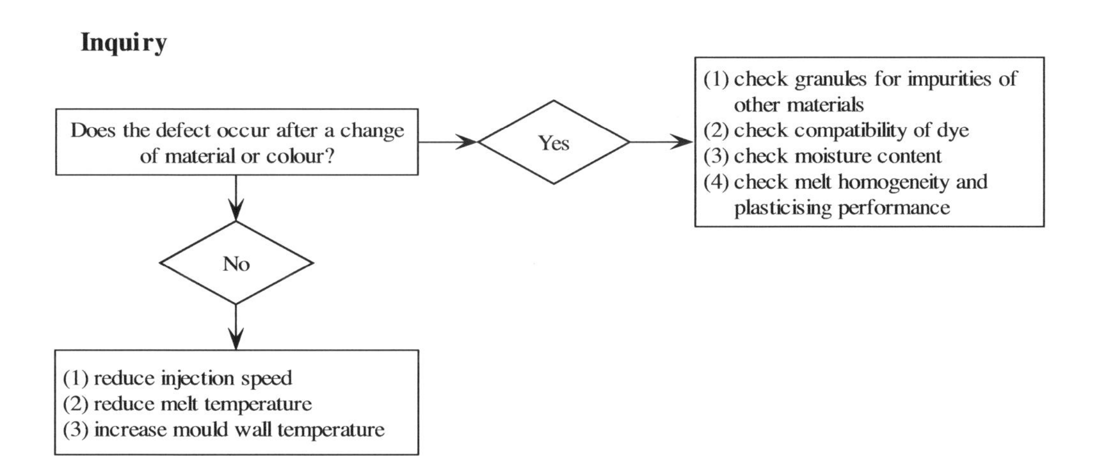

Flow Chart 9.18 Flaking of the surface layer

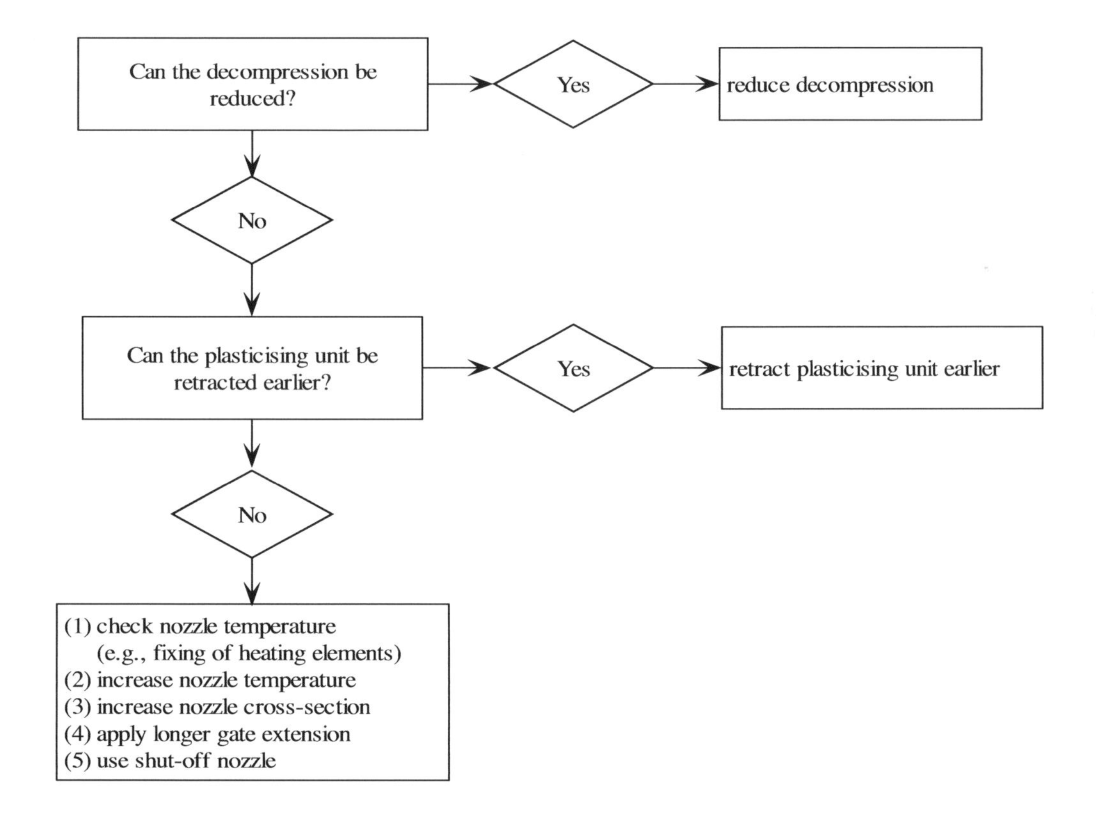

Flow Chart 9.19 Cold slugs/cold flow lines

Inquiry

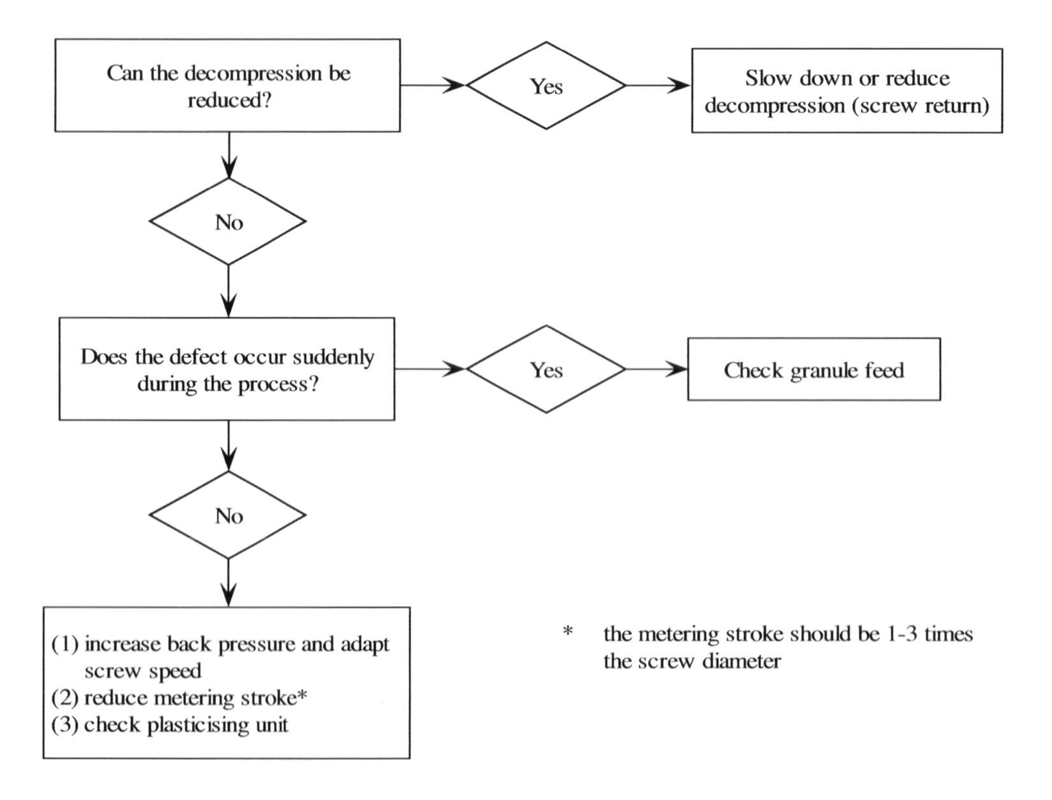

Flow Chart 9.20 Entrapped air

Inquiry

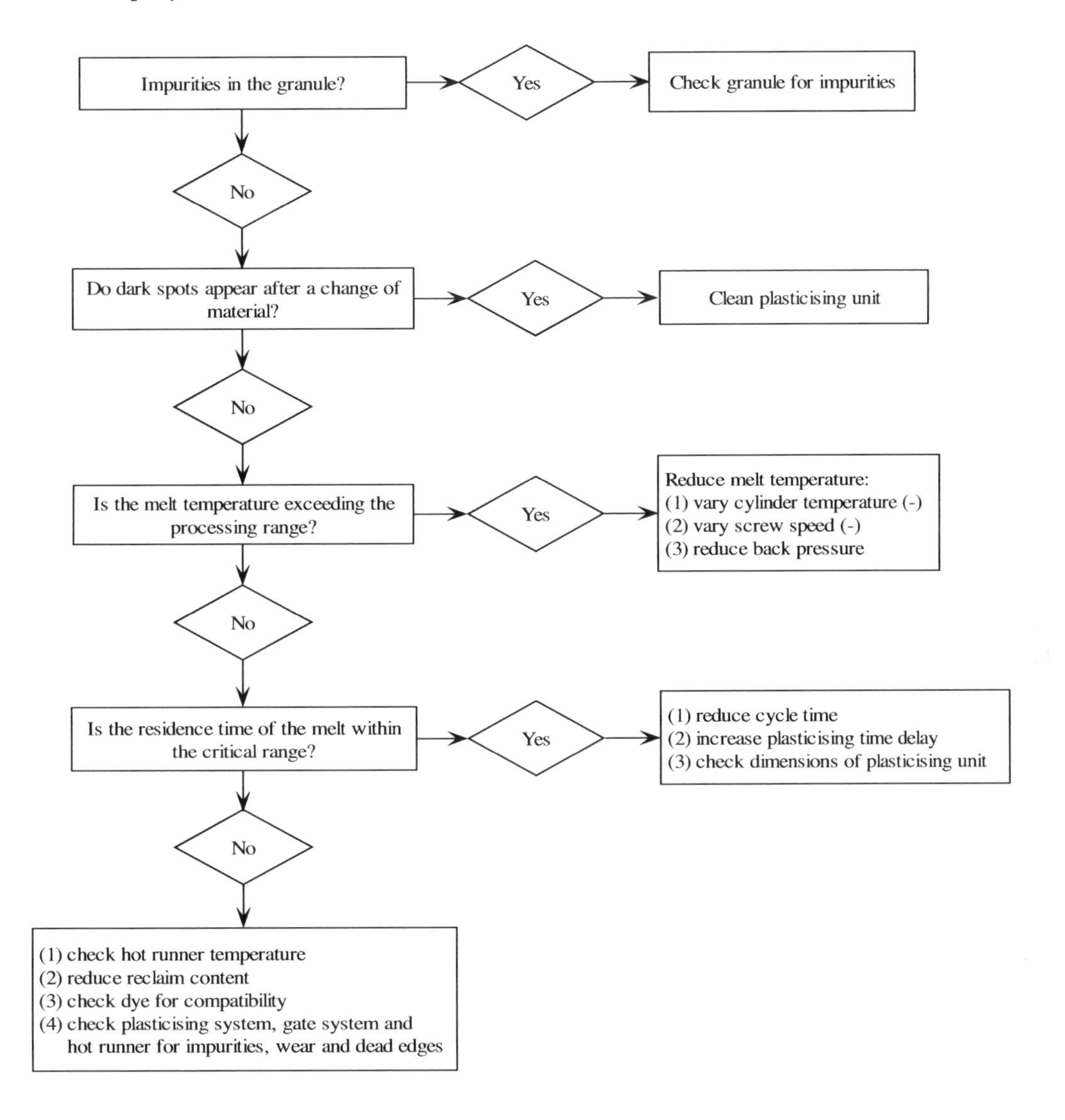

Flow Chart 9.21 Dark spots

Inquiry

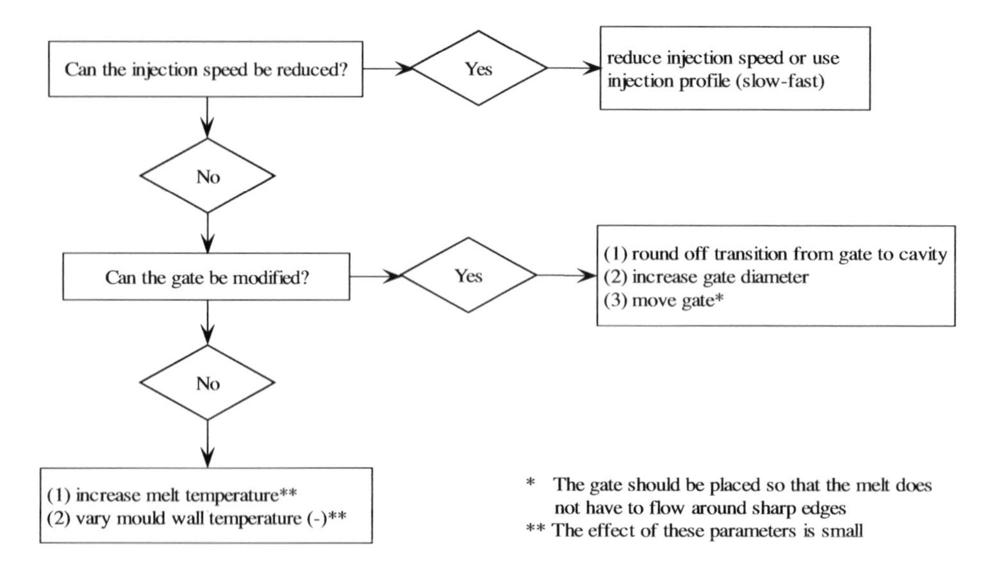

Flow Chart 9.22 Dull spots near the sprue

9.2.3 Sink Marks

Physical cause

Sink marks occur during the cooling process, if the thermal contraction (shrinkage) of the plastic cannot be compensated in certain areas. If the outside walls of the moulded part are not stable enough, due to insufficient cooling, the outer layer is drawn inside by cooling stresses as shown in Figure 9.1.

There are three fundamental cases:

- Solidification too slow
- Effective holding pressure time too short
- Not enough holding pressure transfer; because flow resistances in the mould are too high.

Note: For optimum holding pressure transfer the moulded part should be gated to the largest cross-section. In order to avoid premature solidification of the sprue and gate system, sufficient dimensioning is necessary.

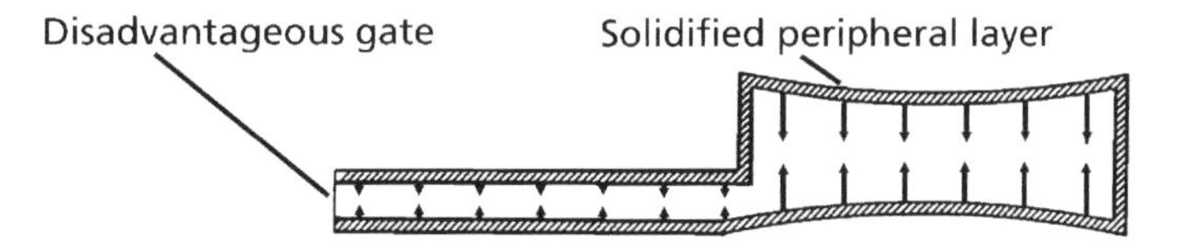

Figure 9.1 Moulded part with sink marks (gating at the thin wall)

Sink marks appear for example near material accumulations as depressions on the surface of the moulded part, if the thermal contraction (shrinkage) cannot be compensated as illustrated by Figures 9.2 and 9.3.

Figure 9.2 Sink marks due to wall thickness variations

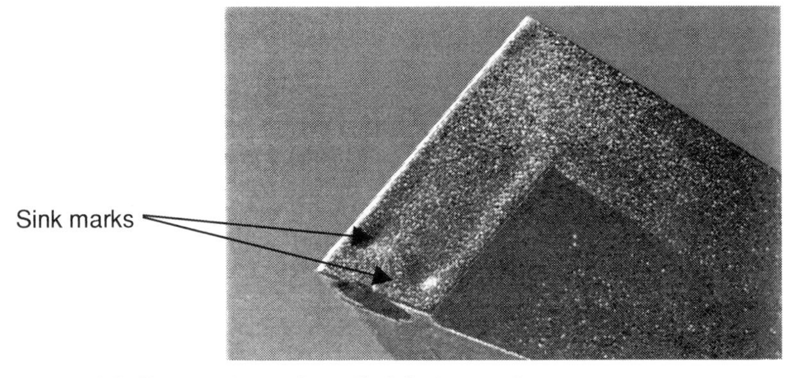

Figure 9.3 Sink marks on the cylindrical core whose temperature was not controlled correctly

Correcting sink marks

Check and/or change machine settings. Change mould or moulding compound. Start new cycle and go through Flow Chart 9.1.

9.2.4 Streaks

Streaks caused by burning, moisture or air can look very similar making classification difficult if not impossible. The signs listed here do not have to appear, they only give reason to suspect a certain type of streak.

Signs for burnt streaks

- The streak appears periodically
- The streak appears behind narrow cross-sections (shear points) or sharp edges in the mould
- The melt temperature is near the upper processing limit
- Lowering the screw advance speed has a positive impact on the defect
- Lowering the melt temperature has a positive impact on the defect
- Long residence time in the plasticising unit or the space in front of the screw (due to, e.g., cycle breaks or low shot volumes)
- High reclaim content, or a part of the material has already been melted several times before
- The mould is equipped with a hot runner
- The mould is equipped with a shut-off nozzle.

Examples of mouldings with burnt streaks are shown in Figure 9.4.

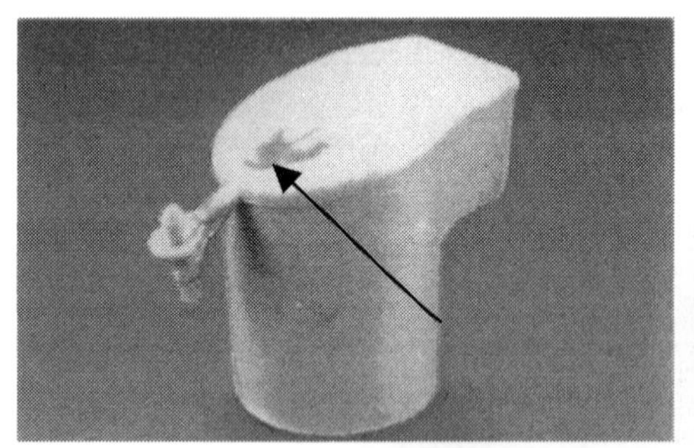

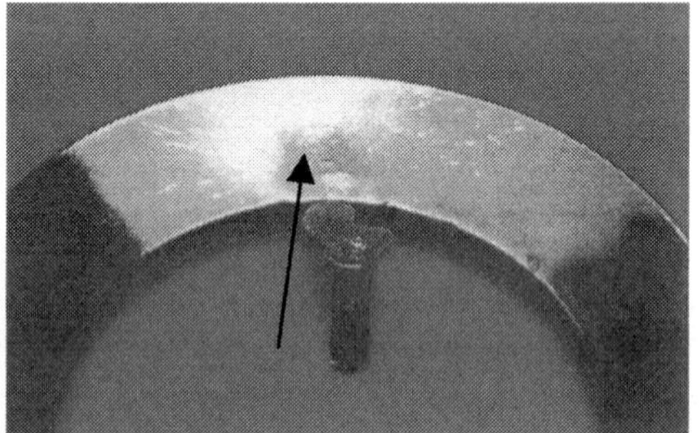

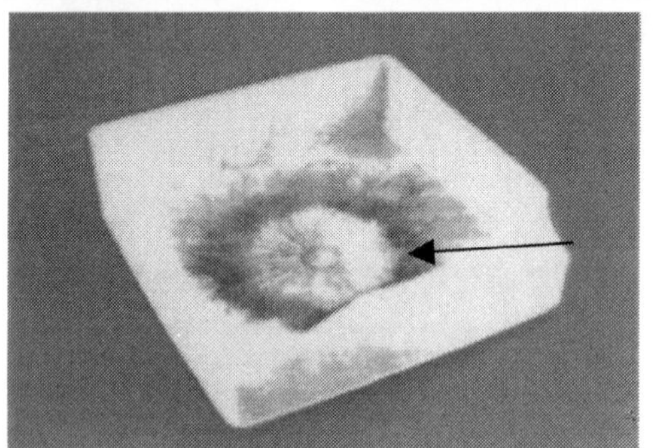

Figure 9.4 Examples of burnt streaks
Top left: Burnt streaks due to excessive residence time in the plasticising cylinder
Top right: Burnt streaks due to high shearing heat in the gate
Bottom left: Burnt streaks due to excessive residence time in the plasticising cylinder

Signs for moisture streaks

- The material tends to absorb moisture (e.g., PA, ABS, CA, PBT, PC, PMMA, SAN)

- When slowly injecting into the air, the melt shows blisters and/or is steaming
- The solidified flow front of a partial filling shows crater-like structures
- The moisture content of the material before the processing is very high.

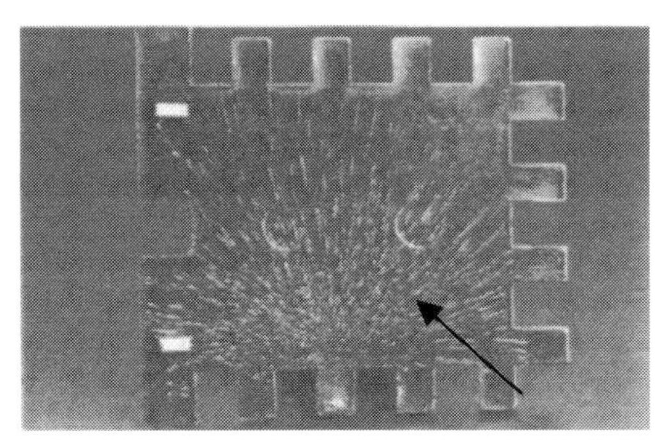
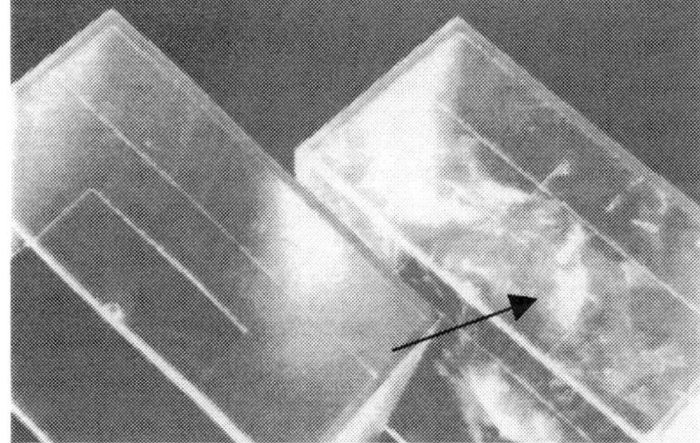

Figure 9.5 Examples of moisture streaks
Left: Streaks due to moist granules
Right: Streaks due to moisture on the mould surface

Signs for air streaks

- The moisture content in the environment is very high (especially in combination with cold moulds and cold granules)
- The defect becomes smaller with lower decompression
- The defect becomes smaller with lower screw advance speed
- Blisters are visible in the injected material
- The solidified front flow of a partial filling shows crater-like structures.

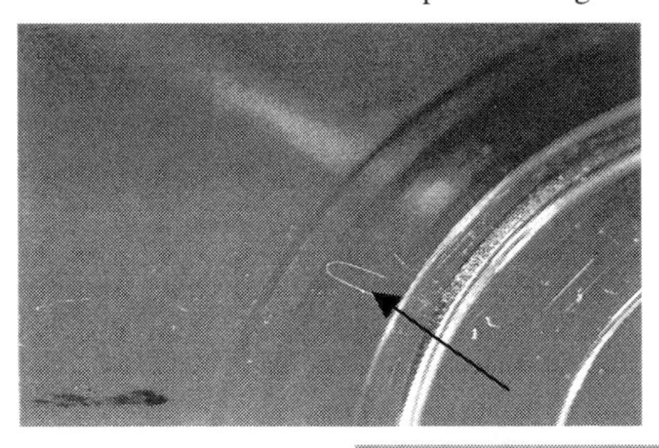
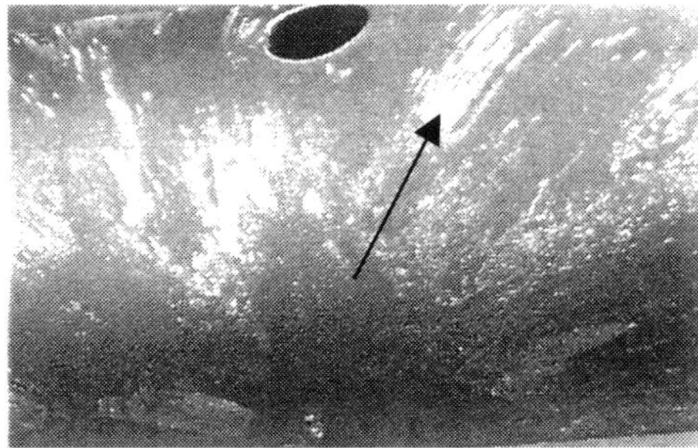
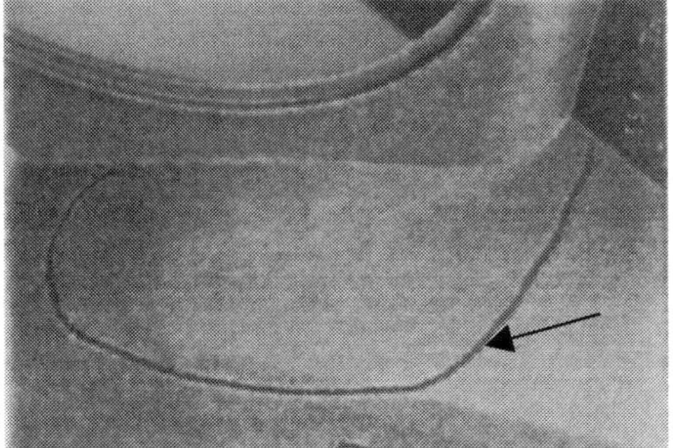

Figure 9.6 Air streaks/air hooks
Top left: Air streak behind a wall thickness variation
Top right: Air streak (near the sprue) due to sucked in air during decompression
Bottom: Air streak due to entrained and stretched air near rib

9.2.4.1 Burnt Streaks (Brown or Silver)

Physical cause

Burnt streaks are caused by thermal damage to the melt. The result can be a decrease of the length of the molecule chain (silvery discoloration) or a change of the macromolecules (brownish discoloration).

Possible causes for thermal damage:

- Temperature too high or residence time too long during predrying
- Melt temperature too high
- Shearing in the plasticising unit too high (e.g., screw speed too high)
- Residence time in the plasticising unit too long
- Shearing in the mould too high (e.g., injection rate too high).

Note: Inject 'into the air' in order to check the melt temperature. Measure temperature with a needle thermometer.

Thermal degradation of the plastic has a negative impact on its mechanical properties, even if no damage is visible on the surface.

Correcting burnt streaks (brown or silver)

Check and/or change machine settings, change mould or moulding compound, start new cycle and go through Flow Chart 9.2 reducing melt temperature.

9.2.4.2 Moisture Streaks

Physical cause

During storage or processing, moisture is absorbed by the granules, forming water vapour in the melt. Due to the velocity profile at the flow front, gas blisters are pushed to the surface of the melt as shown in Figure 9.7. Since they want to compensate the pressure, the blisters burst, are deformed by the moving flow front and freeze at the mould wall.

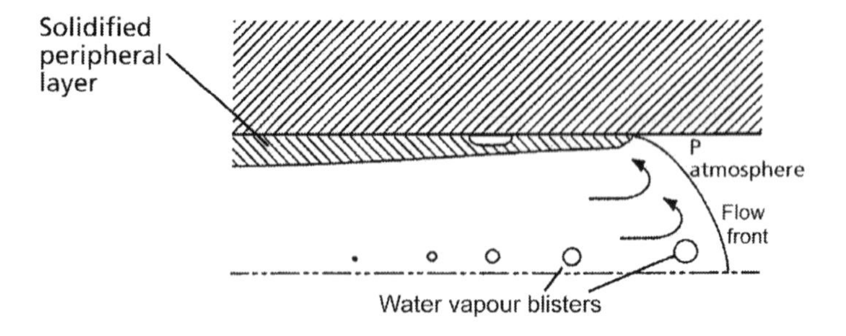

Figure 9.7 Flow of water vapour blisters near the flow front

Possible causes for moisture streaks:

(1) Moisture on the mould surface
   - leaky mould temperature control system
   - condensation water on the mould walls

(2) Moisture in/on the granules
   - insufficient pre-drying of the material
   - wrong storage of the material

Correcting moisture streaks

Check and/or change machine settings, change mould or moulding compound, start new cycle and go through Flow Chart 9.3.

9.2.4.3 Colour Streaks

Physical cause

During pigmentation, pigment agglomerations can lead to differences in the concentration. To some extent this can be mitigated by an increase in shearing as shown in Figure 9.8 and increases in back pressure can be applied during the plastication stage to increase mixing. This kind of poor distribution can be caused by the plastic, the processing parameters, adhesives and other additives. With in-plant colouring using dyes, the defect can occur due to uncompleted solution of the dye particles in the melt.

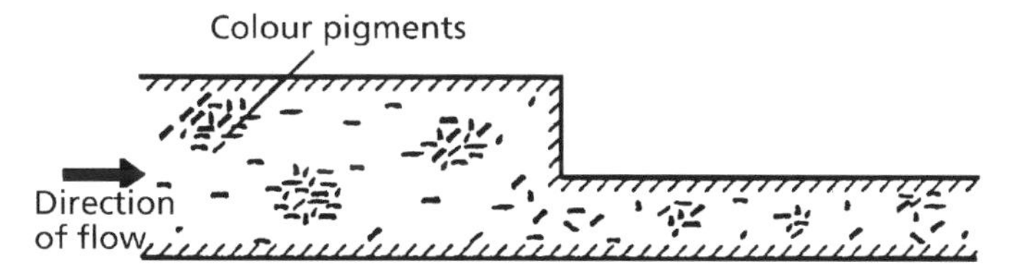

Figure 9.8 Smaller differences in pigment concentration due to higher shearing

Similar to thermoplastics, pigments and dyes are sensitive to excessive processing temperature and residence times. If thermal damage is the reason for colour streaks, they should be considered as burnt streaks.

Extensive stress or warpage can also cause colour differences. The deformed areas break the light in a different manner than other areas.

Note: If using master batches for colouring, make sure the substrate is compatible with the plastic to be coloured. The effect of the use of an incompatible masterbatch is shown in Figure 9.9.

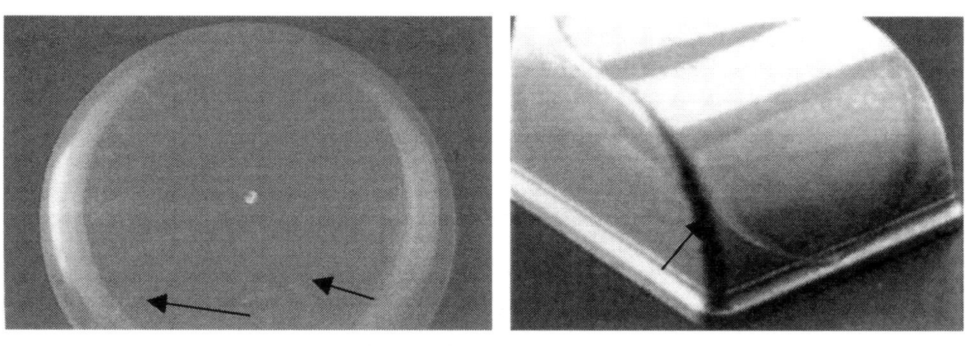

Figure 9.9 Colour streaks
Left: Colour streaks due to an incompatible masterbatch
Right: Orientation of metal-effect pigments caused by the flow

Correcting colour streaks

Check and/or change machine settings, change mould or moulding compound, start new cycle and go through Flow Chart 9.4.

9.2.4.4 Air Streaks/Air Hooks

Physical cause

Air which cannot escape in time during the mould filling, is drawn to the surface and stretched in the direction of the flow. Especially near writings, ribs, domes and depressions, the air can be rolled over and thus entrapped by the melt as shown in Figure 9.10. The result is the formation of air streaks or air hooks.

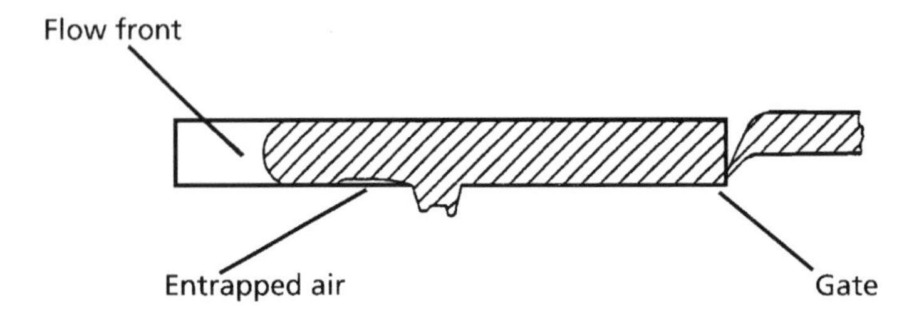

Figure 9.10 Formation of an air streak behind an engraving

If air is sucked into the area in front of the screw during decompression, air streaks will appear near the gate. Here, air is transported into the cavity during the injection, and is then pushed towards the mould wall where it freezes.

Correcting air streaks/air hooks

Check and/or change machine settings, change mould or moulding compound, start new cycle and go through Flow Chart 9.5.

9.2.4.5 Glass Fibre Streaks

Physical cause

Due to their length, glass fibres orientate themselves in the direction of flow during injection. If the melt suddenly freezes when touching the mould wall, the glass fibres may not yet be sufficiently surrounded with melt.

In addition to that, the surface can turn rough because of the big differences in shrinkage (glass fibre: plastic = 1: 200). The glass fibres impede shrinkage of the cooling plastic, especially in the longitudinal direction of the fibre, thus producing an uneven surface as shown in Figure 9.11.

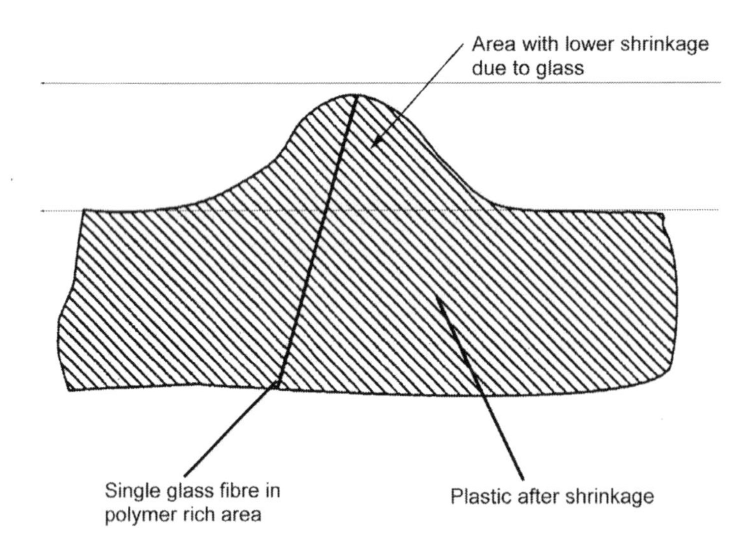

Figure 9.11 Formation of a rough surface due to different shrinkage

Figure 9.12 illustrates the effects that fibre orientation can have on moulded parts.

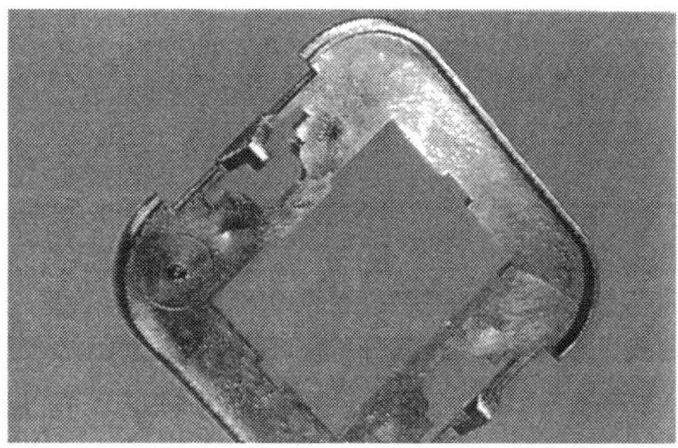

Figure 9.12 Glass fibre streaks
Top left: Glass fibre streaks: clearly visible weld line
Top right: Moulded part with rough silvery surface
Bottom: Glass fibre streak due to orientation near sprue

Correcting glass fibre streaks

Check and/or change machine settings, change mould or moulding compound, start new cycle and go through Flow Chart 9.6.

9.2.5 Gloss/Gloss Differences

Physical cause

The gloss of a moulded part is the appearance of its surface when exposed to light.

If a ray of light hits the surface, its direction will change (refraction of light). While one part of the light will be reflected on the surface, another part will reflect inside the part or penetrate it with different intensities. The impression of gloss is at an optimum, the lower the surface roughness. To achieve this, a polished mould wall should be as good as possible, a textured mould wall would not. This is illustrated in Figure 9.13. Gloss differences are caused by different projection behaviours of the plastic at the mould wall, due to different cooling conditions and shrinkage differences.

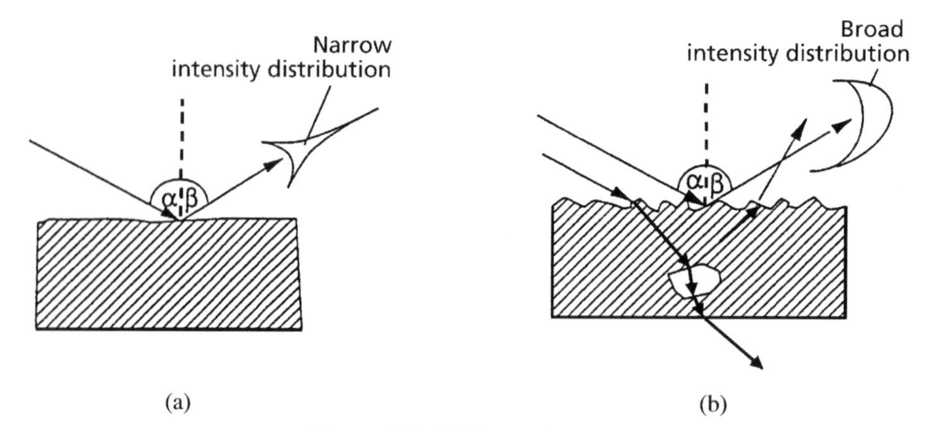

Figure 9.13 Different glosses
(a) Very glossy impression due to reflection on the polished surface
(b) Only slightly glossy impression due to reflection on a rough surface and on filler materials

Stretching of already cooled areas (e.g., due to warpage) can be another reason for gloss differences. Various examples of gloss related defects are shown in Figure 9.14.

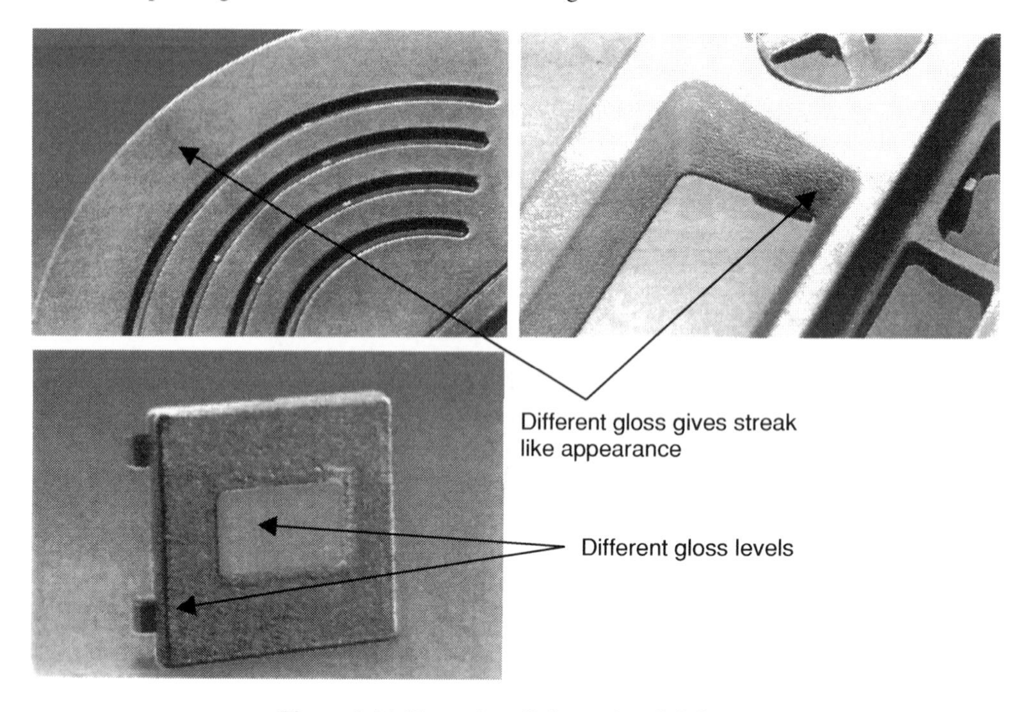

Figure 9.14 Examples of gloss related defects
Top left: Gloss differences near weld lines
Top right: Gloss differences due to wall thickness variations
Bottom left: Glass differences near ribs

Correcting gloss/gloss differences

Check and/or change machine settings, change mould or moulding compound, start new cycle and go through Flow Chart 9.7 and/or 9.8.

9.2.6 Weld Line (Visible Notch or Colour Change)

Physical cause

Weld lines are created when two or more melt flows meet. The rounded flow fronts of the melt streams are flattened and bonded when touching each other. This is shown in Figures 9.15 and 9.16 This process requires stretching of the already highly viscous flow fronts. If temperature and pressure are not high enough, the corners of the flow fronts will not completely develop, creating a notch. Furthermore, the flow fronts no longer melt together homogeneously, possibly producing an optical and mechanical weak spot as shown in Figure 9.17. If moulding compounds containing additives (e.g., colour pigments) are used, strong orientations of these additives near the weld line are possible. This can lead to colour changes near the weld line. Notches are particularly visible on dark or transparent parts with smooth, highly polished surfaces. Colour changes are particularly visible on parts with metallic pigments. Note: significant improvements can only be reached by high mould wall temperatures. Increasing the mould wall temperature increases the cycle time by approximately 2% per °C.

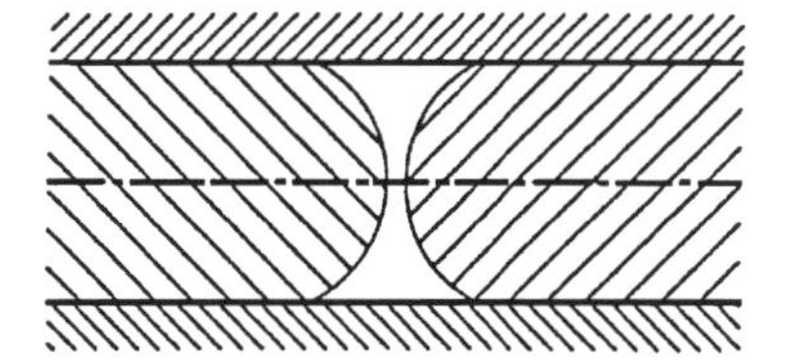

Figure 9.15 Flow fronts before touching each other

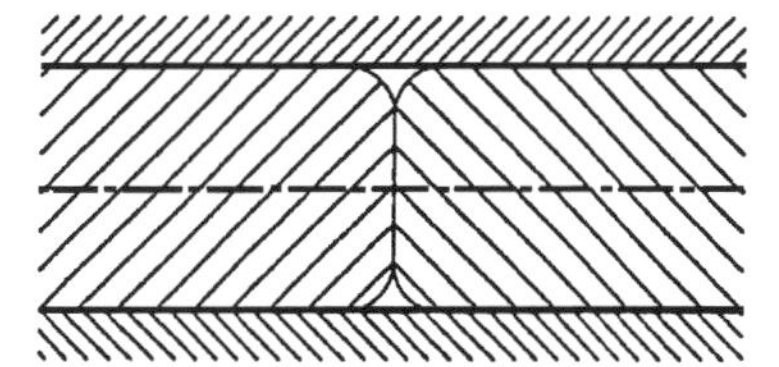

Figure 9.16 Stretching of the rounded flow fronts

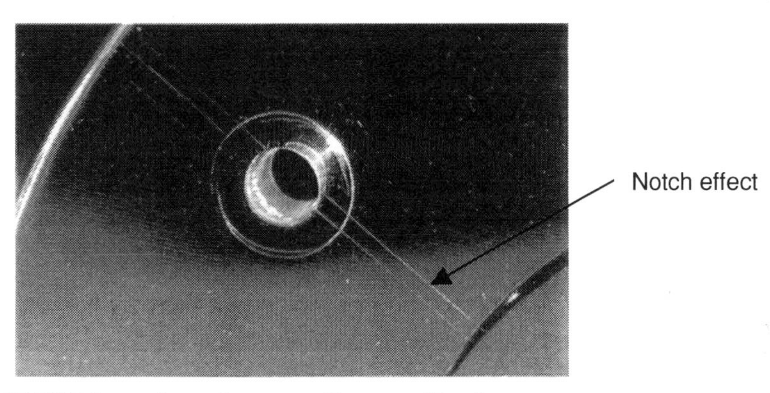

Figure 9.17 Visible notch on the top and bottom side of a transparent part

Improving a weld line (visible notch or colour change)

Check and/or change machine settings, change mould or moulding compound, start new cycle and go through Flow Chart 9.9.

9.2.7 Jetting

Physical cause

Jetting is caused by an undeveloped frontal flow of melt in the cavity. A melt strand is developed which, starting at the gate, enters the cavity with uncontrolled movements. During this phase the melt strand has cooled down to such a degree that it cannot be fused homogeneously with the rest of the moulding compound. This often happens with discontinuously increasing cross-sections of the moulding part in conjunction with high injection speeds. Jetting is illustrated in Figures 9.18 and 9.19. Often jetting causes differences in colour and gloss. In some cases there are similarities to the record grooves effect. Note: jetting can also be influenced by the position of the mould. In order to avoid defects, the cavity should not be filled from top to bottom.

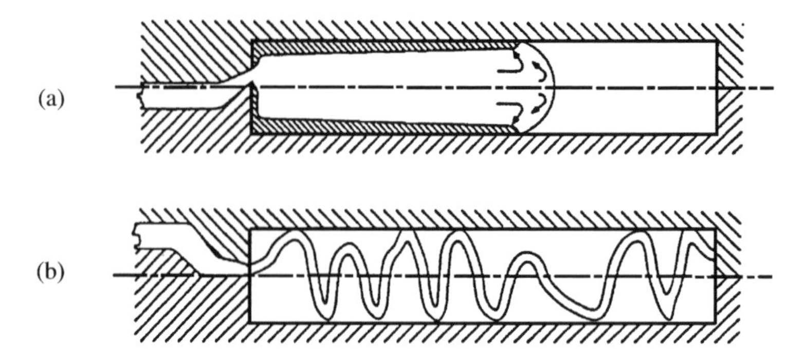

Figure 9.18 Mould filling (a) frontal flow and (b) jetting

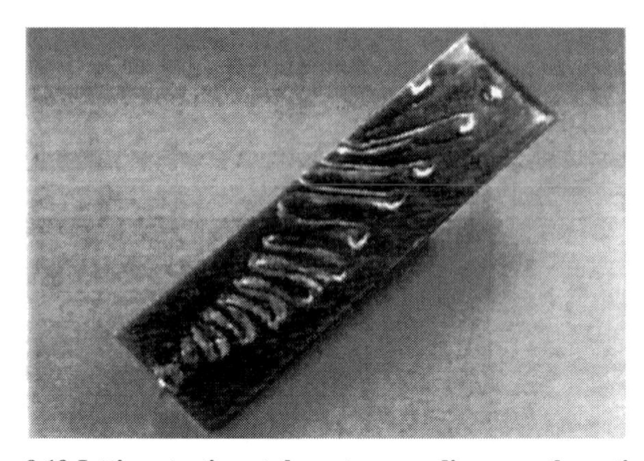

Figure 9.19 Jetting starting at the gate, spreading over the entire part

Correcting jetting

Check and/or change machine settings, change mould or moulding compound, start new cycle and go through Flow Chart 9.10.

9.2.8 Diesel Effect (Burns)

Physical cause

The diesel effect is a pure venting problem. It can occur near blind holes, fillets, the end of flow paths and near points where several flow fronts fuse. It happens whenever the air cannot escape or not quick enough via commissures, venting channels or ejector fits. Towards the end of the injection process, the air is compressed and thus heated to a high degree. The result is very high temperatures which can cause burn marks on the plastic as shown in Figures 9.20 and 9.21.

Note: Due to the burning of the plastic, aggressive decomposition products may be created, which often attack the mould surface, leading to destruction.

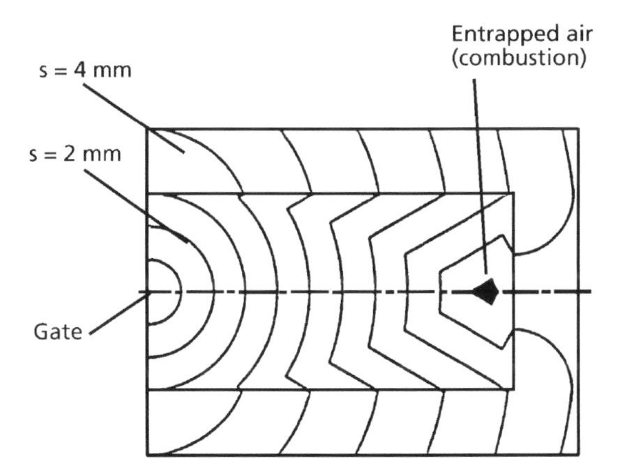

Figure 9.20 Sheet with diesel effect (filling pattern)
(s indicates the two different flow fronts and their distance from the sprue)

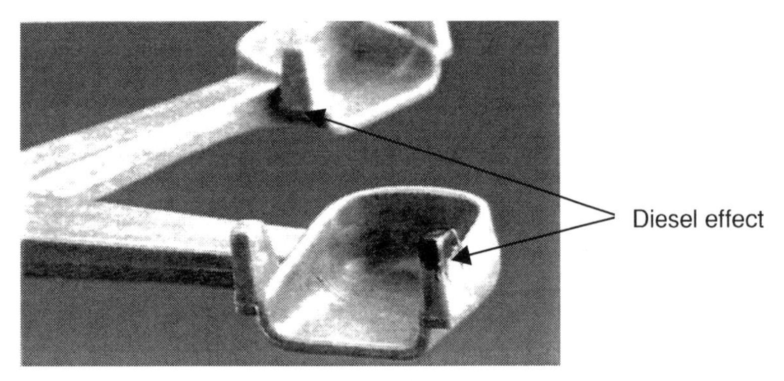

Figure 9.21 Diesel effect (burns) due to merging of several flow fronts with entrapped air at the end of the flow path

Correcting diesel effect (burns)

Check and/or change machine settings. change mould or moulding compound, start new cycle and go through Flow Chart 9.11.

9.2.9 Record Grooves Effect

In this effect very fine grooves show up on the moulded part, which are very similar to those of records. Concentric rings appear near pin-point gates, while markings are parallel towards the end of the flowpath and/or behind the gate. This is shown in Figure 9.22.

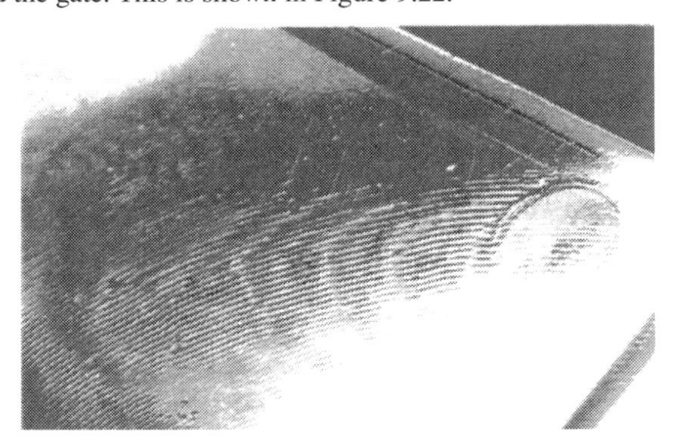

Figure 9.22 Concentric record grooves

Physical cause

- High cooling velocity
- Melt temperature too low
- Injection speed too low
- Mould wall temperature too low.

When injecting the moulding compound into a cold mould, a solidified peripheral layer will be formed behind the flow front due to the high cooling rate. The cooling of the peripheral layer also causes cooling of flow front areas near the mould wall. If this cooling is very high (especially with low injection speeds) these very high viscosity or frozen flow front areas can impede the direct frontal flow of the melt to the wall. Thus the following hot melt will not be pushed towards the wall as usual, but it will cause an elongation of the flow front in the middle. From a certain pressure the flow front will again touch the wall. The cooled down peripheral areas of the flow front have no contact with the wall (see Figures 9.23-9.25).

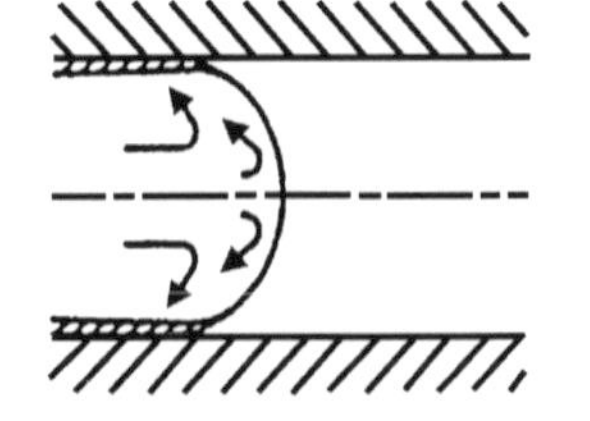

Figure 9.23 Flow front has cooled down near the wall

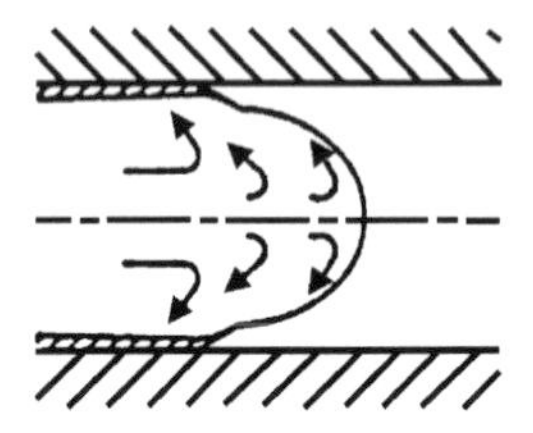

Figure 9.24 Cooled down peripheral layer impedes direct frontal flow to the wall

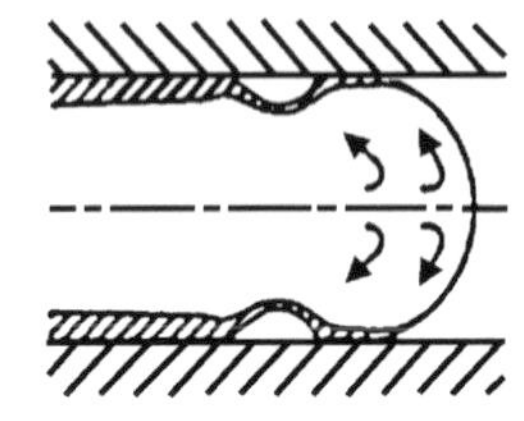

Figure 9.25 Flow front touches mould wall again

Correcting record grooves effect

Check and/or change machine settings, change mould or moulding compound, start new cycle and go through Flow Chart 9.12.

9.2.10 Stress Whitening/Stress Cracks

Physical cause

Stress whitening or stress cracks occur when exceeding the maximum deformation (e.g., due to external stress or warpage). The maximum deformation depends on the type of material used, the molecular structure, the processing and the surrounding climate of the moulded part.

The strength against external and internal stresses can be drastically reduced through physical processes depending on time and temperature. In this case, the linkage forces of the molecules are reduced through wetting, diffusion and swelling processes. This may especially favour stress cracks. Besides internal cooling stresses and stresses due to flow, internal stresses due to expansion are another main reason for internal stresses. External expansion stress is created by demoulding under residual pressure, when the moulded part suddenly shifts from residual pressure to atmospheric pressure. Thus the inner layers of the moulded part put stress on the outer layers. The main reasons for demoulding under residual pressure are insufficiently dimensioned moulds and/or high cavity pressures. The formation of stress is shown in Figure 9.26 and the physical manifestations on the moulding in Figure 9.27. Note: if aggressive substances are used (e.g., alkali solutions, grease, etc.) stress whitening and stress cracks often appear after a very long time of operation.

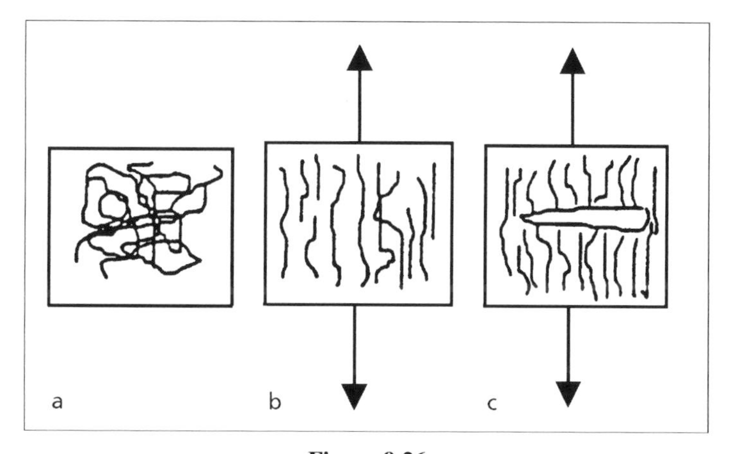

Figure 9.26
(a) Unstressed, felted molecule structure; (b) Orientation of molecules due to force;
(c) Destroyed molecules due to additional force

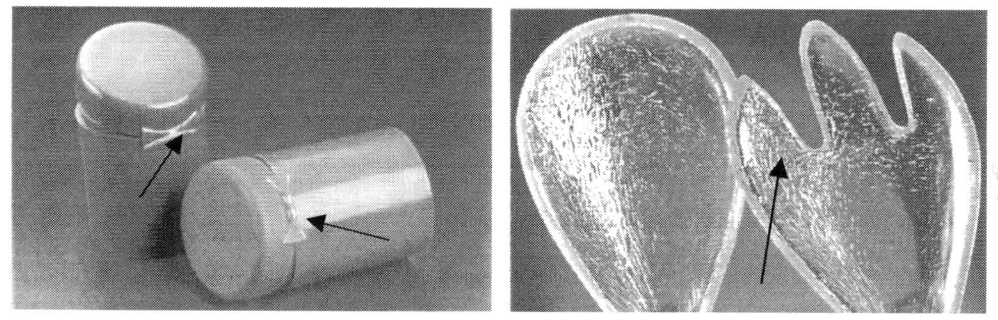

Figure 9.27 Stress whitening
Left: Stress whitening on an integral hinge
Right: Stress cracks on salad servers (damage visible several weeks after purchase)

Correcting stress whitening/stress cracks

Check and/or change machine settings, change mould or moulding compound, start new cycle and go through Flow Chart 9.13.

9.2.11 Incompletely Filled Parts

Injection moulded parts with incompletely developed outer profiles are called incompletely filled parts (or short shots). An example is shown in Figure 9.28. This kind of defect often appears far from the gate if there are long flow distances, or on thin walls (e.g., ribs as shown in Figure 9.29). Due to insufficient mould venting, this defect can also occur in other areas.

Figure 9.28 Housing with incompletely demoulded lattice

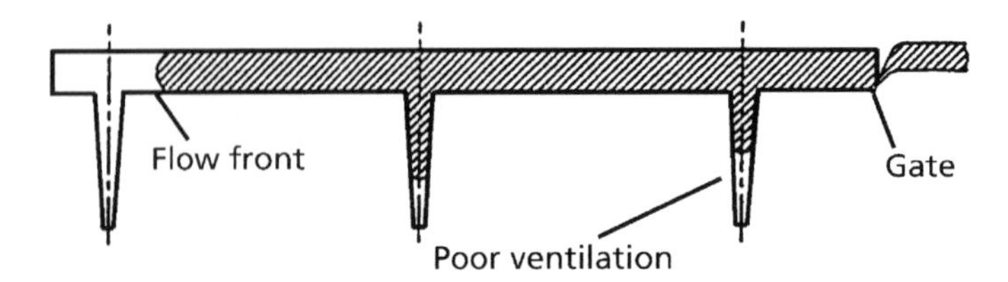

Figure 9.29 Filling problems near thin ribs

Physical cause

There are several physical causes for incomplete filling:

- Injected compound volume too small (e.g., shot volume too small)
- Melt flow impeded due to venting problems
- Injection pressure not sufficient
- Premature freezing of a channel cross-section (e.g., low injection speed or wrong temperature control in the mould).

Note: Incomplete filling due to venting problems does not necessarily cause the diesel effect (Section 9.2.8). Therefore the cause for the defect is often hard to determine.

Correcting incompletely filled parts

Check and/or change machine settings, change mould or moulding compound, start new cycle and go through Flow Chart 9.14.

9.2.12 Oversprayed Parts (Flashes)

Flashes are often created near sealing faces, venting channels or ejectors. They look like a more or less developed film-like plastic edge. Fine flashes are not often immediately visible. Large area thick flashes on the other hand sometimes stick out several centimetres over the nominal profile as shown in Figure 9.30.

Figure 9.30 Large area overspraying (flash)

Physical cause

The different possibilities can be divided into four main groups:

- Allowed gap widths exceeded (mould tightness insufficient, production tolerances too large or damaged sealing faces)
- Clamping force of the machine insufficient or set too low (mould opening force higher than clamping force, mould cannot be kept closed; clamping force deforms platens and mould)
- Internal mould pressures too high (shaping pressure at the gap is so high that the melt is pushed even into very small gaps)

- Viscosity of moulding compound too low (high internal mould pressures and low flow resistance favour flash formation).

Note: Flash formation can very quickly (few cycles) and damage the sealing faces (parting surface).

Correcting oversprayed parts (flashes)

Check and/or change machine settings, change mould or moulding compound, start new cycle and go through Flow Chart 9.15.

9.2.13 Visible Ejector Marks

Ejector marks are depressions or elevations on the ejector side of the moulded part surface. These wall thickness variations can cause gloss differences and depressions on the visible surface of the moulding as shown in Figure 9.31.

Figure 9.31 Gloss differences near the ejector

Physical cause

The different possibilities can be divided into four main groups:

- Process-related causes (e.g., premature demoulding or high demoulding forces due to unfavourable machine settings)
- Geometric causes (e.g., wrong fitting or wrong ejector length)
- Mechanical or strength-related (e.g., faulty dimensioning and design of the mould, the moulded part or the demoulding system)
- Thermal causes (high temperature differences between ejector and mould wall, e.g., Figure 9.32).

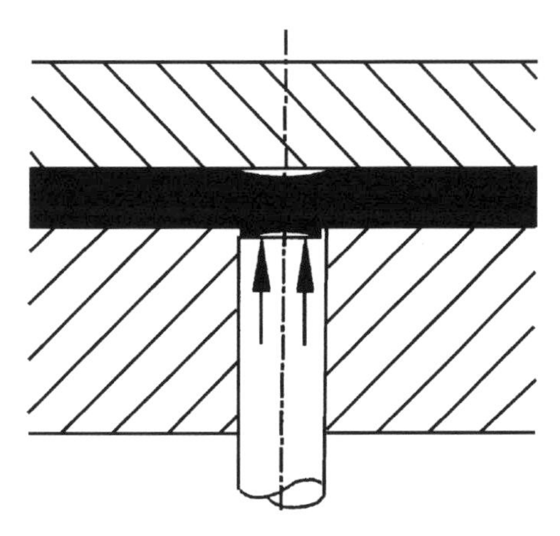

Figure 9.32 Shrinkage near an overheated and poorly fitted ejector

Correcting visible ejector marks

Check and/or change machine settings, change mould or moulding compound, start new cycle and go through Flow Chart 9.16.

9.2.14 Deformation During Demoulding

Depending on the degree of damage there is a classification into extraction marking, cracks, fractures, overstretched areas and deeply depressed ejectors. Critical are moulded parts with undercuts, which are demoulded without movable parts (e.g., slides). Examples of two defective mouldings are shown in Figure 9.33.

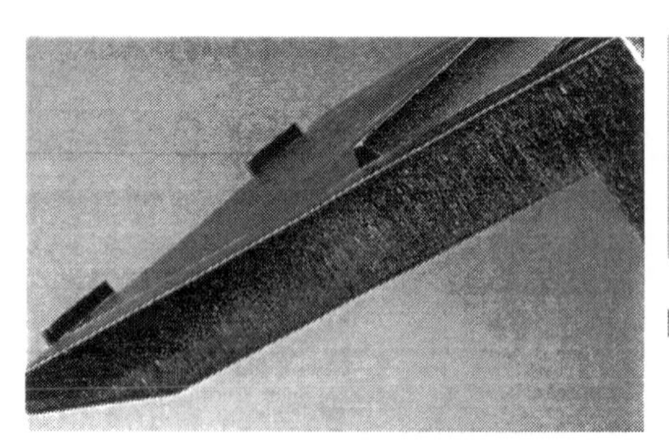

Figure 9.33 Deformation during demoulding
Left: Demoulding grooves on a textured surface
Right: Defomation due to forced demoulding at an undercut

Physical cause

The causes of deformations can be classified as follows:

- The forces necessary for demoulding cannot be applied to the moulded part without damaging it
- The demoulding movement is disturbed.

The amount of demoulding force applied is a crucial criterion and should thus be kept small. Beside other factors the shrinkage of the moulded part has a direct impact on the demoulding forces. Shrinkage and demoulding force can be influenced considerably by varying the process parameters. It is to be considered though that the geometry of the moulded part is a very important influencing factor.

In general, low shrinkage is desirable for sleeve and box-shaped parts, since these parts tend to shrink onto the core (==> increase holding pressure or reduce cooling time).

Near ribs, the shrinkage retroacts on the demoulding force, because the ribs are being detached from the mould walls (==> decrease holding pressure or increase cooling time).

Correcting deformation during demoulding

Check and/or change machine settings, change mould or moulding compound, start new cycle and go through Flow Chart 9.17.

9.2.15 Flaking of the Surface Layer

The layers of material are not homogeneously joined together and start flaking. This can occur at the gate or on the moulded part, and can be either large or very small and thin, depending on the intensity. Examples of both are shown in Figure 9.34.

Figure 9.34 Flaking

Physical cause

Flaking of surface layers is due to insufficient bonding of adjacent surface layers. The different layers are formed by different flow effects and cooling conditions over the cross-section. Shear stresses and inhomogeneities can reduce the bonding of these layers to such a degree that single surface layers start flaking off as shown in Figure 9.35.

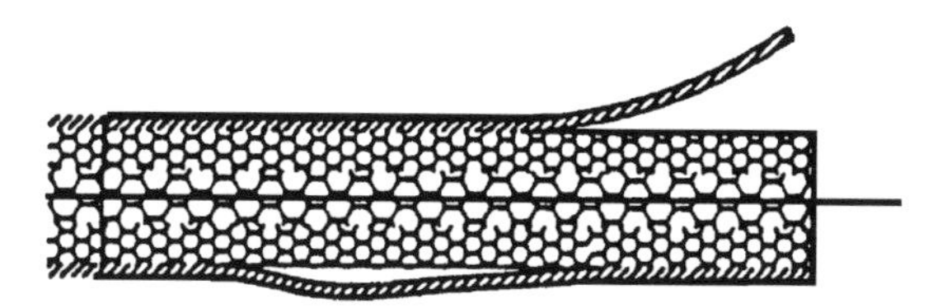

Figure 9.35 Flaking on a cross-section of a moulded part with different structure formation

High shear stresses and thermal damage can be caused by:

- High injection speeds
- High melt temperatures.

Inhomogeneities can be caused by:

- Impurities or other materials among the granules
- Incompatible dye or master batch
- Moisture in/on the granules
- Poorly melted moulding compound.

Correcting flaking of the surface layer

Check and/or change machine settings, change mould or moulding compound, start new cycle and go through Flow Chart 9.18.

9.2.16 Cold Slugs/Cold Flow Lines

Physical cause

Cold slugs are formed when melt solidifies in the gate or in the nozzle, before the compound is injected, and is transported into the mould with the following shot sequence. If the cold slugs do not melt again, they will cause markings which look like comet tails. They can be spread all over the moulded part. The cold slug can also jam a runner, forcing the melt to part. The results are surface defects similar to weld lines. An example is shown in Figure 9.36. Cold slugs are often caused by a wrong nozzle temperature or belated retraction of the plasticising unit. Small nozzle diameters can also have a negative effect. An illustration of the mechanism is shown in Figure 9.37.

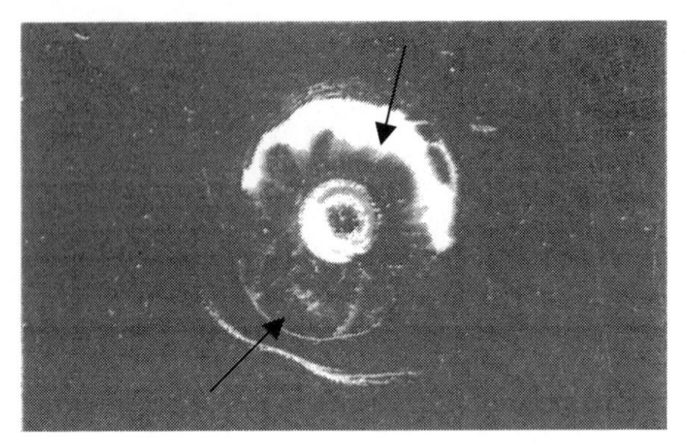

Figure 9.36 Markings caused by cold slug near the sprue

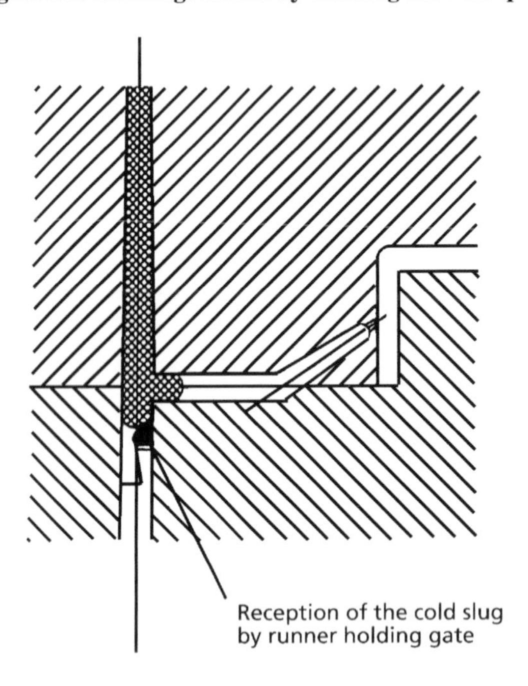

Figure 9.37 Cold slug is transported into the mould by the flow front

Correcting cold slug/cold flow lines

Check and/or change machine settings, change mould or moulding compound, start new cycle and go through Flow Chart 9.19.

9.2.17 Entrapped Air (Blister Formation)

Physical cause

During the injection, air is entrapped in the melt and appears as a hollow (air blister) on the moulded part. Primarily there are two factors responsible for this defect:

- Decompression too high or too fast
- Plasticising performance too low.

Note: There are two types of hollows, entrapped air and voids. Voids are vacuole hollows, formed by the shrinkage of the moulding compound (see 'sink marks'). Distinguishing between the two is very hard, because of their similar appearance. The following hints might be helpful:

- When opening the hollow in a fluid, a void (vacuum) shows no gaseous bubbles

- Entrapped air defects can be reduced by using no decompression
- Changing the holding pressure or the holding pressure time has no effect on the size of the hollows.

Moulded parts with hollows are usually not as strong as parts without. Non-transparent parts should be randomly tested by randomly opening them, if suspicious. Examples of trapped air are shown in Figure 9.38(a) and (b).

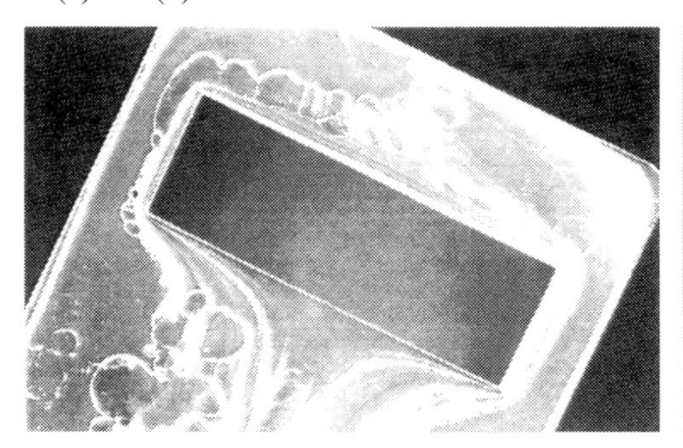

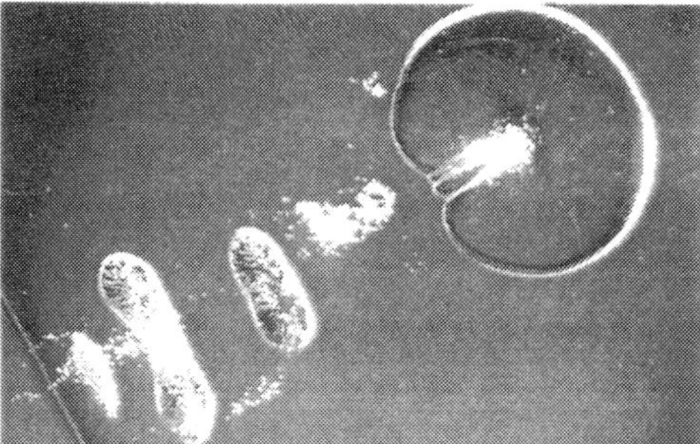

Figure 9.38 Entrapped air

Correcting entrapped air (blister formation)

Check and/or change machine settings, change mould or moulding compound, start new cycle and go through Flow Chart 9.20.

9.2.18 Dark Spots

Physical cause

Black or dark spots appear on the surface due to wear, thermal damage or dirt.

Different factors can cause the formation of dark spots or speckled parts:

- Process-related causes, e.g., melt temperature too high or residence time in the plasticising unit too long. Wrong temperature profile in the hot runner system.
- Mould-related causes, e.g., dirty gate system or wear (dead edges) in the hot runner system.
- Machine-related causes, e.g., dirty plasticising unit or worn screw and cylinder.
- Caused by polymer or dyeing, e.g., impurities in the granule, high reclaim content or unsuitable dye/masterbatch.

Figure 9.39 shows dark spots resulting from thermal damage to the polymer.

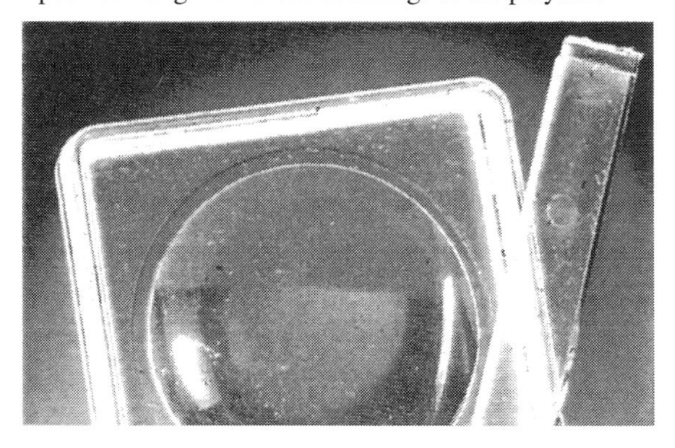

Figure 9.39 Dark spots due to thermal damage

Correcting dark spots

Check and/or change machine settings, change mould or moulding compound, start new cycle and go through Flow Chart 9.21.

9.2.19 Dull Spots Near the Sprue

Physical cause

Dull spots near the sprue are mainly caused by:

- Small gates
- High injection speeds.

Due to high injection speeds, small gate cross-sections and bypasses behind the gate, extremely strong orientations of the molecule chains are formed during injection. There is not enough relaxation time directly behind the gate, so the peripheral layers of the melt are frozen while still strongly oriented. Such peripheral layers can only be stretched to a minimum degree and crack under the impact of the high shear stresses.

The hot melt inside flows to the mould wall and forms very small notches as shown in Figure 9.40. The dull appearance is caused by the widespread reflection in this area.

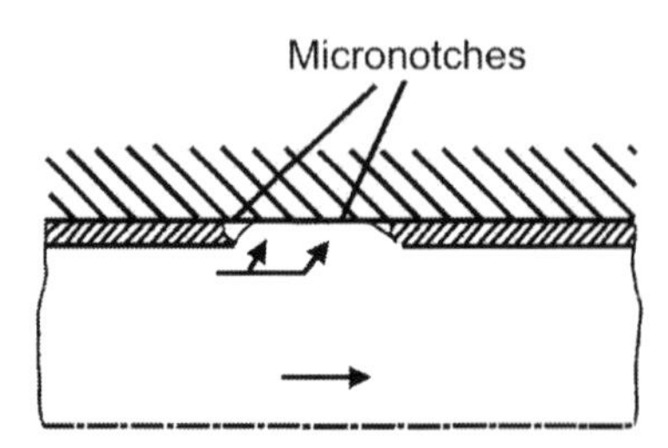

Figure 9.40 Melt flows into the cracked peripheral layer (formation of micronotches)

Correcting dull spots near the sprue

Check and/or change machine settings, change mould or moulding compound, start new cycle and go through Flow Chart 9.22.

9.3 Data Acquisition Record

Should surface defects appear during the proving of the mould, the person setting up the machine will try to eliminate or at least minimise the defect by changing the processing parameters. Recording the proving and optimisation process helps to understand the action taken and to evaluate the effects on the quality of the moulded part. Furthermore, the record can be used as a basis for a discussion, because it also helps other people to understand the process quickly.

9.3.1 Using the Data Acquisition Record

Recording the moulding process should be started as soon as a defect comes up. In the first part of the record sheet, general data about moulded part, material, machine and mould can be entered, if necessary.

When changing processing parameters, the producer can use the flow charts in earlier sections of this chapter. Note that only one parameter is to be changed per cycle. Defects can be evaluated using the recommended key for each trial. It proved to be useful to mark the parts produced during the trials for later reconstruction of the optimisation. This record sheet only contains a selection of the large amount of processing parameters.

9.3.2 Data Acquisition Record for Optimising Moulded Parts

9.3.2.1 Moulded Part Data

Name:	______________________
Moulded part number:	______________________
Special remarks:	______________________
Compound data:	______________________
Trade name:	______________________
Type:	______________________
Batch number:	______________________
Type of colouring:	______________________
Type of dying:	______________________
Special remarks:	______________________
Machine data:	______________________
Type of machine:	______________________
Machine number:	______________________
Special remarks:	______________________
Mould data:	______________________
Mould number:	______________________
Number of cavities:	______________________
Special remarks:	______________________
	______________________

9.3.2.2 Machine Settings and Defect Evaluation

Temperatures (°C):	**Basic settings**	**Test 1**	**Test 2**	**Test 3**	**Test 4**	**Test 5**	**Test 6**	**Test 7**
Cylinder temp. (feed zone):								
Cylinder temp. (middle zone):								
Cylinder temp. (front zone):								
Melt temp. (injected):								
Mould temp. (nozzle):								
Mould temp. (core):								
Mould temp. (defect):								
Hot runner:								
Back pressure in ____:								
Metering stroke in ____:								
Melt cushion in ____:								

Injection profile:	**Basic settings**	**Test 1**	**Test 2**	**Test 3**	**Test 4**	**Test 5**	**Test 6**	**Test 7**
Injection speed in 1 ___:								
Change-over point 1 in ___:								
Injection speed 2 in ___:								
Change-over point 2 in ___:								
Injection speed 3 in ___:								
Change-over point 3 in ___:								
Injection pressure (setting):								
Injection pressure (actual value):								
Holding pressure:								
Holding pressure 1 in ___:								
Holding pressure 2 in ___:								
Holding pressure 3 in ___:								
Holding pressure time:								
Actual values:								
Cooling time:								
Cycle time:								
Defect evaluation:								
Type of defect:								
Evaluation #1:								
(1) moulded part OK (2) defect better, (3) defect unchanged, (4) defect worse, (5) new defect created								

9.4 Case Studies of Injection Moulded Components

The following examples highlight potential errors that can be made in injection moulded components as well as solutions to overcoming them.

9.4.1 Threaded Connecting Sleeves for Ink Drafting Apparatus

With this thin-cavity three-plate mould arrangement, series injection took place in the parting plane. The filling process and the pressure ratios in the two cavities were thus different. The effect of this was that the component did not have a good surface. Moreover, very narrow tolerances could not be maintained. These effects could be overcome by correcting the flow path lengths in the runner. This is shown in Figure 9.41. Identical flow path lengths on multiple equipment in the mould are the basic prerequisite for maintaining narrow tolerances.

This basic principle is generally valid in injection moulding technology.

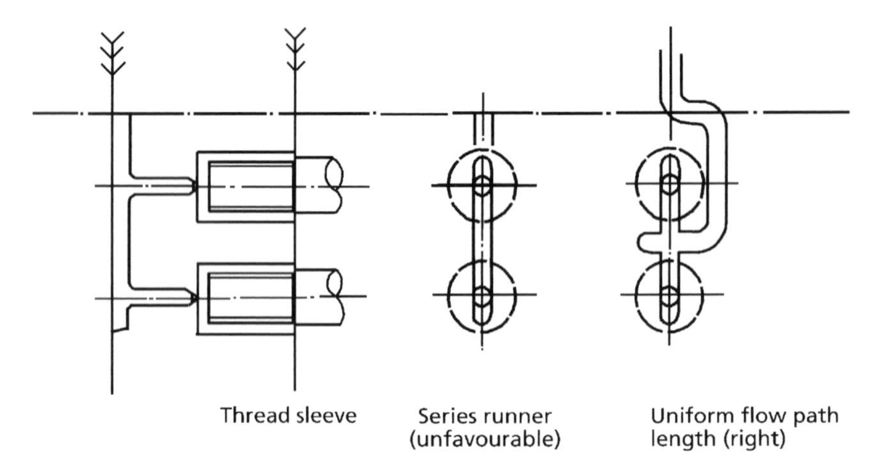

Figure 9.41 Runners for ink drafting apparatus

9.4.2 Meter Cases

On this meter case, there are dovetail guides on the four side faces. The varying wall thickness caused the meeting of flowpaths resulting in an air blister in the vicinity of the gate. This is shown in Figure 9.42. The following measures were adopted to try and remove this air blister:

1. Reduction of initial injection speed; however, a hole now appeared at a new point, where the compound was no longer coalescing.
2. Raising the compound temperature; initially this was a success, in that, by raising the compound temperature step by step, it was possible to reduce the initial injection speed as much as possible.

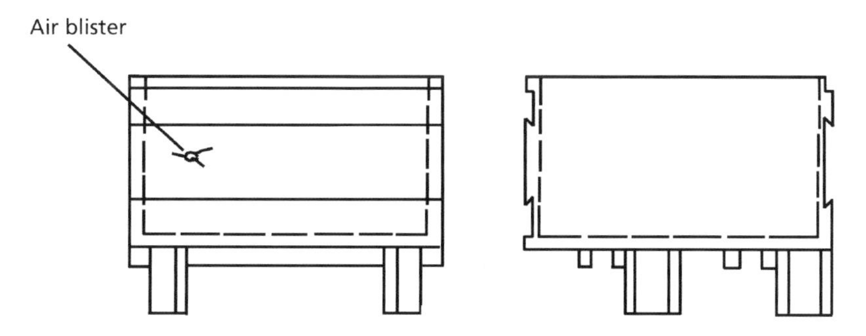

Figure 9.42 Meter case

However, after a certain amount of time warping occurred, which can clearly be traced back to very high internal orientation stresses resulting from excessively slow filling.

In order to remove the air blister even at faster initial injection speeds, the mould clamping force, which had previously been set very high (too high) was reduced until satisfactory running was achieved again. This example shows that a mould clamping force which is not too high, and which is satisfactorily set, guarantees better air venting.

Moreover, the equipment is looked after better: the lower the mould clamping force, the lower the wear.

9.4.3 Wristwatch Glass

A glass for a wristwatch displayed convergence points opposite the runner, i.e., very visible joint lines. It was possible to establish clearly, after a filling sequence with this mould, that no uniform flow front was formed.

When the sample components as shown in Figure 9.43 were measured later, there were thickness variations of 0.1-0.15 mm. The variations in wall thickness were overcome by supplementary work on the core.

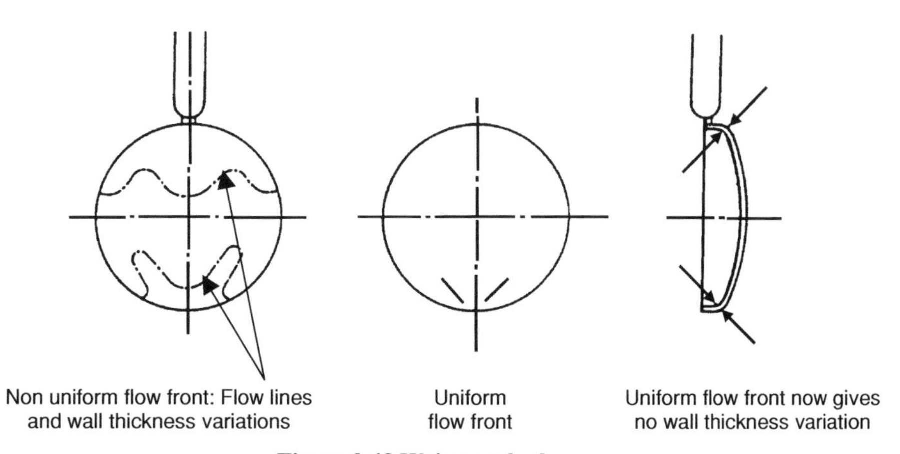

Figure 9.43 Wristwatch glass

Now a uniform flow front could be generated, and the faults caused by the joint lines are a thing of the past.

After the fault referred to above had been eliminated, there was still some dissatisfaction with the surface lustre of the watch glass, although the mould inserts were high-gloss polished. The required surface gloss could not be obtained until new mould inserts had been manufactured from non-porous vacuum steel (in accordance with the vacuum arc refining process).

9.4.4 Alarm Clock Glass

In the manufacture of an alarm clock glass, the following problem arose during sampling. A uniform flow front was not being formed, due to the edges being some tenths of a millimetre thicker, and an air blister was formed on the side of the component opposite the runner. Since there were electroplated inserts in the core and the mould insert, it was not possible to correct the wall thickness by secondary work. The electroplated inserts would have had to be manufactured again.

It was decided to remove the air blister by a simpler method as shown in Figure 9.44. A gas vent was ground, 8.0 mm wide and 0.3 mm thick. The convergence point, with the air, could now be forced out, and it was possible to manufacture good components.

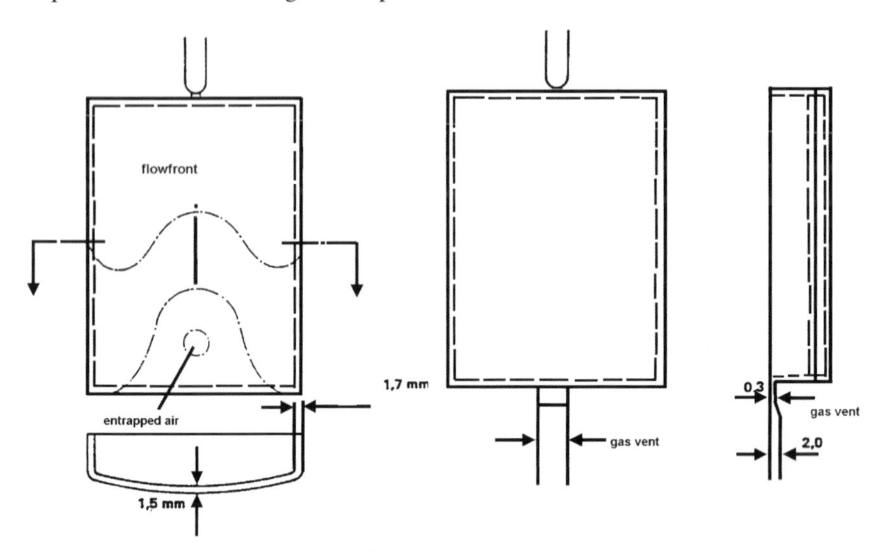

Figure 9.44 Alarm clock glass

9.4.5 Glass Cover for Digital Gauge

This component displayed slight sink marks on the face, above the finning, especially on the side away from the gate. Injection was carried out with a film (fan) gate into the parting plane. However the film runner was produced in such a way that the cross-runner to part space was too large. In addition, the runner cross-section was too small.

By moving the cross-runner closer to the component (shorter gate) and by better formation of the accumulation base, it was arranged that the gate stayed open longer. Before and after designs are shown in Figure 9.45.

This not only made it possible to avoid the sink marks, but it was also no longer necessary to select such high melt temperatures and initial injection pressures. This also improved the quality of the component.

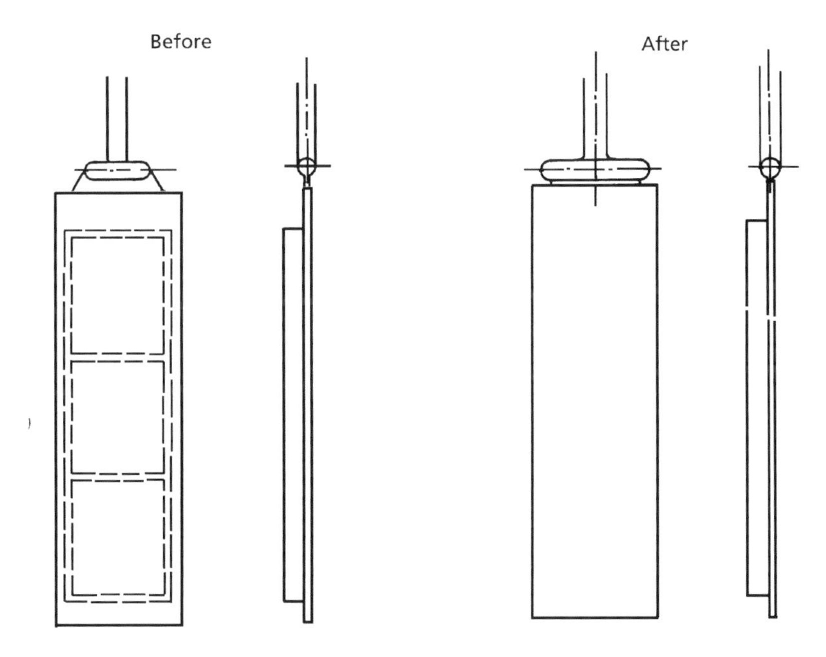

Figure 9.45 Glass cover for digital gauge
(Dashed lines indicate regions where sink marks were occurring)

9.4.6 Plug Boards with Insert Pins

On this component, despite all the technical skills that injection moulding technology can offer, voids were forming in the centre of the moulding. The reason why the voids had formed was because the components had been directly connected to the cross-runner. Since a gate point of this kind very quickly freezes, the holding pressure was not able to act for long enough.

It is generally true in injection moulding technology that gates should start in the centre of the runner or the runner system, so that a plastic flow can form properly from the gate outwards. The design is shown in Figure 9.46.

Direct connections from the runner are very unfavourable.

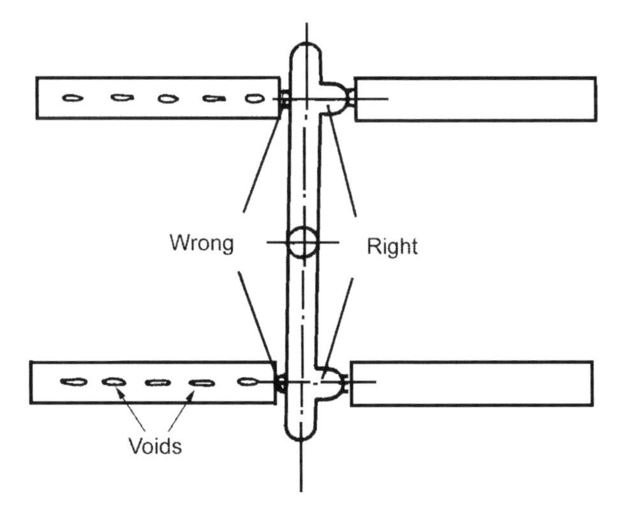

Figure 9.46 Plug boards with insert pins

9.4.7 Hair Slides

It was not possible to manufacture the components at the point away from the gate in this sextuple hair slide mould without sink marks. The root cause lay in the pencil-type transfer from the runner to the gate, where, owing to the lack of an accumulation base, highly viscous material was carried along from the marginal area, thus leading to an over-early closing (sealing) of the gate. The sink marks then occurred, due to the lack of holding pressure. However, the sink marks had to be avoided at all costs, as the component was subsequently heat-stamped with film and the film did not make contact at the sink marks. Due to the subsequent incorporation of a well-formed accumulation base at the gate, it was possible to manufacture the components without sink marks.

For pin-point gates, the transfer from the runner should always take the form of an accumulation base. The accumulation base prevents the highly viscous compound at the edge from being carried along, the plastic core of the runner is thus retained for longer (the gate does not freeze until later).

In spite of the gate being the same size, the holding pressure is now effective for longer, and the risk that sink marks will form is reduced. Both gating types are illustrated in Figure 9.47.

On three-plate moulds too, pencil-form connections should be avoided for this reason.

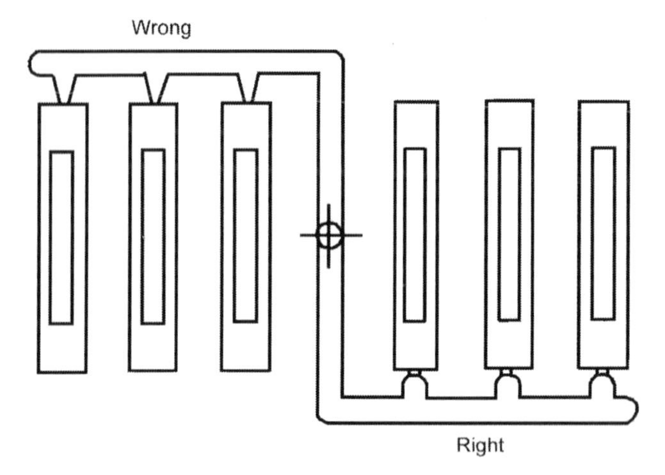

Figure 9.47 Gating in hair slide tool

9.4.8 Toothbrush Components

These components were injected into the parting plane. A so-called free jet arose through the massive wall thickness because of the injection into the free space. This free jet cools relatively quickly on its top face, and thus can no longer be adequately attached to the compound which follows. As filling goes on, it is compressed together like a snake (sausage injection moulding) and leaves creases behind on the top face of the component.

It was possible to prevent a free jet here by fitting a sprung baffle strut, which ensured a flow head of the compound moving away from the gate. When the filling phase ends, the sprung baffle strut is retracted, and leaves only a mark behind on the component. The design is illustrated in Figure 9.48.

Guide values for setting a sprung baffle strut:

- Minimum diameter: 4 mm
- Distance from gate: 2.5 mm
- Insertion depth into component: approx. 2/3 of wall thickness.

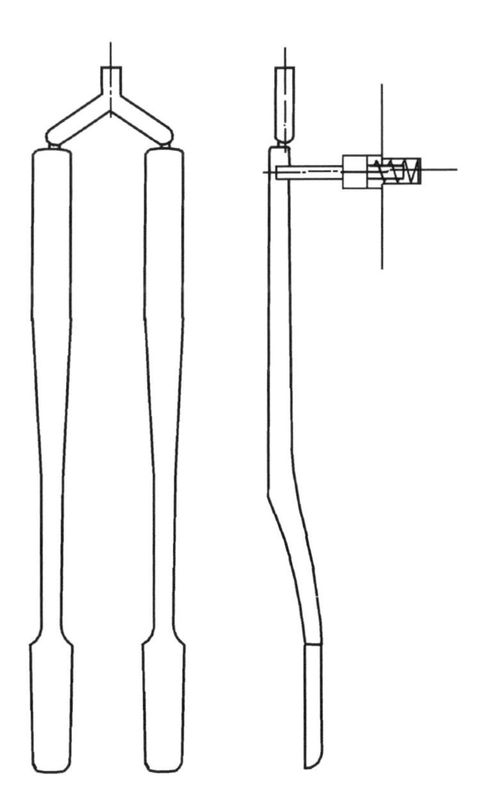

Figure 9.48 Toothbrush component

9.4.9 Screw Cap with Conical Nipple

The screw caps are manufactured in pairs, using a three-plate mould.

During filling, free jet formation (sausage injection moulding) occurred, since the compartment had a larger cross-section at the connection point, and the jet of compound was therefore initially able to form without hindrance. It was possible to fit a baffle or a sprung baffle strut. It was also possible to avoid free jets here by rounding off the sharp transfer from the gate towards the component. Thus the desired flow head is formed when the compound flows in. This is illustrated in Figure 9.49.

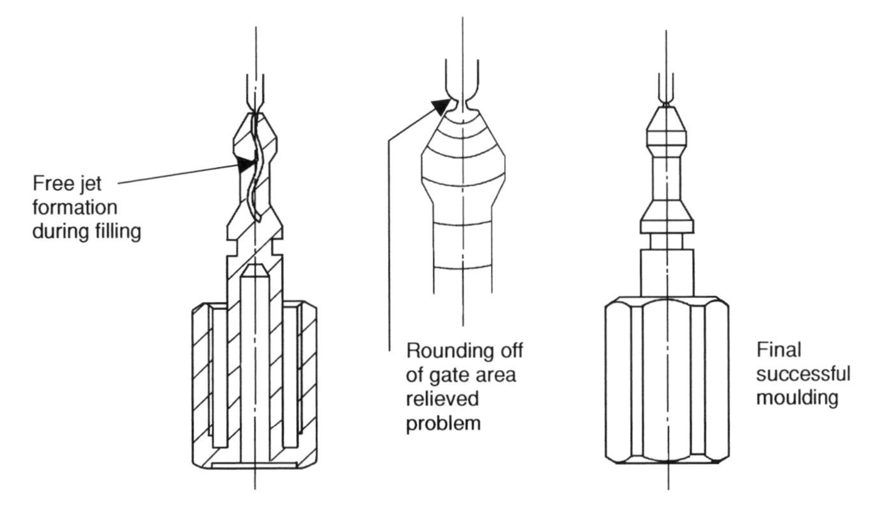

Figure 9.49 Screw cap with conical nipple

It would be generally favourable to make this edge less sharp, but it cannot be done in every case. For example, with tunnel runners, it might be that the runner could not be de-gated at the desired point (directly at the component).

9.4.10 Switch Housing

A switch housing was injection moulded using a rod runner. Strong flow line formation took place on the top face of the component. In spite of mould heating, no improvement could be made.

It was not until the transfer from the rod runner toward the component was rounded off that a good top face was produced. This is shown in Figure 9.50.

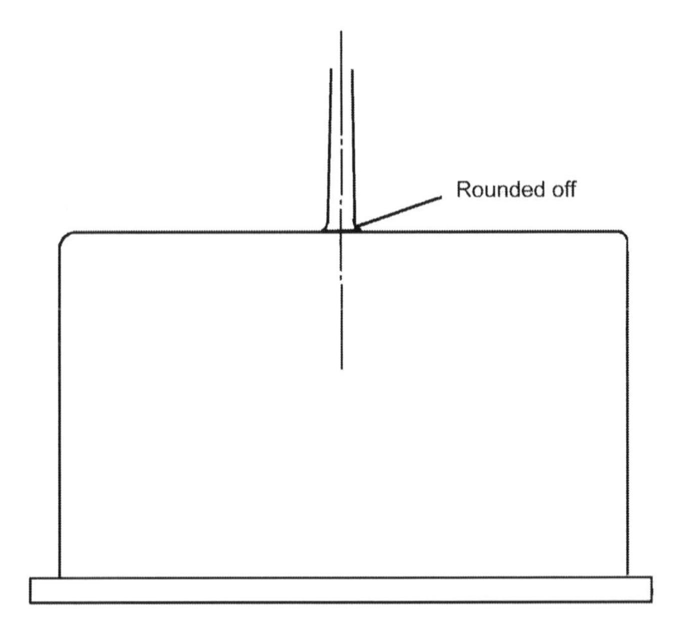

Figure 9.50 Switch housing

9.4.11 Battery Housing

A twin-compartment battery housing was centrally injected using a rod runner. The individual wall thicknesses and the intermediate wall were about 1.5 mm, and the walls were approximately 80 mm high. During injection, a strong wedge effect occurred, due to the compound moving forward in the intermediate wall. This is actually in the middle of the component, with the cores pressed right against the external mould wall, although they were manufactured from a single unit.

To be able to manufacture this component at all, the injection had to be displaced into the parting plane. Because the compound now came in at the base of cores, they were no longer pressed away, but rather were established through the uniform flow front. This is shown in Figure 9.51.

Now, so as to extract the enclosed air at the housing base, it was, of course, necessary to insert some air venting plungers.

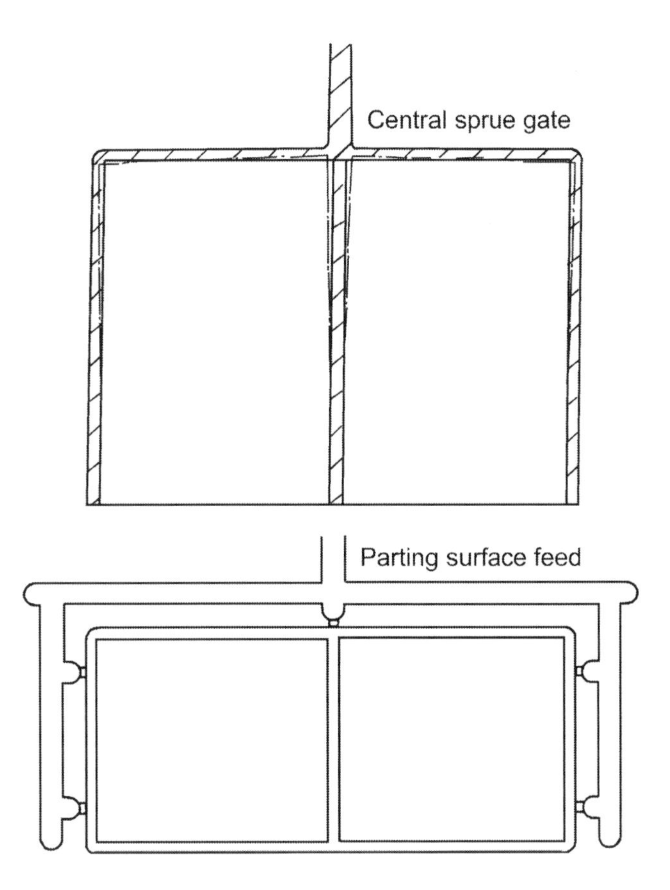

Figure 9.51 Battery housing

10 Advanced Processing Techniques

10.1 Introduction

Whilst injection moulding can be considered a relatively young manufacturing technology, many variants of the basic process have been developed. Examples include processes such as injection-compression, gas assisted, water injection, co-injection and over-moulding. Each of these processes are suitable for specific market applications. An overview of the technologies covered in this chapter is shown in Figure 10.1.

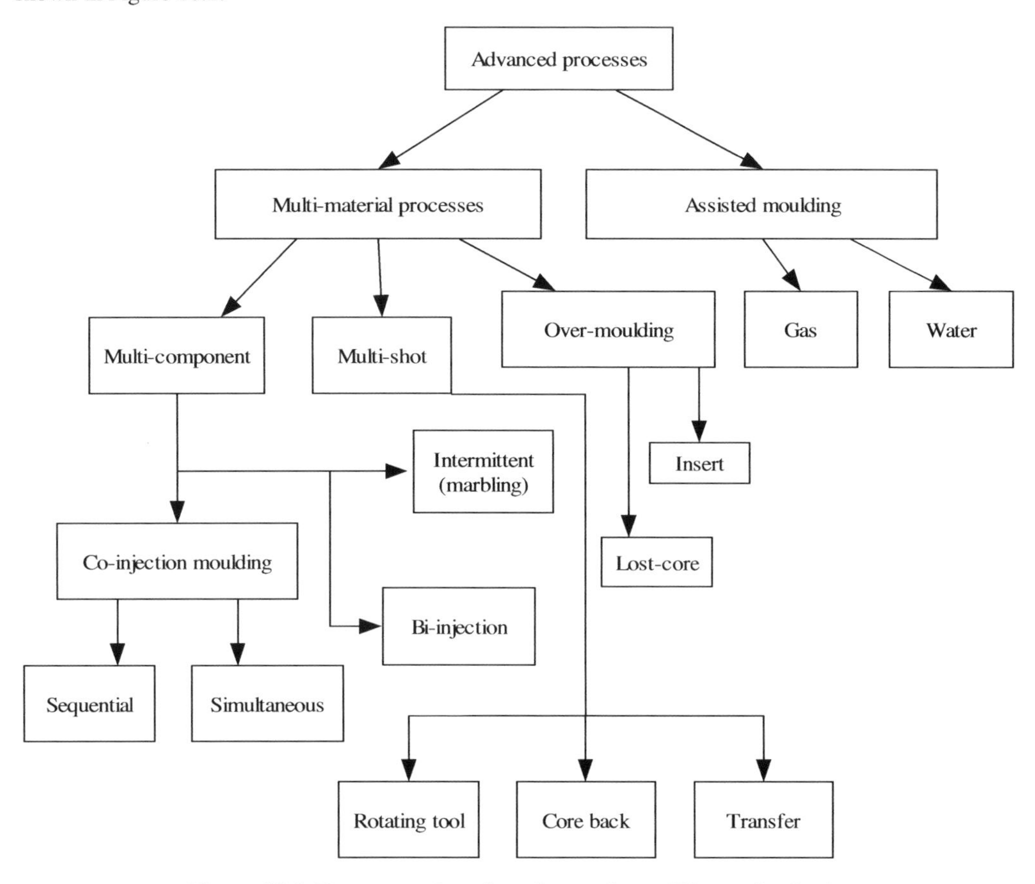

Figure 10.1 Process options for advanced moulding technologies

The processes are split into two categories, multi-material moulding and assisted moulding, the next section will introduce each of these in turn.

Various terminologies are used in multi-material moulding to describe particular process routes. These have been split into three categories here: multi-component, multi-shot and over-mould. Assisted moulding techniques include both gas and water. A brief description of each of these technologies and some of the names commonly used for them in the literature now follows.

10.2 Multi-Component Moulding

10.2.1 Co-Injection Moulding

Also known as dual injection, sandwich moulding, 2K (or 3K). It can be recognised by the production of a defined skin and core structure. The core is fully encapsulated, to produce a sandwich like moulding.

10.2.2 Bi-Injection Moulding

This is simultaneous injection of different materials through different gates as opposed to through the same gate as in co-injection moulding.

10.2.3 Interval Injection Moulding

Also known as marbleising, this is simultaneous injection of different materials through the same gate, with limited mixing. This produces a pattern of either regular or irregular colour distribution such as marble type effects.

10.3 Assisted Moulding

This is in many ways analogous to co-injection moulding as the resultant moulding has a skin and core configuration. However in this case the core injection process is used to selectively hollow out parts of the moulding. This process is primarily used for weight savings and cycle time reduction. The gas or water takes the place of polymer to produce a hollowed out moulding with a reduced section thickness.

Gas injection moulding has no standardised designations. Some common process terminology/trade names include: gas injection moulding technology (GIT), gas assisted injection moulding (GAIM), gas assisted injection moulding (GAIN™), gas interior pressure (German: Gas Innen Druck), gas injection process (GIP) and Koolgas™, which is a variation of the standard gas injection technique using pre-cooled gas injection.

Water injection moulding (WIM) or water assisted injection moulding (WAIM) is much newer and less well established.

10.4 Multi-Shot Moulding

Multi-shot can be used to describe any process whereby distinct multiple material shots are applied to produce a single final component. This includes processes whereby preforms are moulded and then transferred to different cavities on the same machine. It can also be used to describe processes where multiple shots are made into the same tool without the tool opening between shots.

10.5 Over-Moulding

This technique is where components are placed in an injection mould and are then moulded over with another material. The term over-moulding covers both insert moulding and lost core moulding. This technique is not confined to plastics and over-moulding of metal inserts, such as to produce scissor handles and plastics on ceramics is commonplace.

10.6 Business Trends

Multi-material and assisted moulding are two of the fastest growing sectors of the injection moulding industry and all major machine manufacturers such as ARBURG, Battenfeld, Demag, Engel and Ferromatik Milacron, offer multi-material machine capability. Gas and water assisted moulding tend to be less readily available, and more specialised. These technologies very often require licences. The development of gas technology being heavily patented by the industry.

The trend towards multi-material moulding has been pushed by the growing importance of added value manufacture to give moulders commercial advantages over their rivals. An increase in the importance of material recycling has also aided the growth in co-injection technologies, where encapsulation enables recyclate materials to be used.

A further factor pushing the popularity of advanced moulding techniques in Europe has been the transfer of much of the region's traditional trade moulding market to the cheaper manufacturing sectors of the Far East. To combat the increased competition, trade moulders have sought more technically challenging markets as a means of survival, as much of the traditional standard injection moulding market has been and continues to be contracted out to the Far Eastern moulders. The more technically advanced processes however, can often be done more efficiently and with higher quality by established multi-material toolmakers and moulding shops in Europe.

Whilst German manufacturers have been quick to invest in multi-material moulding technologies, moulders in the UK, the rest of Europe and the USA have taken longer to realise the potential advantages in these technologies. However, the versatility these processes offer to manufacturers, and the current highly competitive moulding market is seemingly finally overcoming the resistance generated by the larger capital equipment investment often required to produce these mouldings. This is a positive step forward since a well designed multi-material component can often repay the initial investment quickly and easily. This is achieved by combining distinct moulding jobs, such as when producing mouldings, that then require assembly. By removing the need for the assembly step or perhaps a further finishing stage, a single multi-material moulding can offer a considerable unit cost reduction. Of equal importance is that a more attractive end product is often the result. Consider for example the case of the humble toothbrush. Application of multi-material moulding to this product has transformed this simple and standardised design. The placement of combinations of hard and soft feel materials and a myriad of multiple colour combinations have filled endless supermarket shelves with a variety of attractive designs for the consumer, and in the process transformed the toothbrush market place.

Assisted moulding can likewise bring a competitive advantage to its user. As well as material savings, these processes can reduce cycle times due to both the cooling effect of the fluids and the reduced section thickness of the polymer.

A general introduction to each of the technology areas has now been presented and each will be discussed in more depth later. However before this an introduction to some material selection issues will be presented.

10.7 Material Selection

In selecting material combinations for multi-material moulding applications, consideration must be given to the combination of properties required in the final product. In many cases, there must be a certain level of adhesion between skin and core in order to maintain mechanical integrity. This can be achieved in two ways:

1. The materials are compatible and offer some degree of bonding at the interface.
2. A method must be found to mechanically interlock them. In the case of over-moulding and multi-shot this can be done with clever usage of material properties and tool design.

Not all parts require adhesion. In fact in some cases the requirement may be the exact opposite. If joints are to be produced, it is necessary that the mouldings can move freely at the interface. Examples include over-moulding to produce what will be the moving arms and legs on dolls and other similar toys or to produce ball and socket joint mouldings. In cases like these, materials must be selected for their immisciblity to ensure smooth regions of movement. Where adhesion is required, good interfacial bond strength is a pre-requisite, otherwise the properties may come from a significantly reduced section thickness. For good adhesion, a certain amount of interdiffusion is required between the melts. This can be achieved when there is a high compatibility, or solubility between the melts.

Tables of compatible and incompatible material combinations are readily available from both machine and materials suppliers, such as the one shown in Table 10.1. However caution is required when using such tables, since it has been shown that changing from one particular grade of material to another can affect the bond strength. Occasionally manufacturers may also seemingly disagree on the adhesion properties of materials. Since processing conditions also affect adhesion, experimentation may be required to ascertain optimum conditions for any given material combination. Additives are available to aid the compatibility of materials and not surprisingly, they are called compatibilisers. Through the addition of these materials it is possible to chemically bond some non-adherent materials. These substances usually contain a third polymer that bonds to, or is soluble in, the two materials. Numerous compatibilisers are commercially available to bond various immiscible materials together. Take for example PA 6 and PP. These materials can be bonded by the addition of a maleic anhydride grafted polypropylene. However, compatibilisers tend to be expensive and both the number of chemical sites available at the interface (a result of the number of grafts per chain) and the molten contact time are limiting factors in the final bond strength as there is generally little interfacial mixing.

Table 10.1 Material compatibility table

	ABS	ASA	EVA	PA 6	PA 66	PBT	PC	HDPE	LDPE	PET	PMMA	POM	PP	PPO mod	PS-GP	PS-HI	SAN	TPU
ABS	+	+	+			+	+	-	-	+	+	-	-	-	*	*	+	+
ASA	+	+	+			+	+	-	-	+	+	-	-	-	*	-	+	+
EVA	+	+	+					+	+				+		+	+	+	
PA 6				+	+	*	*	*	*			-	*	-	-	-	+	+
PA 66				+	+	*	*	*	*			-	-	-	-	-	+	+
PBT	+	+		*	*	+	+	-	-	+	-	-	-	-	-	-	+	+
PC	+	+		*	*	+	+	-	-	+		-	-	-	-	-	+	+
HDPE	-	-	+	*	*	-	-	+	+	-	*	*	-	-	-	-	-	-
LDPE	-	-	+	*	*	-	-	+	+	-	*	*	+	-	*	-	-	-
PET	+	+				+	+	-	-	+	-	-		-	-	-		+
PMMA	+	+				-		*	*	-	+		*	-	-	-	+	
POM	-	-		-	-	-	-	*	*	-		+	-	-	-	-	-	
PP	-	-	+	*	-	-	-	-	+		*	-	+	-	-	-	-	-
PPO mod	-	-		-	-	-	-	-	-	-	-	-	-	+	+	+	*	-
PS-GP	*	*	+	-	-	-	-	-	*	-	-	-	-	+	+	+	-	-
PS-HI	*	-	+	-	-	-	-	-	-	-	-	-	-	+	+	+	-	-
SAN	+	+	+	+	+	+	+	-	-		+	-	-	*	-	-	+	+
TPU	+	+		+	+	+	+	-	-	+			-	-	-	-	+	+

Key (-) : No adhesion, (*) : Poor adhesion, (+) : Good adhesion
PS-GP = general purpose polystyrene
PS-HI = high impact polystyrene

As well as adhesion, there are other material characteristics that also need to be considered when moulding with materials of different generic families. Examples are the levels of relative shrinkage and thermal expansion values, these may need to be matched or careful consideration given to the requirements before final material selection takes place.

Certainly in the case of many co-injection techniques, differences in mould shrinkage and thermal expansion can lead to problems such as sink marks, warpage and residual stresses. With over-moulding techniques, differences in shrinkage or the coefficient of linear thermal expansion (CLTE) can produce high stresses between restrained materials. The result in both cases can be the same, premature failure.

Consideration is also required of the long-term properties of the various plastic materials making up a multi-material component, especially if the moulding is to be put under stress. Stress can be produced during the processing stage, especially if unsuitable processing conditions are used. In service factors such as mechanical stress, chemical attack or high temperature may be found. All relevant design parameters should be considered as well as those of shrinkage and CLTE. Depending on the application these may include:

- Time-temperature effect: many properties are temperature dependent. The stiffness, ductility and impact strength may vary considerably with temperature. Therefore, testing should be carried out at conditions representative of the service conditions
- Fatigue: dynamic fatigue can occur when stress is applied periodically in applications such as bearings.
- Creep: where parts are exposed to constant loading there may be a change of stiffness with time. This is termed creep.
- Environmental stress cracking: plastics may be embrittled by exposure to water, light, temperature and oxygen as well as chemical attack.

This section has given an overview of material selection that can be applied not only to multi-material mouldings but also in many cases to standard plastic components. If all relevant process and design criteria are considered at an early stage, conversion to multi-material processes should be a relatively painless transition. Further material considerations for individual processing techniques along with current applications and markets can be found under the relevant technology section.

10.8 Process Technology

This section is split into four main areas:

- Multi-component injection moulding
- Assisted moulding
- Multi-shot moulding
- Over-moulding

Each of these areas will be now be discussed in turn.

10.8.1 Multi-Component Injection Moulding

This method describes a process whereby plastics are injected into one mould during a single injection cycle. The most common process in this category is co-injection moulding. Other less common variants are bi-injection and intermittent techniques.

10.8.2 Co-Injection Moulding

Co-injection moulding is a variant of the standard injection process and has been in use since the early 1970s. A number of terminologies are used that can encompass this process such as sandwich moulding, 2K (2-component) or 3K (3-component) and dual-injection or multi-component. For the purpose of this chapter, co-injection moulding is the preferred terminology. This technique offers the advantages of combining two or more material properties to produce a 'sandwich' structure. This is achieved by making sequential injections into the same mould with one material as the core and another as the skin. This is illustrated in Figure 10.2.

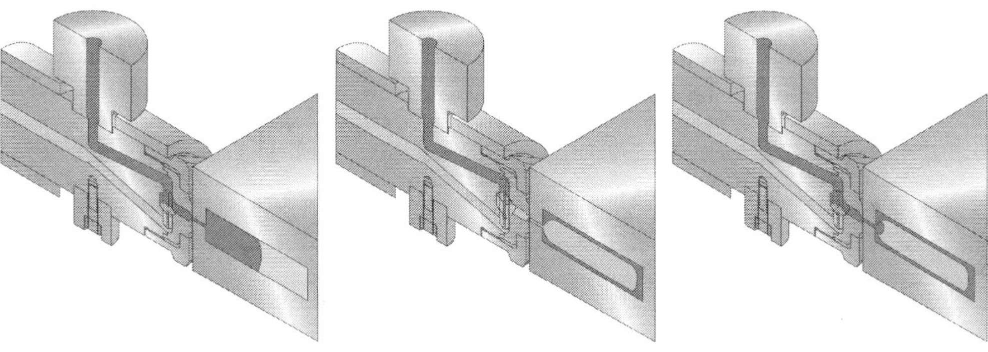

1. Injection of skin material only
2. Injection of core material begins pushing skin material onto mould walls
3. Injection of skin material only to cap moulding and clear sprue for next shot

Figure 10.2 Encapsulation of core material by skin during co-injection moulding

The result is the distinctive sandwich structure of skin and core as shown in Figure 10.3.

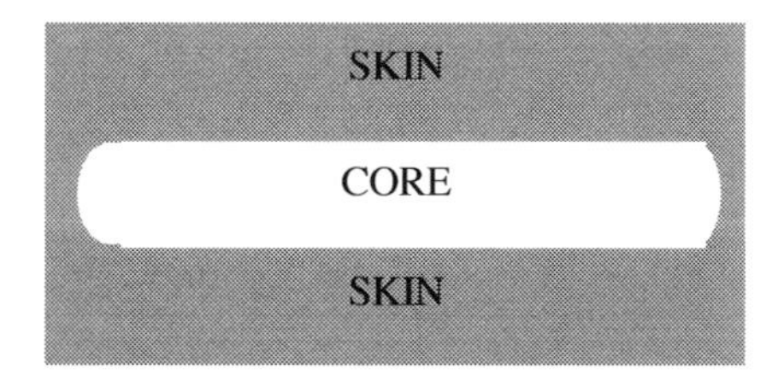

Figure 10.3 Cross-section of sandwich structure of a co-injection moulding

Co-injection moulding offers numerous possibilities in terms of a variety of material combinations, some of which are shown in Table 10.2.

Table 10.2 Current applications for co-injection moulding

Material combination	Properties	Application
Soft feel skin/hard core	High strength core with soft feel skin	Door handles, gear lever
Unfilled skin/core with conductive filler	Electromagnetic interference shielding (EMI)	Computer housings
Virgin skin/recycled core	Environmentally friendly production, cost saving	Garden furniture, automotive bumpers and fascias
Unfilled skin/reinforced core	High surface finish, structural performance	Automotive door handles
In mould paint, variable core	No finishing of product required after moulding	Wheel trims
Unfilled skin/foamed core	Good surface finish, low density, high rigidity	Automotive body panels
Pigmented skin, uncoloured core (or reverse)	Reduced pigment cost, aesthetics	Yoghurt pot

10.8.2.1 Material Selection for Co-Injection Moulding

One of the most difficult technical problems with co-injection moulding is that the core material must be prevented from large scale mixing with the skin material, in order to retain a consistent skin layer thickness and resultant properties. It must also be prevented from penetrating the skin of the moulding. There are also limitations in the variation of flow characteristics between the two materials that are permissible. Rheology plays a large part in the relative skin/core distribution. Therefore the viscosity of the materials is of the utmost importance as it affects the process dynamics and resultant core distribution. As a general guide, in order to keep the sandwich configuration and layer thickness consistent, the skin should have the same or preferably a slightly lower viscosity than the core. If the skin viscosity is too high, the core melt will flow through the skin and form the surface layer. Core distribution can also be controlled by adjustments to the speed and time of injection, polymer melt temperatures and mould temperature. However, the use of similar moulding temperatures for both materials is recommended, since they are processed simultaneously. The rheology of the materials will be discussed in more detail later in this chapter.

The process does have limitations. There is a need for the skin and core materials to be compatible with each other in terms of adhesion and shrinkage. Adhesion of the layers is necessary to prevent the core material becoming detached from the skin especially if the moulding is likely to be exposed to mechanical loads. Therefore materials must be compatible or a suitable compatibiliser used in the core component. The use of compatibilisers in the core component of co-injection moulding was developed and patented by the Rover Group in collaboration with University of Warwick [1]. Researchers from Warwick have also developed and reported methods to mechanically interlock immiscible materials for co-injection moulding but these are currently in the early development stages [2].

10.8.2.2 Process Sequence

In co-injection moulding two compatible melts are injected either sequentially or simultaneously into the mould thus forming a layered structure. The melt injected first forms the skin, whilst the melt injected afterwards forms the core.

Using two polymers with different properties makes it possible to obtain unique property combinations that are not possible in ordinary injection moulding. A number of these including commercial applications are shown in Table 10.2.

The co-injection process was first described and developed by ICI in 1970 and was developed to overcome the surface finish limitations inherent in the structural foam process. Foam mouldings have a rough, irregular surface finish which will vary in quality from moulding to moulding. By using a solid skin with a cellular foamed core, it is possible to obtain a surface finish as good as for a solid part but with the added rigidity of the foam core. There is also the reduction in material cost associated with foaming. This requires fewer raw materials and is therefore cheaper to produce. In products with thickness above 4 mm, for example some automotive body panels, this method is often used.

10.8.2.3 Co-Injection Moulding: Different Techniques

There are a number of variations of the co-injection moulding process which have been developed. They can be split into two types, sequential injection and simultaneous injection. These methods will now be introduced and the advantages and disadvantages of both discussed.

Sequential injection: single channel technique

The single channel method was patented by ICI in 1970 and was the first commercial co-injection technique. An injection moulding machine with two cylinders is used, one for the skin material and one for the core. The polymer melts are injected sequentially into a mould. First the skin and then the core. A specifically designed valve is used, which allows a first injection of skin material only. At a pre-set point, the flow is stopped and the core is injected. At the switch point from one extruder to another there is a pressure drop in the mould. This is the major limitation in this method. This switching of polymer flows can cause the flow to stop, giving surface defects such as shadow marks or gloss marks on the mouldings. In terms of skin/core structure, as with all co-injection mouldings, changing injection moulding parameters or the relative material viscosity can be used to control material. This process is mainly used for thick sectional parts with foamed cores.

Sequential injection: Mono-Sandwich technique

The Mono-Sandwich technique was developed by Ferromatik Milakron. It uses two materials layered in a standard cylinder and is shown in Figure 10.4. This is achieved by melting the skin material in a separate side extruder that plasticates the material. A special hot runner system then leads the molten material through to the front of the screw in the main cylinder. The resultant melt pressure pushes the screw backwards. When the specified amount of melted material has accumulated in front of the screw, the screw starts rotating and feeds the core material. The injection is then done in the same way as for normal injection moulding by pushing the screw forward.

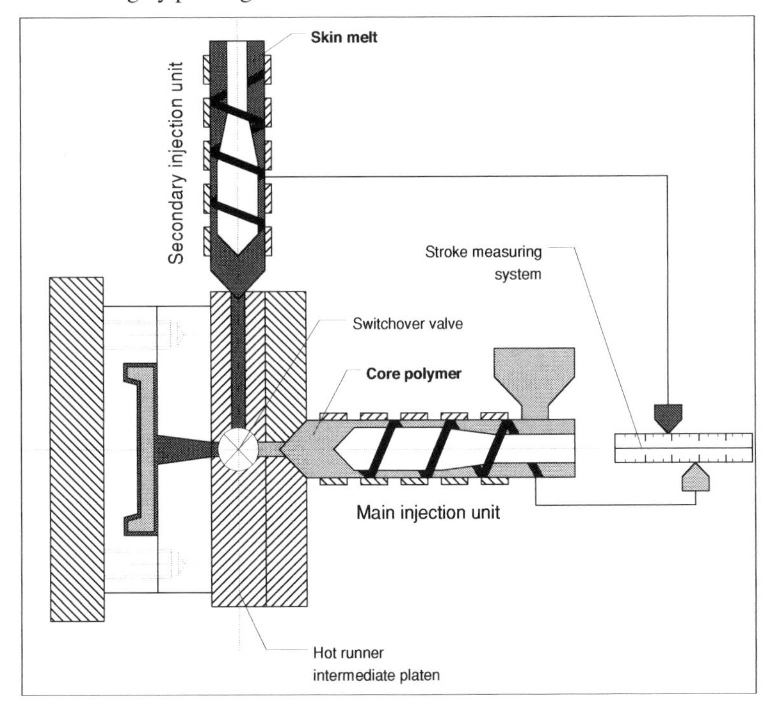

Figure 10.4 Ferromatik Mono-Sandwich technique

There is little or no mixing of skin and core melts in the screw as the dynamics of the screw push the material forward rather than mixing the melts together at the material interface. The maximum amount of core material that can be encapsulated will be entirely dependent upon the mould geometry. For simple symmetrical parts this value will be around 65-75% by volume. The main advantage with this method is that a standard injection moulding machine can be rebuilt to a sandwich machine simply by connecting a side extruder to the main injection unit. This method is also particularly good where very thin-walled parts are required. A further advantage reported by the manufacturer is the speed of colour and material change compared to other sandwich techniques due to the relatively simple construction. Since injection is made in the same way as for normal injection moulding, the process control is fairly simple and similar to that of standard injection techniques. The drawback to this method is that by feeding through one injection unit only, there is a lack of detailed control which is required when moulding complex shapes to control skin/core configuration.

Addmix originated a modified version of the Mono-Sandwich injection technique. It is a similar technique to the Ferromatik one but without an ancillary extruder. The second material feedstock is regulated and fed to the screw from a separate hopper controlled by a stroke measurement system, with injection through a special nozzle. In this way the polymers were layered into one injection cylinder. The moulding then proceeds as per conventional techniques.

In summary, sequential injection techniques provide a cheap and useful method for producing sandwich mouldings. It is a technique more suited to simple geometries since the main problem with sequential injection is the lack of control in the skin/core distribution. With the melt stream injection of both skin and core controlled together and with just one velocity profile, skin thickness cannot be adjusted in various parts of the moulding. To overcome these problems, the simultaneous method was developed. Here skin and core velocities can be controlled separately giving enhanced control.

Simultaneous injection: two channel technique

The two channel method, developed by Battenfeld in the mid-1970s, includes a phase of simultaneous injection. The process sequence can vary but a typical example would be:

1. Injection of skin to a pre-set switch point
2. Injection of core material begins so both skin and core flow together
3. Injection of just core
4. Injection of just skin
5. Part packing and cooling followed by ejection.

This co-injection moulding process entails the injection of molten plastic for the skin layer into the mould cavity. After a certain pre-set time, usually in the region of 0.1-0.3 seconds, a second plastic that will make up the core is injected and, for a period, there is simultaneous injection of both materials. This simultaneous period of flow is where this process gets its name and also how it differs from the sequential techniques described earlier. The injection of core material pushes the moving layer of skin material against the cavity walls where it cools and solidifies. The final stage of mould filling is injection of the core material only, although sometimes the mouldings are 'capped' with skin layer to complete encapsulation of the core.

Two injection units are used in this method, which are joined through a specially designed nozzle. In the Battenfeld design, the nozzle is equipped with two separate concentric channels that can be independently, operated, opened, and closed hydraulically. This allows the process sequence to be carefully controlled.

A phase of simultaneous injection of skin and core avoids the problems inherent in the single channel technique by maintaining a constant flow front velocity. This can be seen by looking at Figure 10.5 and comparing the pressure profiles and screw velocity of the simultaneous and single channel methods. In the sequential method, the period of stagnation after injection of skin (A) but before injection of core (B) can be clearly seen, resulting in a drop of cavity pressure and a period where there is no movement of material in the screw. This demonstrates many of the limitations of sequential injection discussed earlier. This pressure drop is less apparent in the simultaneous method.

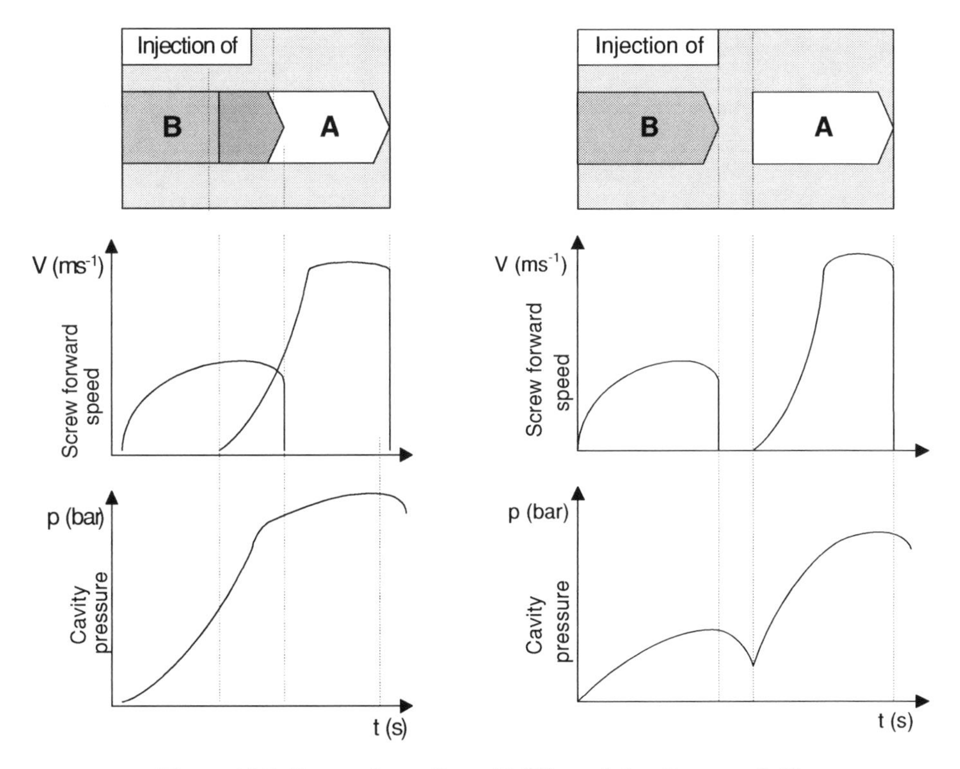

Figure 10.5 Comparison of mould filling of simultaneous (left) and sequential (right) injection moulding

The length of the simultaneous injection phase depends on both the material and the mould geometry. A typical duration is 25% of the injection time for the skin component. Mould filling dynamics dictate that optimum distribution of core material is obtained if the skin viscosity is kept slightly lower than that of the core. This is due to rheological factors that will be explained in the next section. Separate operation of two injection units makes it easier to control skin thickness in various parts of a moulding. This is because, by separate control of the velocity profile of the skin and core, the skin thickness can be adjusted in various parts of the moulding. Extra amounts of skin material can also be injected during holding time to seal the gate area. This also ensures that the nozzle is clear of core material and prepared for the next shot.

Again, due to the dynamics of mould filling, material entering the tool and cooling on the walls near the gate can get re-melted and flushed away due to frictional heat generated by the incoming molten flowing melt. This can lead to variations of skin thickness and leave the skin near to the gate region much thinner than that on the rest of the moulding. This effect is generally more pronounced on the opposite side of the gate due to the higher shear experienced in this region. In order to overcome this the three channel technique was proposed.

Three channel technique

With the three channel technique, an extra channel is used for the skin in the centre of the gate in order to retain skin thickness in this area. An example of a system developed commercially by Kortec is shown in Figure 10.6. This is an example of a hot manifold system. Systems are also marketed by other companies including Kona, Incoe and Battenfeld. The extra channel can reach the opposite side of the moulding, enabling the two surfaces to be regulated separately and surface thickness controlled. This special design can only be used with a central gate, otherwise skin/core distribution will be irregular as detailed in Figure 10.8. For other gate geometries or multi-cavity mouldings, the two channel or single channel technique is preferred.

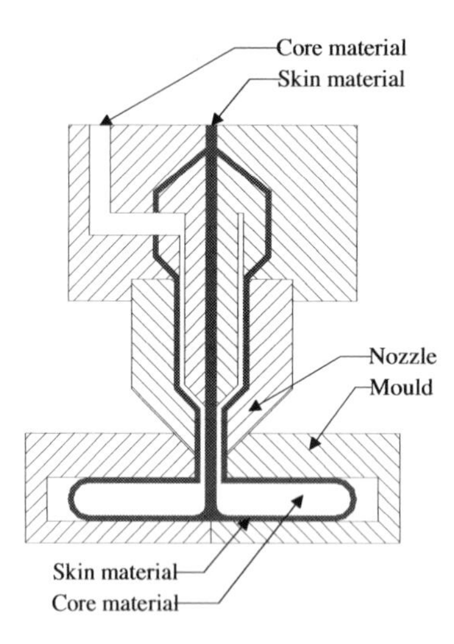

Figure 10.6 Kortec three channel nozzle technique

A three layer technique to combine immiscible material combinations was provided by the Billion Corporation of France. Their solution to polymer incompatibility for sandwich injection moulding used the third intermediate polymer layer as a binder adhesive, this is analogous to methods used in extrusion blow moulding. However, there are obvious machine cost disadvantages here, because the runner system is complex and a third injection unit is required.

The Battenfeld solution to three channel moulding is shown in Figure 10.7. This utilises machine configurations also used in multi-shot techniques as will be described in later sections. The vertical unit can be used to feed the third material into a combined nozzle system based on their two layer technology described in Section 10.8.2.3.

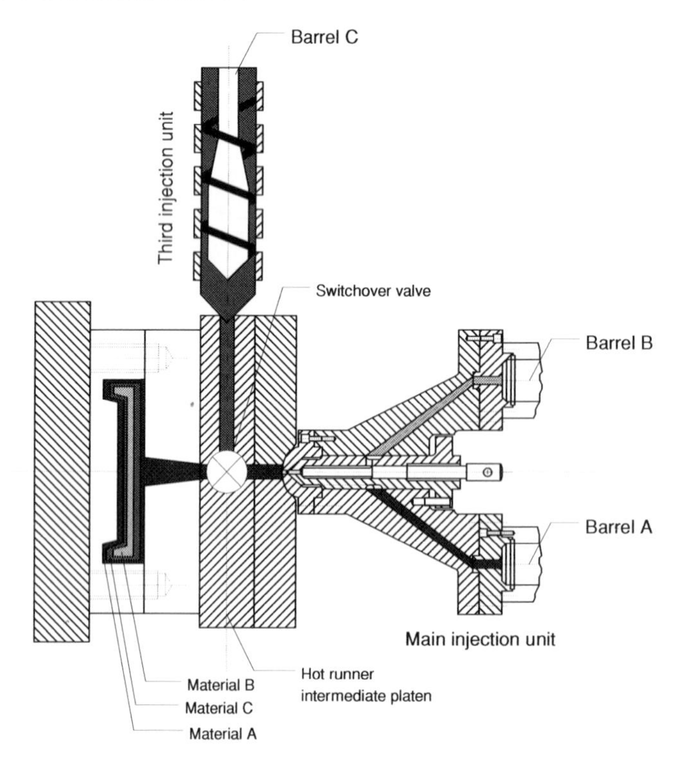

Figure 10.7 Battenfeld three channel technique

10.8.2.4 Part Design and Tooling Requirements for Co-Injection Moulding

Normal injection moulding tooling can be used for both sequential and simultaneous co-injection moulding providing the following factors are considered. Mould filling and the resultant skin/core distribution is strongly dictated by the gate location as shown in Figure 10.8. The injection gate must be designed with consideration for the resultant skin/core distribution otherwise unsatisfactory mouldings may result. If more than one gate is required, or a break in the flow front is produced, the weldline will be skin rich, as the core materials will not completely meet due to the filling dynamics of co-injection moulding. The skin will always pack these areas before the core material can reach them. This is illustrated in Figure 10.8.

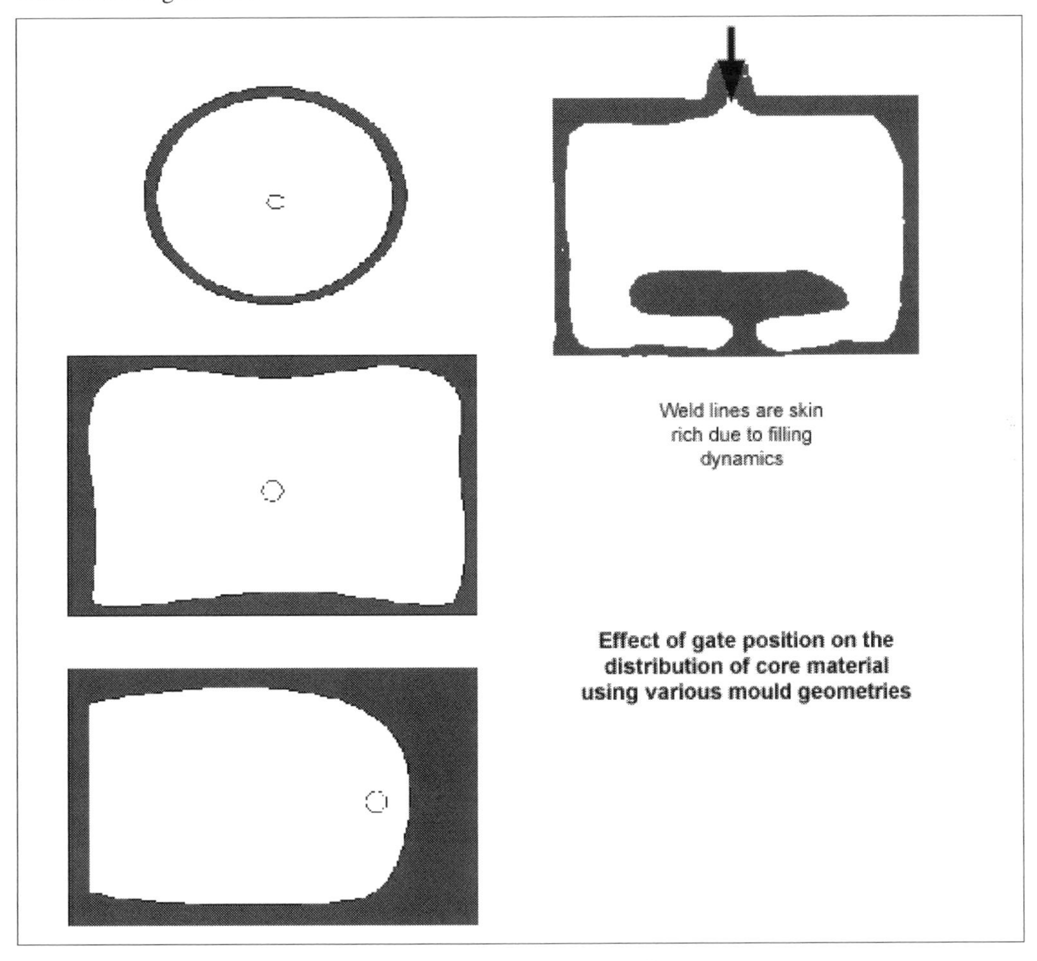

Figure 10.8 Gate position effects

10.8.2.5 Rheology and Mould Filling: Why and How Co-Injection Moulding Works

For readers who have not seen co-injection moulding in action, it can be very difficult to understand how co-injection moulding can produce the distinctive skin/core structure. It should be remembered that the skin material is injected first to cool and form the skin against the tool wall. The core material will then push against the skin to cause it to penetrate deeper into the mould cavity. In this process therefore the core remains encapsulated, providing that the viscosity of the skin and core materials are similar. If they are mismatched, this filling pattern is affected. Before this is discussed further, a recap on rheology basics and the effects of various parameters on viscosity is required. How a material responds to changes in temperature, shear and pressure can greatly affect processing methods and strategies. Other material properties such as density, elasticity, thermal expansion and thermal conductivity can also affect processing behaviour through, for example, decisions on the cooling rates required, shrinkage tolerances and die swell. However, discussions will begin with the basics of rheology.

Rheology basics

Rheology deals with deformation and flow and examines the relationship between stress, strain and viscosity. Most rheological measurements measure quantities related to simple shear such as shear viscosity and normal stress differences. Material melt flows can be split into three categories, each behaving differently under the influence of shear as shown in Figure 10.9: Dilatent (shear thickening), Newtonian and Non-Newtonian pseudoplastic (shear thinning) behaviour.

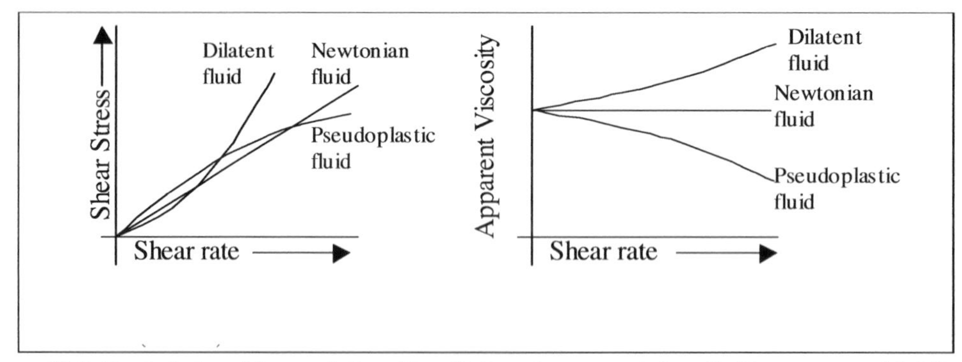

Figure 10.9 Typical stress/shear relationships (left) and apparent viscosity/shear curves (right) for dilatent, Newtonian and pseudoplastic fluids

In an ideal Newtonian liquid flow, the viscosity is independent of shear rate. Plastics fall into the category exhibiting shear thinning behaviour (pseudoplastic). This means that they respond to increased shear by a drop in viscosity. It can be seen from Figure 10.9, however, that all melts approximate to Newtonian fluids at very low shear rates. Unfortunately injection moulding tends to occur at high shear rates. Temperature and pressure also affect the viscosity of the polymer melt. At higher temperatures the viscosity drops, whilst as pressure increases viscosity increases. A complex picture now emerges, as during injection moulding there are both steep gradients of temperature and pressure. Therefore during co-injection moulding materials will be subject to viscosity change throughout the moulding cycle with the skin and core undergoing different shear, temperature and pressure histories.

A further consideration is the response of the skin and core polymer flows to the deformation processes present during injection moulding. This is a complex mixture of shear, elongation and bulk deformations. One way to try to model such effects is to use an element model. A representation of a theoretical polymer element is shown in Figure 10.10. This indicates how stresses can occur in a number of directions. Representations such as these are often used to show the force balance on an element. There are three components of normal stress (xx, yy and zz on Figure 10.10) and six components of shear stress, giving a total of nine stresses (on the three visible planes) using this method of analysis.

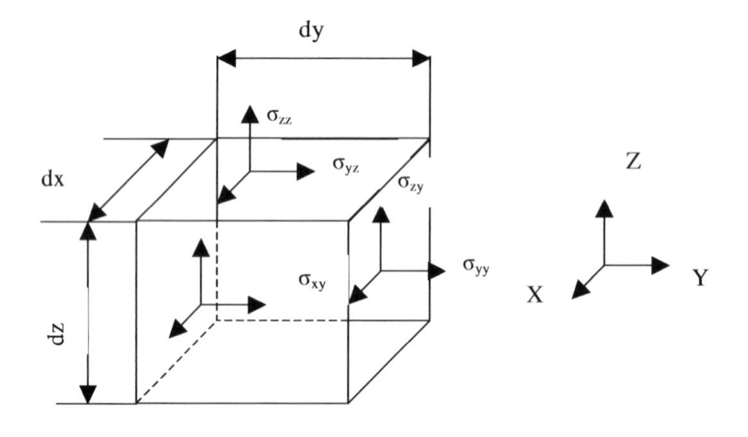

Figure 10.10 Possible stresses on an element

Shear flow is produced when stress is applied tangentially. Extensional flow is the result of stress applied normal to the surface of the material and bulk deformations result when stress is applied normal to all faces.

During shear flow, polymer chains are deformed and orientate in the direction of flow. (The results of this orientation in standard injection mouldings was discussed in Section 2.1.) The tension of the polymer in the flow direction is called the first normal stress. In contrast to this, flow in other directions is relatively small. First normal stress can be defined by the difference between the normal stress component in the flow direction and the normal component in the direction of the shear plane. Combinations of some or all of these forces are applied to the molten polymer material during the moulding process depending upon where on the moulding they are. Analysis using methods such as these, allows material interactions to be better understood.

Impact of rheology on the dual injection process

Now that some of the factors affecting viscosity have been introduced, the effect of viscosity on skin/core formation can be studied. This is illustrated in Figure 10.11.

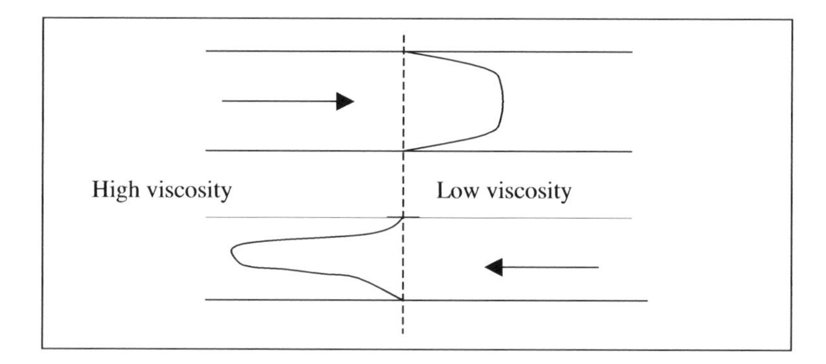

Figure 10.11 Interface behaviour of melts of different viscosity

High viscosity melts flowing into low viscosity melts will force the low viscosity material in front. Low viscosity melts flowing into high viscosity melts will jet through areas of least resistance, giving an effect termed 'melt fingering'. This type of effect can also be seen in gas assisted injection moulding where the core component, gas, has an effective viscosity of zero. The injected gas follows the path of least resistance, making channels in the hottest, thickest and least viscous parts of the melt stream. Breakthrough of the core component in co-injection moulding can cause unwanted surface defects, whereby the core material can be visible in the corners of the mouldings. Therefore, a high percentage of co-injection studies have investigated the relationship between the relative viscosities of the materials used within the process and the resultant skin/core distribution to reinforce these findings.

Previous studies on the effects of viscosity ratios

It was reported as far back as 1974 that two-phase flows such as those found in co-injection moulding are sensitive to differences in the rheological properties of the melts. Different rheological combinations produced different skin/core ratios. Later the simultaneous system driven by viscosity encapsulation phenomena was proposed as shown in Figure 10.11.

It has been found that the most uniform core distributions are achieved at ratios of skin to core of between 0.8 and 1.8. Within these ratios about 60% core could be concealed within a square flat plaque moulding without breakthrough, and about 70% core in a disk mould.

Apart from initial viscosity, the moulding parameters found to have most effect on the skin/core distribution are melt temperature, injection rate and the length of the simultaneous phase. Too short an interval between the injection of skin and core can result in breakthrough, too long and too much skin is forced to the outer edges of the moulding. In terms of other processing effects it has also been found that as in injection moulding, there is considerable melt flow in the region of the gate during the post-filling process of co-injection moulding. This can have implications for the skin thickness around the region of the gate.

From injection moulding it is well known that both the temperature of the mould and the speed at which the materials are injected into the mould cavity also determine the end structure. The temperature of the mould will affect the rate at which the skin material will freeze, the greater the difference in temperatures of the skin and mould surface the more quickly the material will solidify. The rate at which the skin material is injected will determine how much time the leading edge of the skin material will have in contact with the mould surface. A greater speed would lead to the skin material penetrating further into the mould. In relation to these parameter effects in co-injection moulding increasing injection speeds of skin material can help prevent breakthrough. Slowing down the core gives a similar, though not exactly the same effect. This is because as skin speeds are increased, especially at low mould temperatures, instabilities in the flow can also increase. These instabilities can result in a better mixing in the interfacial region between skin and core material. However, taken to extremes these instabilities could also cause breakthrough in the mouldings.

Since viscosity is shear dependent it is worth considering what kind of shears are operating within a mould tool. This is in order to ensure that the viscosity data being used is representative or at least a close approximation of what is actually occurring. It can be seen how the viscosity changes in relation to both shear and temperature. The method used to measure these materials is called capillary rheometry and unlike other methods used to quantify melt flow properties such as MFI and cone and plate rheometry, this method can actually measure the response of a material at the high shears associated with injection moulding. This becomes of even more importance when different generic families are being used as skin and core. A plot of the viscosities of polycarbonate (PC) and acrylic (PMMA) against shear rate is shown in Figure 10.12. This combination of materials has generated considerable interest for possible future automotive glazing applications. It can be seen that not only does viscosity change with shear, different material families also respond differently to changes in shear. A second consideration is temperature, as viscosity is also a function of temperature, so this data must also be representative of the moulding conditions, in this case mouldings were produced with a 20 °C difference in skin and core. The data reflects this. In Figure 10.12 at shears below 1000 s^{-1}, the polycarbonate (PC) material has the lowest viscosity. At shears higher than this the trend is reversed. Therefore if mouldings were carried out on identical injection moulding cavities but one tool had a higher shear at the gate region, the viscosity interaction would be different. Examples such as these highlight some of the possible pitfalls inherent in co-injection moulding. In checking for viscosity matching it is therefore essential that the data used is relevant to the processing temperatures and shear rates within the tooling.

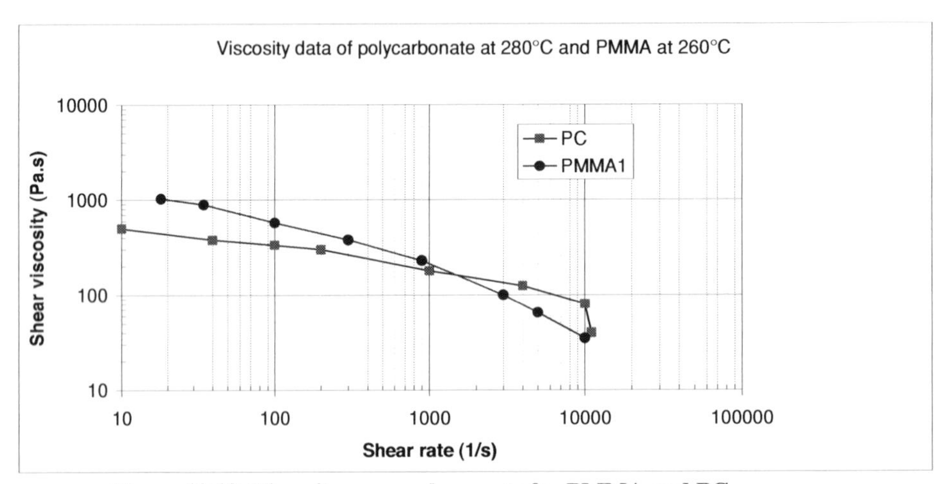

Figure 10.12 Viscosity versus shear rate for PMMA and PC

Whilst the viscosity ratio between the core and skin materials has a major impact on the interface of the materials and the core distribution, it is not the only factor. With mixed material studies complex mechanisms have been found to occur at the interface. A rheological explanation for such effects is insufficient on its own to explain the interfacial effects that are observed, with increases in injection speed appearing key to these studies. Polymers have both viscous and elastic components. Given the high shear rates employed during injection moulding and the relationship between increases of elasticity with shear, it seems reasonable to assume that elastic effects will occur. An increase in injection speed would bring about an increase in shear and therefore also increase elastic property effects in the mouldings. Therefore the resultant mechanical properties for example, impact strength or tensile modulus, may vary as a result of both viscous and elastic interfacial interactions.

10.8.2.6 Immiscible Materials Research in Co-Injection Moulding

If two immiscible materials are moulded using co-injection moulding, it is possible to peel them apart because there is both no adhesion and no mixing of the two materials. (Obviously the properties of such mouldings will be poor where there is no adhesion.) This is because filling occurs by fountain flow in an organised laminar manner giving little mixing between material being injected and the molten material coming in behind it. Fountain flow was illustrated in Chapter 2, and it is as a result of this mechanism that the core structure will tend to have a concave edge configuration. If laminar flow is disrupted, the normal mould filling pattern is disrupted. Small scale interfacial interactions occur, driven by a complex mixture of elastic and viscous interactions which disrupt the stratification at the interface. A common example of fountain flow failure occurs in standard injection moulding. This is the phenomenon of 'jetting' whereby the melt shoots through the gate and therefore fails to gain adhesion to the tool cavity. The result is commonly seen as snake like markings on the final moulding. In co-injection moulding, jetting of this type can disrupt the filling pattern and make it difficult and sometimes impossible to obtain skin-core configurations, an example of this is shown in Figure 10.13. These tensile bar mouldings were produced using a sprue and runner that generated a very high shear flow and caused mixing of the layers and a disturbance of normal fountain flow filling behaviour.

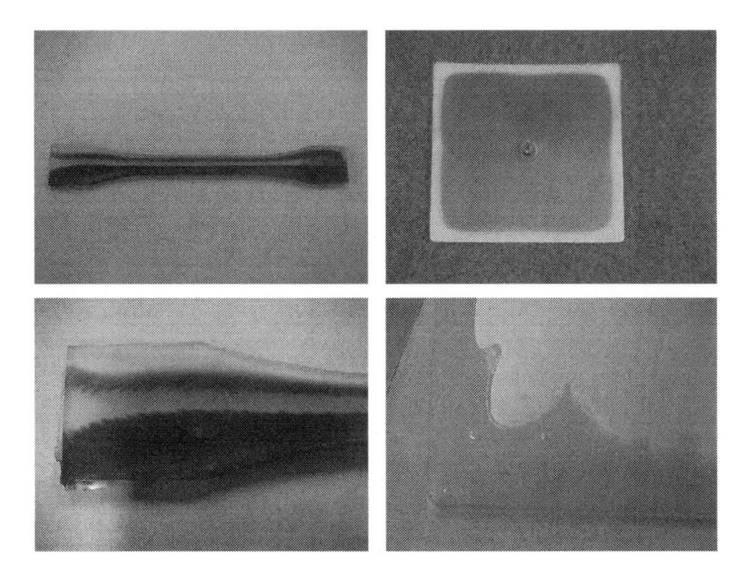

Figure 10.13 Co-injection mouldings produced with non-laminar flow (left upper and lower), with laminar flow (right upper) and melt fingering (right lower)

The moulding parameter of major importance in control of jetting is injection speed with lower speeds resulting in less jetting behaviour. The viscosity will also affect jetting behaviour. However, with adequate tool gating design this should not be an issue unless very high injection rates are used. It does however provide a useful insight into the filling mechanisms at work.

10.8.2.7 Setting Optimum Parameters

In terms of the interface, the main factor in adhesion strength is the thickness of the skin layer. This being due to the increased bonding time available for adhesion with a thicker and therefore slower cooling interface. A thinner layer would also be subject to higher shear from incoming molten material and be more likely to be re-melted and swept away into the melt stream.

Therefore to keep consistent skin thickness across a co-injection moulding, viscosity and moulding parameter effects need to be controlled. It has been found that injection speed is the key parameter in getting uniform skin distribution. Lower speeds being most effective. This will also minimise disturbances in laminar fountain flow especially in the gate region at higher speeds in co-injection moulding. Using materials of different generic families be they immiscible or compatible, there is likely to be a different elastic response to shear and stresses within the mould at the skin core interface, especially in high shear tooling. This instability will affect the mixing dynamics within the interfacial area and in some cases may cause surface defects on the surface of the skin/core interface as a result of stress differences. However, with the development of being able to combine immiscible parts through compatibilisation or through mechanical techniques, it is likely that commercial interest in such material combinations will push further development. There are two obvious areas for this growth, firstly the recycling potential in this technology for mixed materials combinations and secondly the possibility to combine distinct jobs in one in-mould operation. An example of this is to produce in-mould primers for polypropylene to improve paintability or adhesion. This removes the need for the pretreatment stages currently used. Examples of current applications for co-injection techniques now follow.

10.8.2.8 Co-Injection Moulding Application Case Studies

Preforms for blow moulding

Injection blow moulding is commonly used to produce containers such as PET carbonated drinks bottles. In this method, preforms are injection moulded and then inflated in a separate operation. Core bars are used to transfer mouldings from the moulding station to the inflation station. Multi-layer for extrusion blow moulding has been practised for some time with nine or even more distinct layer configurations possible. Multi-layer with injection blow moulding is a newer development.

Thermoplastic olefinic elastomer (TPO) fascia and bumper

Automotive applications require reduced cost and potential recyclability to meet end-of-life vehicle legislation requirements (see next section). Co-injection offers not just the potential to incorporate recyclate in the core but also to utilise post industrial painted recyclates. One such example is the Ford P207 Fascia which uses recyclates in this manner to achieve part recyclability. Once the initial capital machine cost is overcome it is also an economical approach in terms of both material utilisation and recycling potential.

10.8.2.9 Recycling and Legislation

The TPO fascia is an example of the recycling potential inherent in the co-injection moulding technique. Environmental legislation affects a number of sectors of the plastic industry. This legislation varies globally but in Europe packaging manufacturers must consider the Directive on Packaging and Packaging Waste, Directive 94/62/EC. Automotive manufacturers and suppliers must consider the end-of-life vehicle directive (ELV), Directive 2000/53/EC. The electronic and electrical equipment manufacturers must comply with the future demands of the proposed Directive on Waste Electrical and Electronic Equipment (WEEE) (2002/96/EC). Other similar legislation is in place in countries such as the USA and Canada. What this means in real terms is that many current and future plastic applications will demand a consideration of recycling capability. If just worldwide injection mouldings alone are taken into account the quantities of material requiring recycling will be considerable. In this respect co-injection moulding is an ideal solution. Recyclate materials can be buried in the core and, with the exception of transparent materials and tooling of complex geometries or multiple gating, standard injection mouldings can easily be switched to co-injection moulding. With this in mind it is therefore highly likely that this method will become much more widely utilised in the future.

10.8.2.10 Discussion and Conclusions

Co-injection moulding provides processing routes for obtaining property combinations that are, in general, not possible with conventional injection moulding. However, there are a number of factors that have tended to limit the commercial take up of co-injection moulding technologies.

- High capital cost – co-injection machines are around 40% more expensive than an ordinary injection moulding machine.
- Restrictions with regard to mould geometry – parts with sharp corners, changes in wall thickness, ribs or bosses are difficult with co-injection moulding due to the problems of distributing skin and core materials.
- Weld lines – at weld lines there is only skin material, and special solutions, like overflow channels, are necessary if the core material is needed here.
- There is a complicated and poorly understood relationship between rheology and process parameters. Interrelationships exist between melt temperature, viscosity, and temperature differences between the skin and core melts.
- Restrictions with regard to material combinations – even for compatible materials, the choice of a certain grade could influence adhesion and give inferior properties. Since processing conditions can affect the rates of interdiffusion of skin and core, they can also affect the properties of the final component. Again, the effects are complicated and not well understood.

However, with the increased need to meet environmental recycling legislation and targets, in addition to an increased understanding of the process, it is likely that uptake of co-injection technology will continue to rise in future years.

10.8.3 Bi-Injection

In bi-injection, materials are processed simultaneously at different points into a tool. In this manner it has features of co-injection (simultaneous injection of two materials but through one nozzle) and core back moulding (where injection is at two points but in sequence). This method is discussed in detail later.

Moulding in this manner gives a shorter cycle time than core back moulding and it is a method of achieving the usual requirements of multi-shot such as multi-colour or hard/soft combinations. The knit line is also stronger due to the higher temperature at the interface when the flows meet. However, the materials do not maintain good separation and definition at the interface, which can be a problem in potential applications for this technique.

10.8.4 Interval Injection Moulding

This method produces marbling type effects and is generally, though not always, confined to mouldings of different coloured materials of the same type. This can produce a random or regular colour distribution pattern as required. It works by injecting the materials in either a simultaneous or sequential manner from a combined nozzle as shown in Figure 10.14. It is similar to methods used to produce co-injection mouldings. In this case, however, injection is not to produce a skin/core configuration, the material flows together and small scale mixing operations take place but not at a level to completely mix the two materials together. The two injection units are coupled together using a special interval unit, inside which the mixing nozzle is located. This is shown in Figure 10.15.

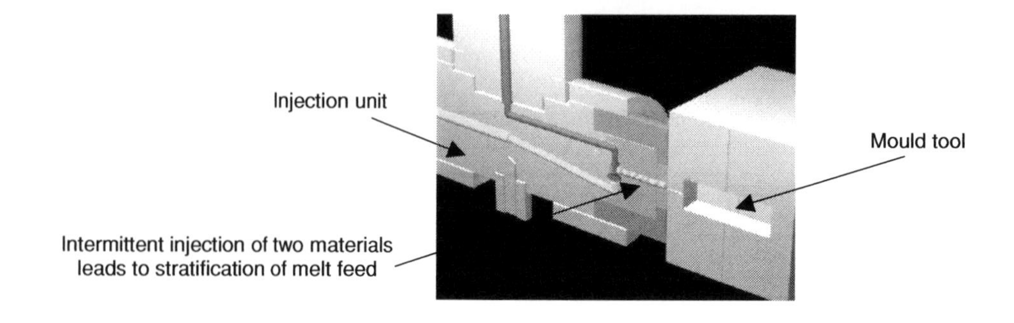

Figure 10.14 Interval injection moulding

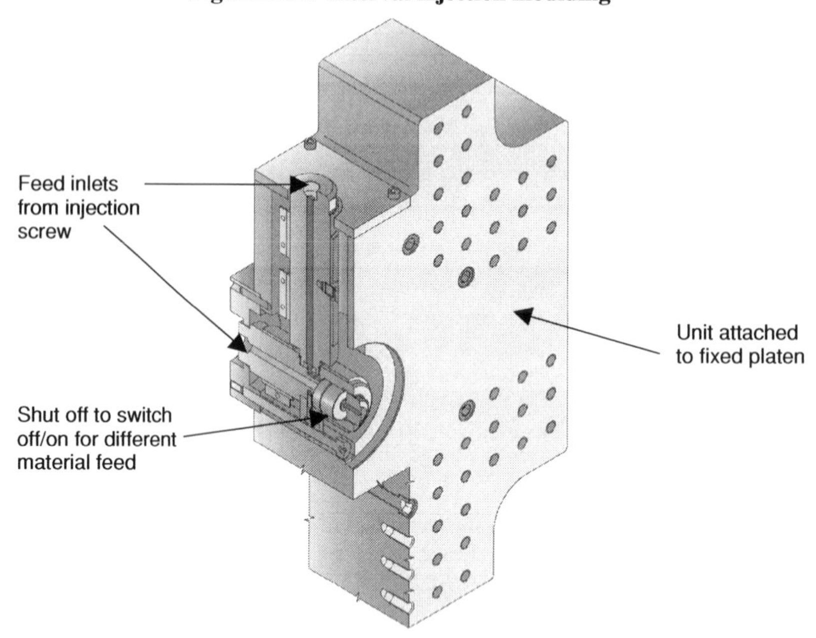

Figure 10.15 Unit for interval moulding

The resulting colouration is controlled by the mixing dynamics of the materials in question and the shear they are subject to, for example, by the size and position of the sprue. Injection can be simultaneous or set on an alternating cycle to give a pulse type effect. The size of the injection steps and the speed of injection will also alter the colouration of the resultant moulding. Applications tend to be for mainly cosmetic applications such as covers and boxes and applications such as buttons where the colouration can be used to decorative effect. An example of a moulding is shown in Figure 10.16.

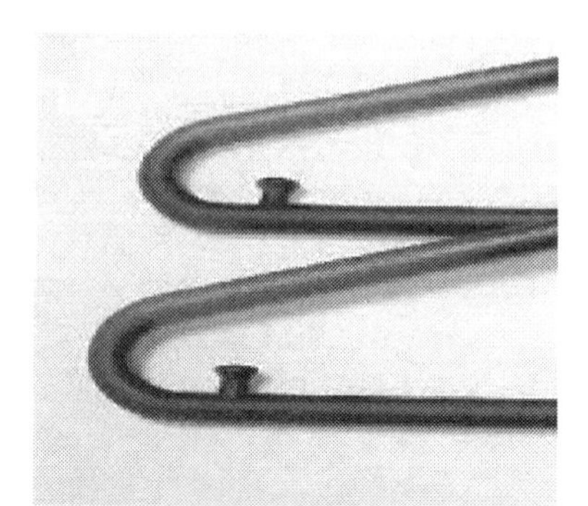

Figure 10.16 Marbling effect on a moulding

10.9 Assisted Moulding

10.9.1 Gas Injection Moulding Technology (GIT)

The characteristic feature of gas injection moulding technology is the filling of a form with two different materials. Plastic materials form the first component. The second component consists of a gas, generally nitrogen (N_2). The two components do not mix. All standard injection moulding machines which are equipped with a device for introducing gas are suitable for the GIT process. The injection of the gas may be performed by a machine nozzle or by a separate injection module in the mould.

10.9.1.1 Process Technology

The process sequence begins with the injection of the first component – the plastic. A dose of approximately 70% ±20% of the volume of the cavity is proportioned and injected.

Shortly before the conclusion of the injection phase, the gas injection phase begins. The brief overlapping of injection and gas injection phases is intended to prevent a speed break of the melt front and the switch over marking which is related to this. The location for introduction of the gas is best in areas with large melt accumulations. The gas fills the cavity and forces the melt forward. Design specific cavities in the moulded part are the result.

Once the cavity is completely filled, the gas-holding pressure phase begins. This pressure phase is applied until the part is dimensionally stable (maximum gas pressure is 400 bar). Gas pressure during this time is constant throughout the entire canal. Because of the relatively low gas pressure, there are correspondingly low interior pressures in the mould, in turn leading to low clamping forces in the injection moulding machine.

After the initial solidification of the melt, the gas pressure is reduced. This takes place either by allowing the gas to escape into the ambient atmosphere, or by recovering a certain proportion (up to 90%) through the machine nozzle or the mould nozzle. The process is illustrated in Figure 10.17.

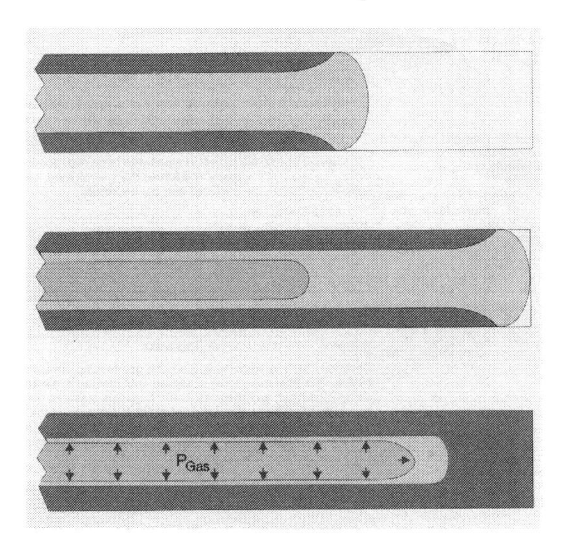

Figure 10.17 The gas injection moulding process

10.9.1.2 Patent Situation

Development in GIT, has been restricted to some extent by the number of patents that exist relating to this process. If a process that is protected by a patent is employed, expensive licence fees must sometimes be paid. This section is written without regard for the patent situation, which must be clarified if considering adoption of this technology.

10.9.1.3 Advantages and Disadvantages of GIT

Depending on the mould construction, the following advantages can be realised:

- Greater range of configuration options in the design of moulds
- Material savings (generally 20-30%)
- Cycle time reduction for moulded parts with thick walls
- Increase in mechanical rigidity with equal weight
- Moulds without sink marks
- More uniform shrinkage, lower residual stresses, significantly less distortion
- Reduced clamping force
- Implementation of long flow paths
- Better surface in comparison to foamed parts
- Simpler mould construction possible in some cases
- Possibility of integrating thick and thin areas

The most significant disadvantages of gas injection moulding technology are:

- Additional costs for gas pressure generation equipment and pressure regulator modules, gas, machine nozzles or mould nozzles, and license fees
- Frequent jetting
- Empirical derivation of setting parameters
- Greater weight deviations
- Hole at the injection point
- Strength/tightness at sealing is frequently inadequate
- Welding seams are typically more frequently visible
- Injector must be cleaned more often in some cases.

10.9.1.4 Process Variations in the Application of Gas Injection Moulding Technology

Gas injection through the machine nozzle

Here, the introduction of the gas is performed directly through the machine nozzle. The process sequence is:

(1) The hydraulic GIT needle shut-off is open. The plastic material is injected. The material cushion for the sealing process remains in the screw pre-chamber as illustrated in Figure 10.18a.

(2) The hydraulic GIT needle shut-off nozzle is closed. Nitrogen is introduced. The melt core is displaced. The pressure holding phase is active. After the conclusion of the pressure holding phase, the recovery phase of the nitrogen gas begins. There is a brief retraction of the nozzle for pressure reduction. This stage of the process is shown in Figure 10.18b.

(3) Finally the hydraulic GIT needle shut-off nozzle is opened and the material cushion is injected for sealing as shown in Figure 10.18c.

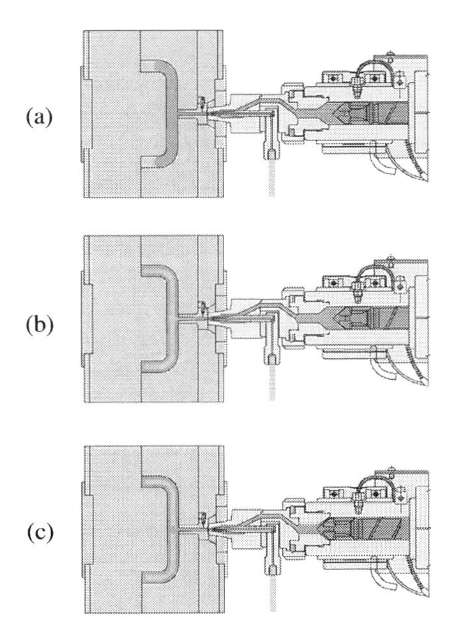

Figure 10.18 Gas injection through the machine nozzle

Gas injection through an injector module in the mould

This can be carried out in two ways. The gas can be introduced either through the sprue or directly into the moulded part. The hydraulic needle shut-off nozzle is open. The mould material is completely injected. No material cushion remains behind. This stage of the process is shown in Figure 10.19.

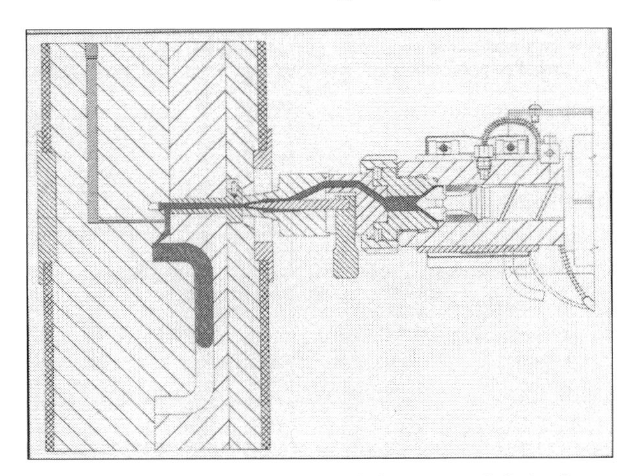

Figure 10.19 Gas injection through an injector module in the mould (1)

The hydraulic needle shut-off nozzle is closed. Nitrogen is introduced into the mould through the injector module. The gas displaces the melt core. The screw can proportion doses, since the holding pressure is provided through the gas pressure phase. Upon conclusion of the gas holding phase, the recovery of the nitrogen is activated. This process stage is shown in Figure 10.20.

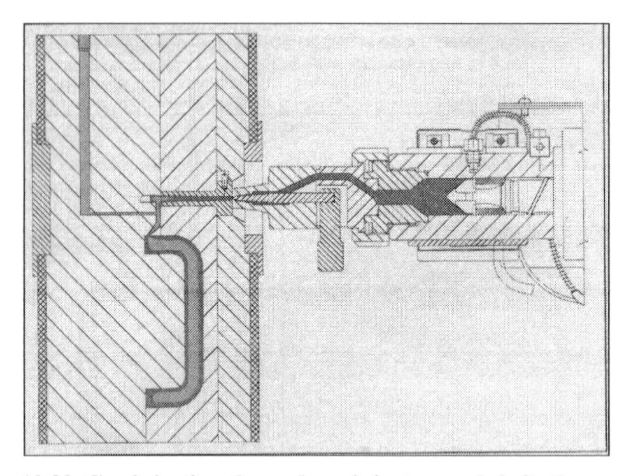

Figure 10.20 Gas injection through an injector module in the mould (2)

Gas injection in the moulded part

In this process, the introduction of the gas takes place directly into the moulded part through an injector module in the mould. The gas introduction point can be at any selected location on the moulded part in this process. The disadvantage here is that the hole that is created in the part cannot be automatically sealed.

In this process, it is possible to work with the nozzle open at intervals. The mould material is completely injected. No material cushion is left behind.

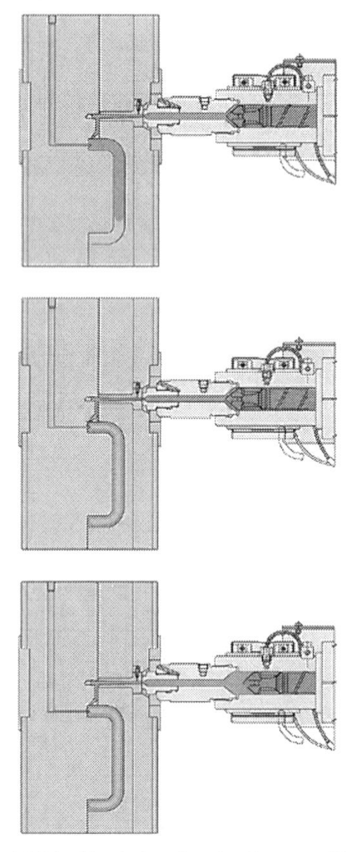

Figure 10.21 Gas injection in the moulded part

Nitrogen is introduced in the mould through the injector module. The holding pressure is realised through the gas holding phase. The delayed dosing phase runs until the sprue is sealed. Upon conclusion of the gas holding phase, the recovery of the nitrogen is activated. The screw can simultaneously begin dosing again. The process sequence is shown in Figure 10.21.

Melt blow moulding technology

Melt blow moulding technology is also divided into two subprocedures, melt back pressure technology and melt extrusion technology in a secondary cavity. The advantage of these processes is the avoidance of switch over markings on the surface of the moulded part.

Melt back pressure technology/gas injection in the moulded part

In melt back pressure technology, the entire moulded part is first filled with material (Figure 10.22a). Nitrogen is then introduced at the flow path end, which forces the plastic melt back into the plasticising cylinder (as shown in Figure 10.22b). With the hydraulic needle shut-off nozzle open, the mould material is injected (the moulded part is filled completely). The hydraulic needle shut-off nozzle remains open. The gas is injected. The screw is withdrawn. After the completion of the gas injection, the hydraulic needle shut-off nozzle is closed. The gas holding phase and recovery of the gas follow; the screw can simultaneously begin dosing again. The process is illustrated in Figure 10.22.

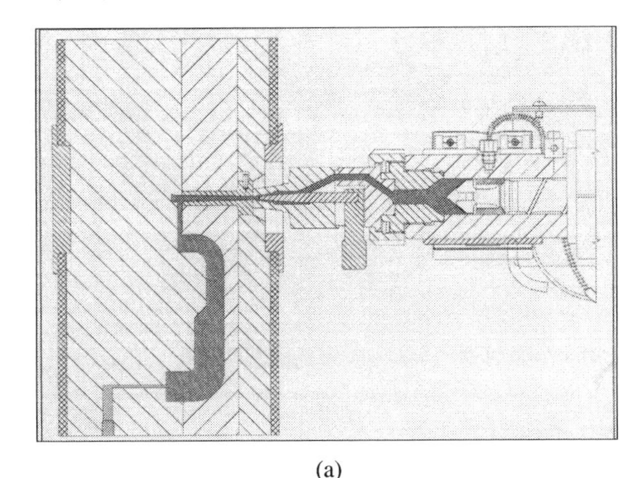

(a)

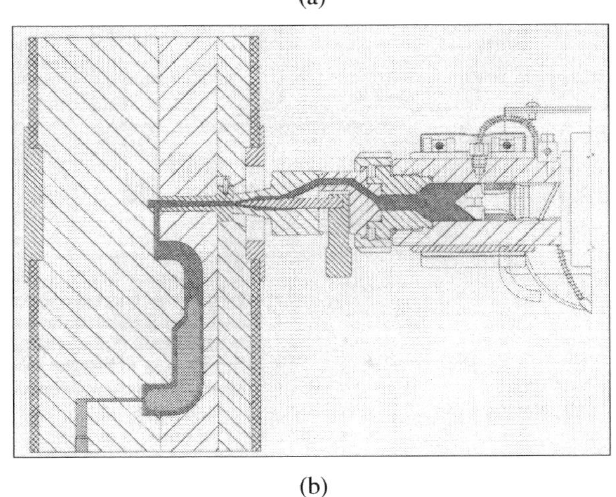

(b)

Figure 10.22 The melt blow moulding technology process

Melt extrusion technology/gas injection in the moulded part

In melt extrusion technology, the cavity is first completely filled with material (Figure 10.23a). Concurrent with the start of the gas injection, one or more secondary cavities into which the excess melt may be displaced are opened (Figure 10.23b).

The process is performed with an open nozzle. The material is injected, and the moulded part is filled completely. The second cavity is closed. A hydraulic slide gate opens the secondary cavity. The gas injection begins concurrently and the excess material is forced into the secondary cavity. The gas holding sequence runs completely through. The recovery of the gas follows after the gas holding phase. The screw can simultaneously begin dosing again. The process is shown in Figure 10.23.

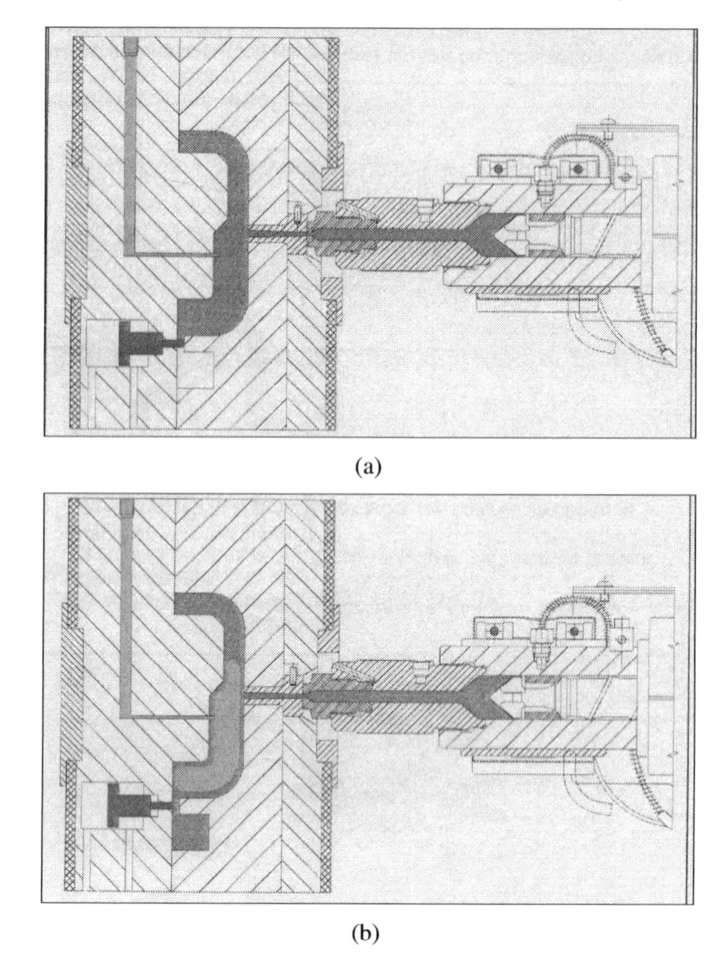

(a)

(b)

Figure 10.23 Melt extrusion GIT

Core pull technology

In core pull technology, the cavity is first completely filled and the moulded part is subjected to melt back pressure (Figure 10.24a). Before or during the introduction of the gas, an additional displaced volume is created in the main cavity by the withdrawal of one or more of the cores (Figure 10.24b).

In core pull technology, work also proceeds with an open nozzle. The core is inserted. The material is injected, and the moulded part is completely filled. The core is withdrawn from the moulded part hydraulically, thus opening a space for additional volume. The mould material is completely injected. The injection of the gas into the moulded part occurs simultaneously. The gas holding phase follows. The recovery of the gas follows after the gas holding phase. After the gate is sealed, the screw can begin dosing again. The process is illustrated in Figure 10.24.

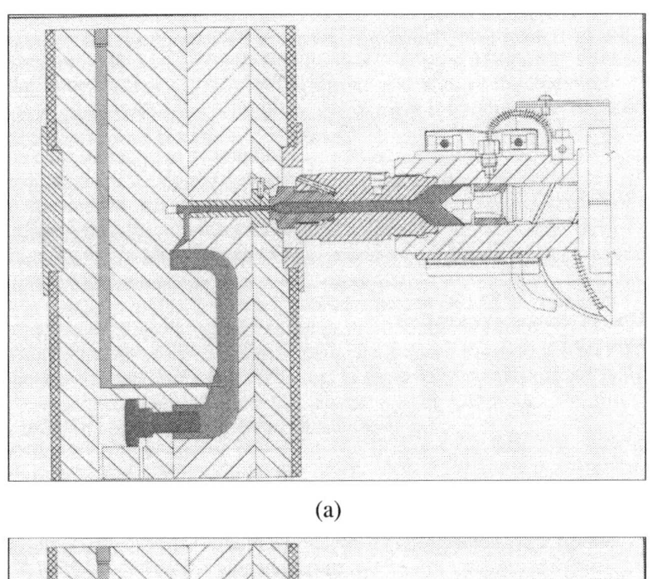

(a)

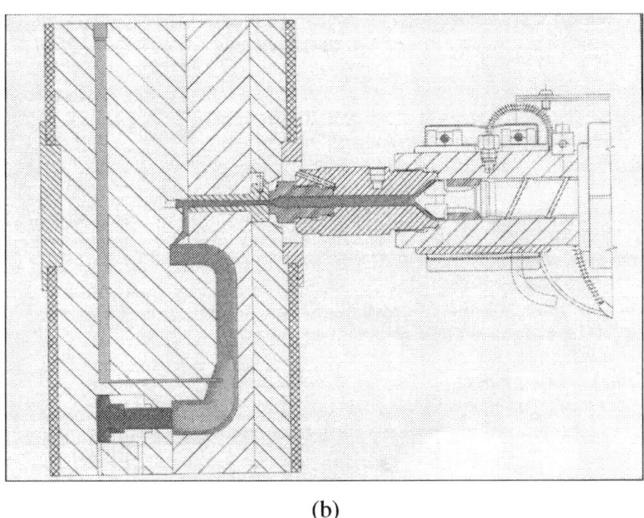

(b)

Figure 10.24 Core pull technology GIT

10.9.1.5 Systems Technology for the Implementation of Gas Injection Technology

Gas pressure regulation

Gas pressure regulation consists of the pressure regulation module and the electrical control unit. The pressure regulation module reduces the established system pressure to the desired gas pressure. The electrical control of the pressure regulation module is provided through either the machine control unit or through an external memory with programmed controls. The switch over from melt injection to gas injection can take place either as a function of time, stroke or pressure. Either the hydraulic pressure of the injection moulding machine or the internal pressure in the mould can be used for the pressure dependent impulse.

Two different design concepts exist for the control unit:

(a) Stationary unit: the stationary unit consists of the pressure regulation modules and an electrical control unit which is integrated in the machine controls. This process method is implemented with the control unit. External operating elements are not necessary here. The simple and direct entry of all necessary parameters through the user monitor of the machine control unit is especially advantageous with this system. In addition, the GIT device can be integrated into the monitoring function of many modern machines.

(b) Mobile control unit: the mobile control unit consists of pressure regulation modules and a separate electronic control unit with a manual programming device. The link to the machine is created through an interface. The mobile gas pressure control unit can be installed in a short period on any other injection moulding machine with the GIT interface.

GIT nozzle systems

Machine nozzles and mould nozzles are available for the introduction of the gas into the moulded part.

Machine nozzle

After the cavity is filled with the plastic melt, the gas is also injected through the same machine nozzle. The injection gate and the gas opening are at the same location of the moulded part.

Mould nozzle

Mould nozzles provide advantages due to the fact that the melt injection gate and the gas injection point do not have to be at the same location. Because of this, several gas introduction points may be arranged on the mould piece to meet specific requirements. Thus, the separate gas runners can be controlled differently, allowing special requirements to be fulfilled. Mould nozzles are thus distinguished by their great flexibility in planning and in possible mould modifications. With regard to the manner of installation, a distinction is made between fixed and movable mould nozzles.

(a) Fixed mould nozzles: these mould nozzles are installed in a fixed position in the direction of mould ejection. They may be easily placed in the mould since they have especially small dimensions.

(b) Movable mould nozzles: when mould nozzles cannot be installed in the direction of mould ejection for design reasons, they must be installed with movement capability (the nozzle can then be inserted into the cavity or withdrawn from it). The system for moving the mould nozzle can also be operated with nitrogen.

10.9.1.6 Configuration Guidelines for GIT Moulding

General

Moulded parts which are especially suited for the application of the GIT process are rod shaped parts, such as handles. These parts are characterised by very high wall strength.

- This makes it possible to produce plastic parts with very high wall strength without sink marks in one process run.
- Material savings up to 50% are possible.
- Significantly shorter cycle times result from shorter cooling times due to the reduced wall thickness.
- Two-piece parts which previously had to be glued or welded together after the injection moulding process may now be manufactured in one process run.

Cross-sections

In the design of the cross-section of moulded parts which are to be produced in the GIT process, it should always be observed that the cross-section form of the gas pocket is as round as possible. For this reason, an attempt should be made to achieve a circular outer contour in order to avoid uneven material accumulations. Design guides are shown in Figure 10.25.

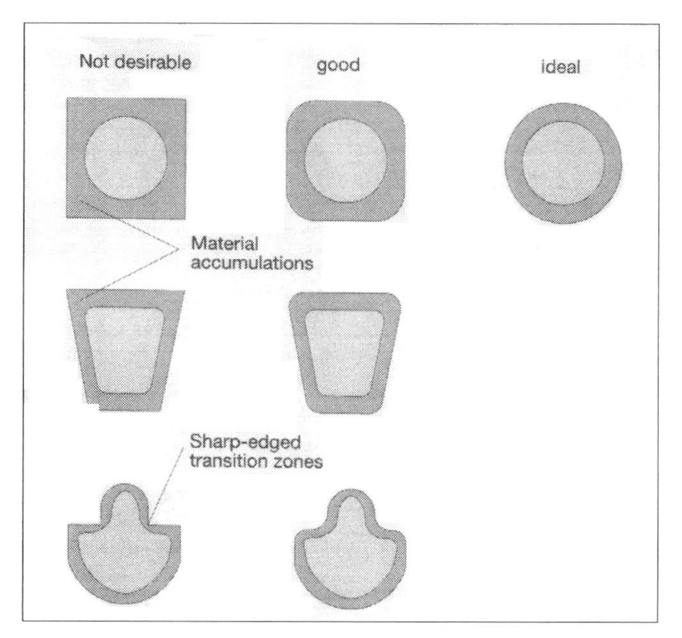

Figure 10.25 Cross-sections in the GIT process

Since rounded external contours are frequently impossible in practical applications, the following configuration guidelines should be observed:

- Implement external contours as close to round as possible
- Avoid sharp edges, round liberally instead
- Avoid material accumulations at the corners
- Provide for uniform wall-thickness for the entire moulded part.

In the configuration of right-angled cross-sections, it is possible that the melt cannot be displaced in the narrow sides. Observe these rules of thumb for this situation:

- Maximum moulded-part width = 3 to 5 times the height of the moulded part
- Minimum moulded-part length = 5 times the height of the moulded part.

Bends and curves

There is always the chance of material accumulations on the exterior surface in the area around bends, as well as reductions of the wall thickness on the interior surface. For this reason, observe the following:

- Avoid sharp edge areas
- Select the greatest possible bend radius.

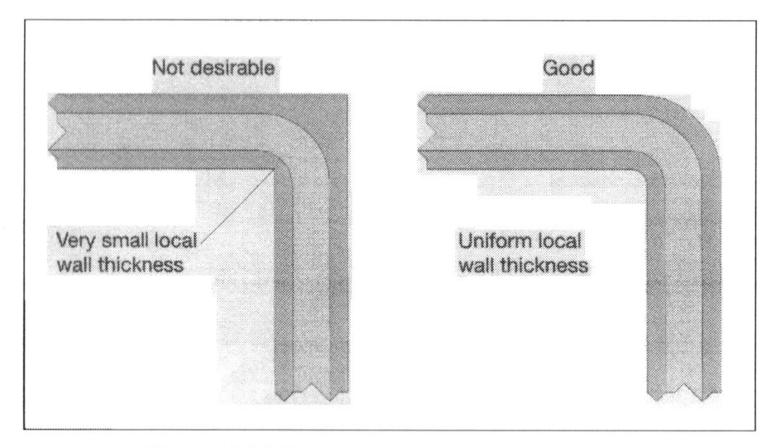

Figure 10.26 Local wall thickness at a 90° angle

Cross-section changes

It occurs frequently that significant changes in the cross-section must be made within a part. For these parts, it should always be observed that the melt flow and the gas flow are guided from the larger to the smaller cross-section. This procedural method leads to the desired origin of flow of the plastic in the cavity and hence to a uniform distribution of the melt. As a reciprocal effect, uniform distribution of the melt creates a uniform distribution of the gas.

Gas introduction/injection location

The injection position and the position of the gas introduction have special significance. Since attempts should always be made to achieve wall thicknesses which are as uniform as possible, it is extremely important to fill the cavity homogeneously with the melt so that the gas bubble can be specifically guided as designed.

The following basic rules must be observed in the selection of the gas introduction location:

- Implement gas and melt injection from the front side of the moulded part in a latitudinal direction whenever possible.
- Plan for only one flow path and one cavity whenever possible.
- Do not establish the injection location in a visible area or in an area of mechanical stress of the moulded part.
- Position the gas introduction location in the vicinity of injection so that the gas bubble can follow the melt flow and only expand in one direction.
- Observe the force of gravity, fill the moulded part from bottom to top in order to achieve the desired swell flow.

If it is nevertheless necessary to establish the gas introduction point at a different location of the moulded part, then it must be observed that:

- The injection point lies exactly in the middle of the moulded part,
- The flow paths are balanced as precisely as possible and
- The mould piece is configured symmetrically.

10.9.1.7 KoolGas™

KoolGas™ is a further development of gas assisted injection moulding, originally developed at the University of Warwick, UK. In this process, cryogenically cooled gas is used to form the moulding core, and also to provide internal cooling. It has several important advantages over the standard GIT techniques including reduced cycle time, smooth bores, balanced thermal stresses and improved morphology across the moulding wall. It can also be used to produce thinner walls than conventional methods. The equipment requirement is similar to conventional GIT except that a special cryogenic cooling module is required to cool the gas to suitable temperatures.

10.9.2 Water Assisted Injection Moulding (WAIM)

Water assisted injection moulding (WAIM) or water injection technology (WIT) was developed by the IKV (Institute for Plastics Processing) in Germany in 1998. Although this was not the first research work in this field, significant barriers to success in previous work were overcome. The aim was to replace the nitrogen gas with water in order to reduce cooling times. The use of water injection presented technical challenges: to generate both high pressures and high flow rates as well as defining a single hollow section. With these problems now overcome WAIM has become a viable fluid injection technique of increasing commercial interest.

The process itself is very similar to GIT except that during the water injection one or more hydro-pumps are used to inject the water at temperatures between 10 °C and 80 °C, and up to 350 bar, into the polymer pre-injected into the cavity. Like GIT, the melt is displaced into the low pressurised areas completely filling the cavity. The water flows through the hollow body to provide cooling until the part has sufficiently solidified for ejection. Finally, the water is removed either by gravity-induced draining or by feeding compressed air through one of the nozzles.

The WAIM process can be used by a short shot technique but current WAIM applications tend to employ the full shot technique and subsequent displacement of the melt into overflow channels to produce the hollow shape.

The main advantage of WAIM is the reduction in cycle time, due to the higher specific heat of water compared to nitrogen gas. Cycle time reductions of up to 70% have been claimed by WAIM, as opposed to conventional GIT. This technology has generally found applications in rod shaped mouldings with large diameters (typically above 40 mm in diameter).

Because of viscosity differences between gas and water (the water is much more viscous), larger injection nozzles are required in water injection in order to achieve suitable flow rates. Therefore, the size of the nozzle places a restriction on its potential applications.

Investment costs are similar to those associated with GIT, including items such as a water pressure generator, pumps and nozzles. There are also safety issues associated with this method as both operators and moulds need to be protected from the high pressure water jets and water damage.

There are a number of machine manufacturers who now offer water injection systems. For example Battenfeld offer AQUAMOULD^{TM}, Engel offer WATERMELT^{TM} and Ferromatik Milacron offer the Aquapress system, to name but three.

10.10 Multi-Shot Moulding

Multi-shot moulding has been around for over thirty years and is used as a method of placing materials either side by side (abutting), one on top of the other within an overlap, or superimposition of one shot onto another. To do this special tooling and machinery is required. Common examples of such mouldings are keypads with the numbers made of one colour and the letters moulded-in using another colour. The advantage of this technique being the elimination of the printing processes which would otherwise be required to mark the keypads. Multi-colour automotive taillights are also made by this technology. Another common application of this technology is to combine hard and soft materials to produce a 'soft feel' component. Handles such as on doors or toothbrushes are common products benefiting from this technology.

Multi-shot moulding techniques are well established, their growth being pushed by the development of thermoplastic elastomer (TPE) materials, enabling rigid and flexible material combinations such as those described previously to be employed. These can also be seen in a variety of other applications from automotive seals to bras.

A number of moulding methods can be employed to produce a multi-shot moulding but whilst mouldings can be produced by a variety of methods, those produced from the same material but in multi-colour multi-shot enjoy the highest market share.

Multi-shot techniques produce not only multi-colour but also multi-material mouldings. The most common methods are the use of tool rotation, the core back technique and transfer tools. Transfer tools can be used to move shots from one cavity to another, this technique is very similar to over-moulding. These processes will be described in detail in later sections. Some applications are listed in Table 10.3.

Table 10.3 Applications for multi-shot mouldings		
Material combination	**Properties**	**Application**
Same material/different colours	Aesthetics	Buttons, mobile phone casings, toothbrush, automotive light casings
Soft feel/hard feel	Increased customer appeal through 'soft touch' properties	Various handles, toothbrush, camera, screwdriver
Transparent/coloured	Viewing panel incorporated into moulding	Light casings
Very flexible TPE, hard substrate	Sealing properties	Lids with moulded on seals
LSR, hard substrate	Sealing properties	Seals

As well as these applications multi-shot also finds use for power tool cases, battery cases and domestic appliances (e.g., kettles handles, vacuums, lawn mowers and electric toothbrushes).

10.10.1 Machine Technology

Multi-shot processes, as the name implies, require multiple shots of material to make a single component. For each one of these materials an injection unit is required. To mould these multiple shots also requires special tooling and equipment. Multi-shot capability can be built either into the injection tool or controlled by the injection moulding machine. To enable multi-shot, multiple injection units can be arranged to feed machines in a number of ways as the next section will explain.

10.10.1.1 Injection Unit Configurations

Multiple injection units can be arranged around the clamping units as combinations of horizontal and vertical units in piggy back or right angle configurations. Some examples are shown in Figure 10.27.

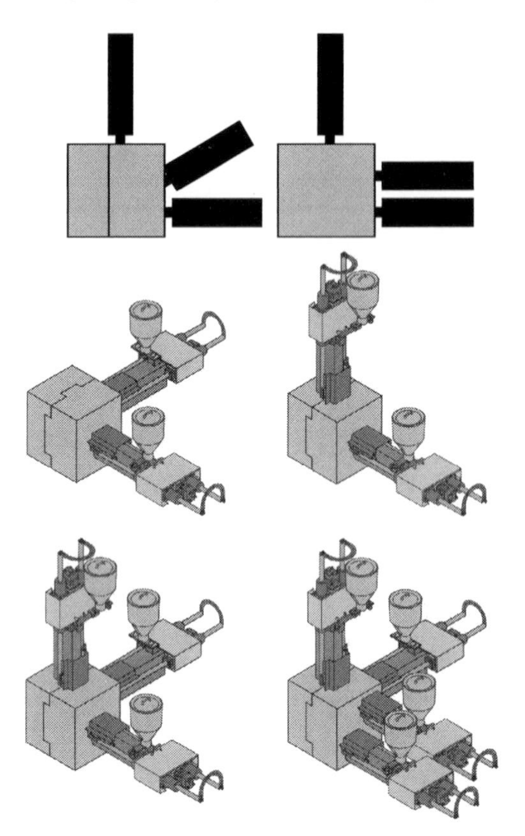

Figure 10.27 Possible arrangements of injection units

Using a vertical injection unit can save space and hence is the most used position for multi-shot moulding at the split-line. For mould changing, the units can be slid towards the nozzle. Where the vertical position cannot be used, perhaps because of lack of factory height, the second unit can be positioned at right angles. The position of the unit can be adjusted both horizontally and vertically, although the former is available as an option rather than as standard, by some manufacturers. Again as in the vertical unit, it can be moved to the nozzle-side to change the mould.

With the piggyback method, units can be angled above the main unit. This method saves floor space like the vertical method but requires less ceiling height. The additional unit can be moved in two ways. Attached to the main unit but with the nozzles thermally separated or with a cylinder which moves separately from the main cylinder.

Plastication design

In terms of the injection screw configurations within the injection units, these tend to have become standardised at 22D (the ratio of screw length (L) to its diameter (D)). For faster production and high quality mixing this is often increased up to 26D. Specialised screws for hygroscopic materials, which enable venting, screws for increased mixing of materials and configurations for temperature sensitive materials such as PVC are all commonly available from suppliers.

Machine type

When robots are required to control insertion and removal of parts, machinery design can be an issue. The use of tie-bar-less machines as opposed to the conventional tie bar models is usually a contentious subject with advantages and disadvantages inherent in both machinery designs. However, when it comes to multi-material technology, tie-barless machines can offer distinct advantages in the ease of which robots can be utilised and the larger mould mounting area these machines offer. Often, this can mean a smaller capital outlay on machinery as a smaller machine may be purchased.

Moving platen support
Large daylight required,
extended tie bars (max. 1250 mm 200 ton machine)

Figure 10.28 Example of clamp unit for multi-shot

Now machinery issues have been briefly overviewed, individual technologies for multi-shot will be introduced. These are:

- Core back moulding
- Rotary method
- Transfer moulding

10.10.2 Core Back Moulding

Also referred to in various literature as composite injection moulding or multi-shot. The manufacturer Battenfeld uses the trade name Combiform for this process. Core back is a tooling controlled process.

Core back moulding, thought of simplistically, is one tool taking multiple shots within a single machine cycle. It allows different areas of the tooling to be opened or closed to specific material feeds. This is achieved through the use of moving slides or inserts and is illustrated in Figure 10.29.

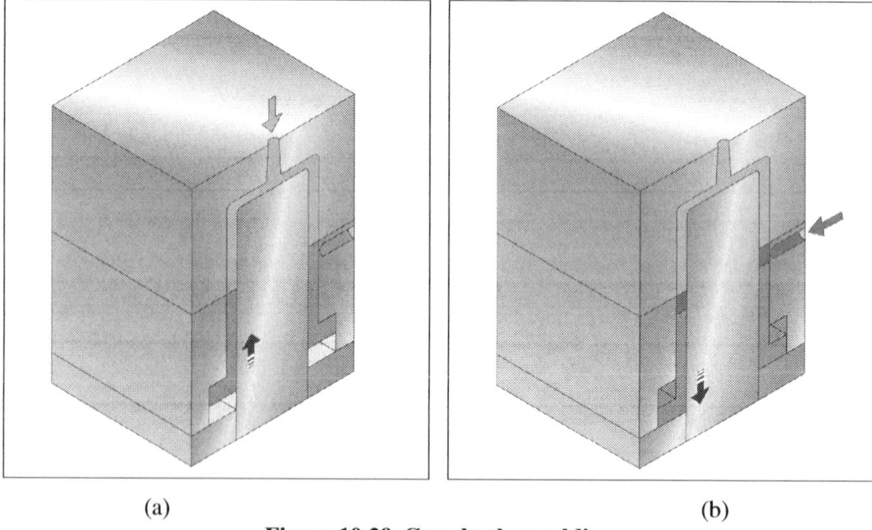

(a) (b)

Figure 10.29 Core back moulding

In Figure 10.29a the insert is closed. This constrains the first injected material to this area of the cavity. When the insert is opened, as shown in Figure 10.29b, the second material can feed into the newly opened cavity area and flow into the material already injected to give the multi-shot component. In this way slides and inserts are used to block and control access to the cavity for particular materials feeds.

For example, a two-component process such as shown in Figure 10.29 may consist of the following stages:

1. The first material is injected into the cavity
2. Using a core puller to activate a slide seal, a further area of cavity is revealed
3. The second material is injected into the cavity
4. The completed multi-shot component is ejected.

The ability to complete the process without mould opening or preform transport are the main advantages to this method when compared to other multi-shot techniques. However, increasing the number of components beyond two will certainly significantly increase the cost of tooling due to the increased intricacy required. The machine must also have the necessary means to actuate all the slides in the tooling.

As in all multi-material injection, attention must be paid to the compatibility of the melts. The use of the core back technique enables greater bond strengths to be achieved than in other multi-shot techniques as the time between injection of the first melt and injection of the second material can be optimised. However, the sequence of injection of the first material then the second material is longer than in other multi-shot techniques, which proceed in parallel. Therefore in components whose design lends it to both techniques, a detailed analysis of the economic implications of the process routes may be required to determine the most appropriate method of production.

However for some parts, other methods may be inappropriate. This could be due to either tooling costs, especially on large parts, or to the nature of the material. For example robotic transfer of a very flexible material may be difficult. Other important factors may be cycle times, for example the cooling times may be very short on a thin moulding. In cases such as these, a core back tooling system may be the best option.

10.10.3 Rotating Tool

In this method, the mould rotates through 180° for a two-shot part or 120° for a three-shot part. Rotational capability can be machine or tool based. There can be an integral rotary capability designed into the tool or the machine can be equipped with a rotary attachment to the moving platen as in Figure 10.30. The choice usually comes down to economics. If rotational capability is to be used regularly it is cheaper to have it on the machine, than to continually buy more expensive tooling. A rotary platen must have an accurate indexing device to control the rotation and the stroke needs to be both fast and cushioned to prevent damage. The platen must also have the facility to mount ejector pins. An example of a mounted electric unit is shown in Figure 10.31. A tool is mounted on the indexing unit and moulding can proceed as shown in Figure 10.32.

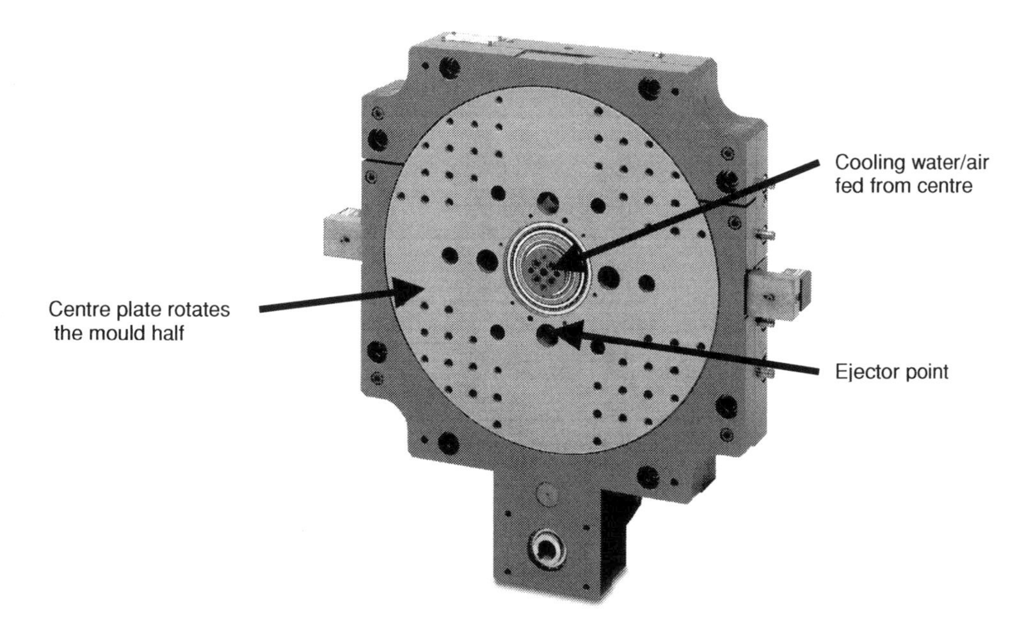

Figure 10.30 Indexing unit

Figure 10.31 Electric indexing unit

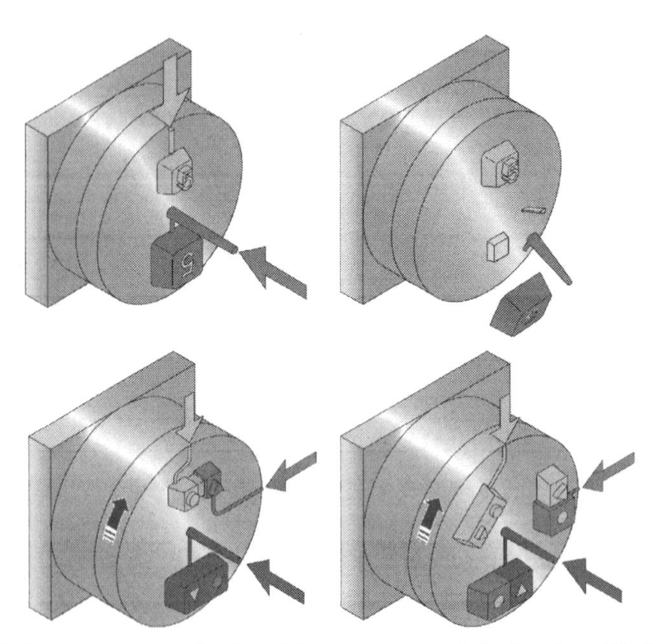

Figure 10.32 Two-component (top) and three-component (bottom) multi-shot moulding

The process proceeds in parallel so at any stage there is a shot being produced by each cavity. This makes the overall cycle time per moulding shorter than the core back technique described earlier. Generally the moulding produced in the first cycle should be expected to melt only on the very surface. This gives the good material separation required but still forms an adequate bond. This does however require good control of the process.

10.10.4 Transfer Moulding

In this method, instead of rotating the mould, a robot is used to transfer the moulding to the next cavity where it can then be over-moulded. This is shown in Figure 10.33. In this example the robot will move the upper moulding to the lower larger cavity as the tool opens after each cycle. Like rotary methods, moulding proceeds in parallel with a moulding produced in each of the cavities during any cycle. Therefore the cycle time will be dependent on the moulding requiring the longest moulding time. As in rotating methods, a good bond is required whilst maintaining distinct separation of materials. High accuracy is required in placing the insert to get good definition and registration on the final component. A means must therefore exist to hold the preform accurately in place before the second material is injected onto it.

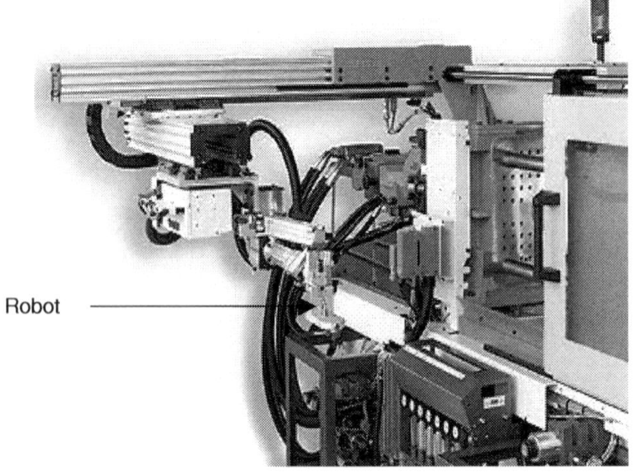

Figure 10.33 Transfer tool with robot

Transfer moulding is not restricted to one machine. Robotics can be used to move the preform to a second machine. However, this would involve investment in both another machine and a second injection tool. If multiple layers are required however, this can be used for example, to produce a four layer moulding: two layers in one machine, robotic transfer to a second machine whereby two further layers can be added. There are several advantages to the use of automation. These include reduced cycle times and the ability to control the process via the machine interface. The robots can also ensure surface scratches or damage is kept to a minimum if an aesthetic component is in production.

10.10.5 Multi-Shot with a Single Injection Unit

The Mono-Sandwich technique for co-injection moulding was described in Section 10.8.2.3 to which the reader should refer for machinery details. This technique can also be used for over-moulding by using the core back technique, again described earlier. In this technique termed the monosandwich 5 process, 'an additional valve is required in the runner system that can release different valves as necessary'. Once the melt is layered, the first component is injected. The valve is switched within the mould to expand the cavity and then the rest of the shot is injected to create a multi-shot component.

10.10.6 Materials for Muti-Shot Moulding

10.10.6.1 Material Selection for Multi-Shot Moulding

This section will introduce material selection issues for multi-shot moulding as well as providing some introductory information on two classes of materials which may not be familiar to the non-specialist moulder namely thermoplastic elastomers (TPE) and liquid silicon rubbers (LSR). This section will begin, however, with consideration of adhesion.

Tables of compatibility have already been introduced, such as Table 10.1. These describe material miscibility and therefore likely adhesion. Other characteristics may also be considered, for example, possible modifications through alloying or addition of compatibilisers. The thermal properties are also important: both softening and melting temperatures and ranges. This is due to the requirements of limited melting to form a bond but ensuring the individual materials are clearly separated. Moulding machine variables will affect this, through changes to melt and mould temperature and injection speed. The molten contact time will drastically affect the final bond strength and the separation achieved. If the first shot becomes molten at too deep a layer, it will flow when the second material is injected onto it. This may also affect the decision as to what type of multi-shot technique can be best employed. The testing of bond strength can be achieved by peel testing or by standard tension testing as required.

10.10.6.2 Material Process Order

The melting and softening characteristics of materials have been mentioned already. They also affect the potential process order. Since the material injected first is already well on its way to cooling when the second material is injected, it is better to mould the material with the lowest melting temperature first. This is where one technique may become preferable over another for certain material combinations. For example, core back moulding can use less than a full cycle between material injection. This means that injection of the second component can occur before full cooling has taken place, therefore increasing the potential to achieve adhesion in some cases. This is not possible with the rotary or transfer methods.

10.10.6.3 Using Thermoset Materials

The combination of thermoplastic/thermoset combinations is a further area of development There are two possibilities for this process.

(1) The thermoplastic is moulded and placed in the thermoset mould
(2) The thermoset is moulded and cured and placed in a thermoplastic mould.

With the first of these processes, the choice of thermoplastic will be much dependent on the temperature required in the thermoset mould. If high cure temperatures are required very few materials may be suitable.

With route (2), it may prove difficult to achieve a bond between a fully cured thermoset and the molten thermoplastic, as there will be no interdiffusion. However clever design, making use of both material properties such as shrinkage, and tooling to promote mechanical interlock, may remove the need for adhesion. This is an area of ongoing development in both Germany and the UK and is likely to hit commercial exploitation in the near future.

The core back technique offers increased opportunity to obtain adhesion between materials. This route would allow the thermoplastic to contact the thermoset before the part had fully cured. This, however, presents problems of its own as controlling the mould temperature and the differing thermoset/thermoplastic flow fronts would be extremely difficult.

Flexible/rigid combinations can be achieved by combining liquid silicone rubber with thermosets. The similar processing temperature requirements and process control is much enhanced with this combination providing that the mould is designed with adequate consideration of flows.

10.10.6.4 Liquid Silicone Rubber (LSR)

Liquid silicone rubber can also be combined with some thermoplastics and is an area of much commercial interest, one current application is in the automotive industry. General Motors produce a thermoplastic/LSR multi-shot air intake manifold on the Northstar model [3]. Other potential applications include shower heads and water resistant mobile phones.

The reason for this interest is the added versatility offered to soft/hard combinations using LSR. LSR materials are thermally stable showing little change in properties with temperature. This can be of special interest at elevated temperatures where the use of TPEs may become limited.

LSRs also have rapid cure cycles and, because post cure is not necessary, finished parts cost less than other thermoset technologies.

Injection mouldable grades are formulated as two-component systems and cure at temperatures of around 170-230 °C. Like moulding with thermoplastic materials, the cycle times are part dependent. Cycle times of 15-60 s are typical. Because mix-meter pump systems are used to feed the injection machines, raw material handling of LSR materials is relatively simple. More details of LSR materials can be found in Chapter 7.

10.10.6.5 Thermoplastic Elastomers (TPEs)

The properties of thermoplastic elastomers that have made them so commercially successful are their low modulus and flexibility. The ability to recover from stress and return to their original shape makes them suitable for applications such as sealing rings. These are the traditional markets of thermosetting rubber materials, which TPE materials are now replacing. TPEs have the added advantage of recyclability over thermoset materials, especially important in markets where environmental legislation is in place.

Some examples of common thermoplastic elastomers are shown in Table 10.4. As well as adhesion considerations, the suitability of a TPE will also depend on properties such as its hardness and compression ratio.

Table 10.4 Examples of TPEs and substrates

Type	Elastomer description	Subgroup	Possible substrates
TPE-O	polyolefin blends	PP/EPDM PP/EPDM crosslinked	PP PA (modified)
TPE-V	polyolefin alloys	various	PS, ABS, PET, ASA, ABS/PC blends
TPE-S	styrene	SEBS, SBS, SEBS/PPE	PP PA (modified)
TPE-A	polyamide	PA 12 based PA 6 based	PA
TPE-E	polyester	polyesterester polyetherester	PA*(poyesterester only) PET, PBT
TPE-U	polyurethane	polyester urethane polyether ester urethane polyether urethane	PA, ABS, POM, PC, PBT, PVC

10.10.7 Multi-Shot Application Case Studies

Now that the major areas of multi-shot have been described, two recent applications of this technology by UK based companies will be described.

10.10.7.1 Trio Knob

An interesting example of a relatively early UK multi-shot application from Sifam, now SMT multi-shot, Devon, England details the cost savings that can be achieved using multi-shot technology. The original costings are from 1996, but the relative savings involved are still applicable today (see Table 10.5). The three production methods are as follows:

(1) Traditional: moulding two components, (a knob and a coloured cap), assembly and printing
(2) 2-shot: knob and cap moulded together, eliminates assembly, still requires printing
(3) 3-shot: part re-designed so that print effects are instead moulded in, thereby eliminates assembly and printing.

Table 10.5 Run cost comparison for Trio Knobs

Production method	Traditional	2-shot	3-shot
Tool cost	£8,500	£12,000	£19,000
Cycle time (s)	21 + 21	21	21
Machine cost (100,000 pieces)	£1,241	£657	£695
Materials	£345	£365	£385
Assembly/print	£1,650	£825	£0
Job cost	£3,236	£1,847	£1,080

It can be seen from Table 10.5 that despite the major increase in tooling costs the following factors reduced job cost:

- Utilising only one injection machine instead of two
- Keeping the cycle time the same due to sequential moulding whereby the three cavities are each moulding at any one cycle
- Removal of assembly and printing costs.

Moulding was carried out on a three-cavity tool moving the rotating side of the mould tool by 120° after each injection cycle. At any one stage there are three mouldings in the tool: a first single shot, a two shot intermediate moulding and a complete three shot component which is then ejected. Since all three cavities fill simultaneously, cycle time is fixed to the cavity requiring the longest fill time.

10.10.7.2 Stanley Screwdriver

Stanley Tools, Engel and Burnett Polymer Engineering, Northampton, UK developed a four shot process to manufacture screwdriver handles. A nylon core is used over which two different coloured layers of PPE are then moulded. These layers give the final part both aesthetic appeal and commercial branding. A final TPE is added for a soft feel grip. Adhesion is achieved in the layers by clever use of material properties and tooling design, to achieve bonding through mechanical interlock. This imaginative design results in an incredibly strong component giving improved impact resistant and torque. An application of this type requires extreme accuracy in location of the various components, to keep definition of the end product and to avoid damage to the tooling. This product highlights what can be achieved with successful collaboration between manufacturer, moulder and machinery supplier.

10.10.8 Limitations to Multi-Shot

Perhaps the most important issue with regard to multi-shot moulding is the cost of tooling. Obviously this varies depending on the part design and the complexity of tooling required. More complex rotating mould tools may also be more prone to breakdown than standard tools. Another problems associated with tooling concerns machine sizes. Over-dimensional, bulky moulds require larger machines, increasing both machine costs and larger space requirements due to the larger machine footprint. One way around this is to use tie-barless machines, the advantages of this machine configuration in relation to multi-shot have already been discussed. The use of two injection machines in a transfer process also involves high investment as two machines and tools are required.

Accurate positioning of the mould cavities must also be achieved. Moving and rotating the tool to achieve precision positioning of parts, often at very high speeds to optimise cycle times, can create wear and inaccurate registration. This also places restrictions on cycle times due to the time required to carry out this change of positioning.

In methods using the same mould and multiple cavities, process restrictions may limit bonding strength with multiple materials, since only one tool temperature can be employed. The choice of tool temperature may be compromised from what would be the optimised bonding temperature.

10.11 Over-Moulding

In this process, a component termed a preform is placed into the tool of an injection moulding machine. A second material is then moulded onto or around the preform. Two methods fall under this category: insert moulding and lost core moulding.

10.11.1 Insert Moulding

Insert moulding with plastics is a two-step process whereby a first preform component is placed into the open mould cavity. Injection then proceeds as with traditional moulding methods with injection of a molten plastic onto the preform. This process is not limited to two material components and the resultant mouldings can be transferred in this way until the required number of layers is achieved. Inserts can be loaded by hand or by the use of robots. Inserts must be accurate in both their dimension and their placement into the over-moulding tool to prevent tool damage and provide accurate registration of one material on another. A means must also exist to hold them in place within the tool. In this way it has similar requirements to that of in mould lamination techniques commonly used to decorate plastics with films or foils, details of which can be found in a specialist Rapra Review Report [4] and will not be covered further here.

10.11.2 Lost Core Moulding

The lost core technique, like insert moulding, is often used in combination with metals as well as plastics. It produces a hollow component similar to those produced by techniques such as extrusion blow moulding or gas assisted injection moulding. This technique overcomes disadvantages inherent in both these processes. It enables high dimensional accuracy, unlike extrusion blow moulding, as well as a defined interior surface not possible with gas assisted injection moulding. For the manufacture of plastic components, the core is first produced either from a low melting metal alloy, usually tin-bismuth, or a soluble plastic material. The core is then inserted either by hand or robot into a tool and over-moulded. The core material can then be melted or dissolved out and the final component cleaned of any residual waste material. There are advantages and disadvantages with this technique, as described in Table 10.6.

Applications for this technology include nylon air inlet manifolds for car engines, canoe paddles, tennis rackets, BMX bike wheels and hot water heating pumps.

Table 10.6 Process attributes of lost core moulding

Advantages	Disadvantages
Mould design relatively simple	Expensive
Complex geometry possible	High unit cost
High quality surface finish	Long development times
Good, seam-free interior surface	Core materials can be expensive
Weight saving	
High cavity pressures permissible	

10.12 The Future?

A whole host of advanced injection moulding technologies exist, a number of which are beyond the scope of this chapter which is far from a conclusive guide to all the technologies currently available. Other developments of note include micromoulding. This enables tiny components to be accurately and cost effectively manufactured, which is of special interest to the medical industry. The use of supercritical fluids to assist in moulding is also a growing field. These materials (generally supercritical CO_2 or N_2), offer a huge number of potential applications in the fields of polymeric foam production, for viscosity reduction (and therefore melt temperature reduction), as contaminant removals in recyclate and as compatibilising agents. Other multi-material injection moulding variations also exist. Some have been described here, others are beyond the scope of their chapter, which has concentrated on plastic material moulding only. It is likely that in the future, commercially successful moulders will require both the understanding and adaptability to utilise a wide range of the process options available. It is also likely that developments will continue to extend beyond the use of just plastic materials. Future components may be hybrids of many different materials such as plastics, metals and ceramics all created into a single component.

Tooling costs will be key to future developments in this area, due to the prohibitive costs often involved in tooling for multiple materials. Developments in both the mouldability and adhesion characteristics of both thermoset and thermoplastic materials, and machinery enhancements in terms of improved speeds and control are likely to further enhance processing and material combination options. With such possibility inherent in these processes it is likely that the future limitations on this technology may extend only to the limits of designer imagination.

References

1. R.A. Easterlow, P.D. Stidworthy, R.J. Coates and G.F. Smith, inventors; Rover Group Ltd., assignee; Painted Plastics Articles, patent EP0816066, 1998.

2. V. Goodship and K. Kirwan, *Plastics, Rubber and Composites*, 2001, **30**, 1, 11.

3. K. Baraw, Proceedings of the Annual Conference, Composites Institute, Society of the Plastics Industry, 1997, S 15-C/1.

4. J.C. Love and V. Goodship, *Rapra Review Report*, 2002, **13**, 1, Issue 146.

Further Reading

J. Avery, *Gas-Assist Injection Molding: Principles and Applications*, Hanser Publishers, 2001.

W. Michaeli, A. Brunswick and M. Gruber, Step on the Gas with Water Injection: Water Assisted Injection Moulding (WAIM): An Alternative to Gas Injection? *Kunststoff Plast Europe*, 1999, **89**, 4, 20.

P. Mapleston, Water-Assist Moulding Nears Debut for Auto Ducts, *Modern Plastics International*, 2002, **32**, 1, 34.

W. Michaeli and A. Brunswick, Manufacture of Conduits for Media by GAIM: Product Oriented Process Development, *Kunststoff Plast Europe*, 1998, **88**, 1, 10.

D. Vink, Getting to grips with Water, *European Plastic News*, 2001, June, 34.

W. Michaeli, T. Juntgen and S. Habibi-Naini, Fluid Injection Techniques Meet Demanding Requirements, *Modern Plastics International*, 2002, **32**, 1, 46.

PRW, Battenfeld Takes the Plunge into Water Injection, *Plastics & Rubber Weekly Magazine*, 2001, March.

Appendix 1 Abbreviations and Acronyms

2K	2-component
3K	3-component
ABS	acrylonitrile-butadiene-styrene
AMMA	acrylonitrile methyl methacrylate
APE	aromatic polyester
ASA	acrylonitrile-styrene-acrylate
ASTM	American Society for Testing and Materials
B	butadiene
BgVV	Bundesinstituts für gesundheitlichen Verbraucherschutz und Veterinärmedizin
BMA	ultra low abrasion
BMC	bulk moulding compound
CA	cellulose acetate
CAB	cellulose acetate butyrate
CAP	cellulose acetate propionate
CD	compact disc
CLTE	coefficient of linear thermal expansion
CP	cellulose propionate
CPVC	chlorinated PVC
CTFE	polychlorotrifluoroethylene
DAP	diallyl phthalate
DMC	dough moulding compound
E	ethylene
ELV	End-Of-Life Vehicle Directive (EU)
EMI	electromagnetic interference
EP	epoxide
ETFE	ethylene-tetrafluoroethylene
EVA	ethylene-vinyl acetate
FDA	Food and Drug Administration (USA)
FEP	tetrafluoroethylene hexafluoropropylene
GAIM	gas assisted injection moulding
GAIN™	gas assisted injection moulding
GIP	gas injection process
GIT	gas injection moulding technology
HDPE	high density polyethylene, PE rigid
HF	high frequency
IKV	Institute for Plastics Processing (Germany)
L/D	length/diameter ratio
LCP	liquid crystal polymer
LDPE	low density polyethylene, PE soft
LIM	liquid injection moulding
LLDPE	linear low density polyethylene
LSR	liquid silicone rubber
MBS	methyl methacrylate-butadiene-styrene
MF	melamine-formaldehyde
MFI	melt flow index
MP	melamine-phenol formaldehyde
PA	polyamide, also known as nylon
PA 11	poly(11-amino-undecanoic acid), polyamide 11
PA 12	poly(laurolactam), polyamide 12
PA 6	poly(ε-caprolactam), polyamide 6
PA 66	poly(hexamethylene diamine/adipic acid), polyamide 66
PA am.	amorphous polyamide
PAA	polyacrylic acid
PAE	polyacrylic ester
PAEK	polyarylether ketone

PAI	polyamide-imide
PAR	polyacrylate
PAS	polyarylsulfone
PBT	polybutylene terephthalate
PBTP	polybutylene terephthalate
PC	polycarbonate
PCTFE	polychlorotrifluoroethylene
PE	polyethylene
PE hard (HD)	unplasticised polyethylene
PE soft (LD)	plasticised polyethylene
PEC	polyester carbonate
PEEK	polyether etherketone
PEEKK	polyether etherketone ketone
PEI	polyetherimide
PEK	polyether ketone
PEKEKK	polyether ketone etherketone ketone
PEKK	polyether ketone ketone
PES	polyether sulfone
PET	polyethylene terephthalate
PETP	polyethylene terephthalate
PF	phenol-formaldehyde
PFA	perfluoro(alkoxyalkane) copolymer
PI	polyimide
PIB	polyisobutylene
PM	polyacrylic acid
PMMA	polymethyl methacrylate
POM	polyacetal, polyoxymethylene
PP	polypropylene
PPA	polyphthalamide
PPE	polyphenylene ether
PPO	polyphenylene oxide
PPS	polyphenylene sulfide
PS	polystyrene
PS-GP	general purpose polystyrene
PS-HI	high impact polystyrene
PSO	polysulfone
PSU	polysulfone
PTFE	polytetrafluoroethylene
PTMP	polytetramethylene terephthalate
PU	polyurethane
PUR	polyurethane
PVC	polyvinyl chloride
PVC hard	unplasticised polyvinyl chloride
PVC soft	plasticised polyvinyl chloride
PVDF	polyvinylidene fluoride
PVFl	polyvinylidene fluoride
RTV	room temperature vulcanisation
SAN	styrene-acrylonitrile
SB	styrene-butadiene
SBS	styrene-butadiene-styrene
SI	silicone
SMC	sheet moulding compound
Tg	glass transition temperature
Tm	melting temperature
TPE	thermoplastic elastomer
TPE-A	thermoplastic elastomer – amide type
TPE-E	thermoplastic elastomer – ethylene type

TPE-O	thermoplastic elastomer – olefin type
TPE-S	thermoplastic elastomer – styrene type
TPE-U	thermoplastic elastomer – urethane type
TPE-V	thermoplastic elastomer – vulcanisate type
TPO	thermoplastic olefinic elastomer
UF	urea-formaldehyde
UP	unsaturated polyester
WAIM	water assisted injection moulding
WEEE	Waste Electrical and Electronic Equipment Directive (EU)
WIM	water injection moulding
WIT	water injection technology

Appendix 2 Trade Names, Specific Weight and Suppliers of Some Plastic Materials

Abbreviation	Trade names	Specific weight (g/cm³)	Plastic type
ABS	Terluran (3), Novodur (4), Lustran ABS (15), Lacqran (31), Cycolac (32)	1.03-1.07	thermoplastic
AMMA		1.17	thermoplastic
CA	Setilithe (1), Cellidor A, S, U (4), Tenite (9)	1.26-1.32	thermoplastic
CAB	Cellidor B (4), Tenite Butyrat (9)	1.16-1.22	thermoplastic
CP	Cellidor CP (4)	1.19-1.23	thermoplastic
DAP	Neonit (5), Moldap (5), Supraplast (26)	1.51-1.78 filled	thermoset
EP	Bakelite (2), Araldite (5), Epikote (24)	1.7-2.0 filled	thermoset
ETFE	Tetzel (7), Hostaflon ET (12)	1.7	thermoplastic
EVA	Lupolen V351 oK (3), Alathon E/VA (7), Evatane (14), Wacker EVA (28), Supraplast (26)	0.92-0.95	thermoplastic
FEP	Teflon FEP (7)	2.14-2.17	thermoplastic
MF	Bakelite (2), Melopas P (5), Ultrapas (8), Resart (20), Resopal (20)	1.5-2.0 filled	thermoset
PA 6 PA 66	Akulon (1), Ultramid (3), Durethan (4), Zytel (7), Grilon (10), ≈ Orgamide (31), Maranyl (14)	1.13 1.14	thermoplastic
PA 11 PA 12	Grilamid (10), Vestamid (13), Rilsan (31)	1.04 1.02	thermoplastic
PA am.	Trogamid T (8)	1.12	thermoplastic
PAE		1.14	thermoplastic
PAS	Astrel 360	1.36	thermoplastic
PBTP (PTMT)	Ultradur (3), Pocan (4), Hostadur B PTMT (12), Tenite (9), Deroton (14)	1.29	thermoplastic
PC	Makrolon (4), Lexan (11)	1.20-1.24	thermoplastic
PCTFE	Hostaflon (12)	2.1-2.12	thermoplastic
PE soft (LD)	Lupolen LD (3), Hostalen LD (12), Alkathene (14), Wacker-PE (28)	0.91-0.93	thermoplastic
PE hard (HD)	Lupolen HD (3), Hostalen HD (12), Vestolen A (13)	0.94-0.96	thermoplastic
PF	Bakelite (2), Trolitan (8), Resinol (18), Supraplast (26)	1.25-2.0 filled	thermoset
PETP	Arnite (1), Crastin (5), Hostadur A, K (12)	1.37 crystalline 1.34 amorphous	thermoplastic
PI	Vespel (7), Gemon (11), OX 13 (14), Kinel (22), Kerimid (22)*)	1.42	thermoplastic, thermoset
PIB	Oppanol (3)	0.91-0.93	thermoplastic
PMMA	Degalan (6), Diakon (14), Tesarit (19), Plexiglas (23)	1.18	thermoplastic
POM	Ultraform (3), Delrin (7), Hostaform (12), Kematal (14)	1.41-1.43	thermoplastic
PR	Novolen (3), Hostalen PP (12), Vestolen P (13), Propathene (14)		
PP		0.9	thermoplastic
PPO	PPO-Copolymer: Noryl (11)		
PPE		1.06-1.10	thermoplastic
PPS	Ryton PPS (16)	1.34	thermoplastic
PS	Polystyrene (100 range) (3), Hostyren N (12), Vestyron (13), Lustrex (15), Lacqrene (31)	1105	thermoplastic

PES	Polyethersulfone (14), Udel (27)	1.24 1.37	thermoplastic
PTFE	Teflon (7), Hostaflon TF (12), Fluon (14)*	2.14-2.20	thermoplastic, elastomer
PUR	Desmopan (4), Hytrel (7), Elastollan (3)	1.14-1.26	thermoplastic, thermoset
PVC soft	Trosiplast (8), Vestolit (13), Vinnol soft (28)	1.2-1.35	thermoplastic
PVC hard	Viniflex (3), Trosiplast (8), Hostalit (12), Vestolit (13), Vinnol hard (28)	1.38-194	thermoplastic
PVDF	Dyflor 2000 (8), Solef (25), Foraflon (33)	1.76	thermoplastic
SAN	Luran (3), Lustran SAN (15), Lacqran (31)	1.08	thermoplastic
SB	Polystyrene (100 range) (3), Hostyren 5 (12), Vestyron (13), Lustrex (15)	1.04	thermoplastic
SI	Silikon (28)*	1.86-1.68	thermoplastic, thermoset
UF	Pollopas (8), Resopal (20)	1.17-2.0 filled	thermoset
UP	Bakelite (2), Keripol (17), Resiopl (18), Harex (20), Supraplast (26)	1.17-2.2 filled	thermoset

*Only moulding compounds or laminating resins or similar (not injection mouldable)

Supplier key:

(1)	Akzo-Plastics	D	(17)	Phonix	D
(2)	Bakelite	D	(18)	Raschig	D
(3)	BASF	D	(19)	Resart-lhm	D
(4)	Bayer	D	(20)	Resart	D
(5)	Ciba-Geigy	CH	(21)	Resopal	D
(6)	Degussa	D	(22)	Rhone-Progil	F
(7)	Du Pont	CH	(23)	Röhm	D
(8)	Dynamit	D	(24)	Shell	CH
(9)	Eastmann	CH	(25)	Solvay	B
(10)	Emser Werke	CH	(26)	Sud-West-Chemie	D
(11)	Gen. Electric	NL	(27)	Union Carbide	CH
(12)	Hoechst	D	(28)	Wacker	D
(13)	Huls	D	(31)	Aquitaine	F
(14)	ICI	GB	(32)	Borg-Warner	B
(15)	Monsanto	B	(33)	P.C.U.K.	F
(16)	Phillips Petroleum	USA	(34)	3M-Comp.	USA

Index

O

P

U

V

W

Printed in Great Britain
by Amazon